Skin Anti-Aging Strategies

Skin Anti-Aging Strategies

Guest Editor

Remo Campiche

MDPI

Basel • Beijing • Wuhan • Barcelona • Belgrade • Novi Sad • Cluj • Manchester

Guest Editor
Remo Campiche
Beauty & Care
Dsm-Firmenich
Kaiseraugst
Switzerland

Editorial Office
MDPI AG
Grosspeteranlage 5
4052 Basel, Switzerland

This is a reprint of the Special Issue, published open access by the journal *Cosmetics* (ISSN 2079-9284), freely accessible at: https://www.mdpi.com/journal/cosmetics/special_issues/L92IQFBN2L.

For citation purposes, cite each article independently as indicated on the article page online and as indicated below:

Lastname, A.A.; Lastname, B.B. Article Title. *Journal Name* **Year**, *Volume Number*, Page Range.

ISBN 978-3-7258-6872-8 (Hbk)
ISBN 978-3-7258-6873-5 (PDF)
https://doi.org/10.3390/books978-3-7258-6873-5

Contents

cosmetics

MDPI

Editorial

Skin Anti-Aging Strategies

Remo Campiche

Beauty & Care, Dsm-Firmenich, CH-4303 Kaiseraugst, Switzerland; remo.campiche@dsm-firmenich.com

The pursuit of youthful, healthy skin remains one of the most compelling drivers in the cosmetics industry. Anti-aging claims dominate consumer expectations, and scientific research continues to evolve to meet these demands. This Special Issue of *Cosmetics—Skin Anti-Aging Strategies*—brings together novel insights, innovative technologies, and modern approaches to address the multifaceted nature of skin aging.

Skin aging is influenced by intrinsic factors like genetic predisposition and cellular senescence, as well as extrinsic stressors including UV radiation, pollution, and lifestyle habits [1]. Climate change and regional variations further complicate this landscape, emphasizing the need for tailored solutions across diverse populations and ethnicities.

Traditional anti-aging strategies have focused on antioxidants [2], anti-inflammatory mechanisms [3], and extracellular matrix protection [4]. Recent advances highlight emerging frontiers: the role of the skin microbiome [5], senolytic interventions [6], exosome-based therapies [7], and neurocosmetics that integrate emotional well-being into skincare [8], as well as skin rejuvenation and longevity claims [9].

This Special Issue features 14 peer-reviewed contributions that span molecular pathways, biochemical mechanisms, and clinical applications. More specifically, this includes:

- Bioactive Ingredients: Studies on naturals as well as peptides demonstrate multifunctional benefits—antioxidant, anti-inflammatory, and anti-melanogenic properties—underscoring the power of nature-inspired actives.
- Advanced Delivery Systems and formulation: Research on composite nanocarriers and innovative formulations illustrates how technology enhances efficacy and consumer experience.
- Nutraceutical Approaches: Clinical trials on collagen peptide supplementation reaffirm the synergy between oral and topical interventions for skin health.
- Device driven skin care and AI: The use of LED-based phototherapy or the integration of artificial intelligence reflect the industry's commitment to innovation.

Beyond scientific evidence, this Special Issue is meant for collaboration and knowledge exchange, driving progress toward effective and sustainable anti-aging solutions.

We invite researchers, formulators, and specialists in the field to explore these contributions and join the dialogue shaping next-generation anti-aging strategies.

Acknowledgments: During the preparation of this manuscript, the author used Copilot Pro for the purposes of generating text. The author has reviewed and edited the output and takes full responsibility for the content of this publication.

Conflicts of Interest: The author declares no conflict of interest.

https://doi.org/10.3390/cosmetics13010002

References

1. Wyles, S.P.; Maredia, H.S.; Ansaf, R.B.; Dweydari, M.R.; Hurt, R.T.; Bonnes, S.L.; Khosla, S.; LeBrasseur, N.K.; Draelos, Z.D.; Davis, M.D. Skinspan(TM): A Healthy Longevity Framework for Skin Aging. *Mayo Clin. Proc.* **2025**, *100*, 1976–1991. [CrossRef] [PubMed]
2. Farris, P.; Krol, Y. Under Persistent Assault: Understanding the Factors that Deteriorate Human Skin and Clinical Efficacy of Topical Antioxidants in Treating Aging Skin. *Cosmetics* **2015**, *2*, 355–367. [CrossRef]
3. Baechle, J.J.; Chen, N.; Makhijani, P.; Winer, S.; Furman, D.; Winer, D.A. Chronic inflammation and the hallmarks of aging. *Mol. Metab.* **2023**, *74*, 101755. [CrossRef] [PubMed]
4. Sparavigna, A. Role of the extracellular matrix in skin aging and dedicated treatment—State of the art. *Plast. Aesthetic Res.* **2020**, *7*, 14. [CrossRef]
5. Sfriso, R.; Egert, M.; Gempeler, M.; Voegeli, R.; Campiche, R. Revealing the secret life of skin with the microbiome you never walk alone. *Int. J. Cosmet. Sci.* **2020**, *42*, 116–126. [CrossRef] [PubMed]
6. Pils, V.; Ring, N.; Valdivieso, K.; Lämmermann, I.; Gruber, F.; Schosserer, M.; Grillari, J.; Ogrodnik, M. Promises and challenges of senolytics in skin regeneration, pathology and ageing. *Mech. Ageing Dev.* **2021**, *200*, 111588. [CrossRef] [PubMed]
7. Thakur, A.; Shah, D.; Rai, D.; Parra, D.C.; Pathikonda, S.; Kurilova, S.; Cili, A. Therapeutic Values of Exosomes in Cosmetics, Skin Care, Tissue Regeneration, and Dermatological Diseases. *Cosmetics* **2023**, *10*, 65. [CrossRef]
8. Rizzi, V.; Gubitosa, J.; Fini, P.; Cosma, P. Neurocosmetics in Skincare—The Fascinating World of Skin–Brain Connection: A Review to Explore Ingredients, Commercial Products for Skin Aging, and Cosmetic Regulation. *Cosmetics* **2021**, *8*, 66. [CrossRef]
9. Klinngam, W.; Chaiwichien, A.; Osotprasit, S.; Ruktanonchai, U.; Kanlayavattanakul, M.; Lourith, N.; Wongrakpanich, A.; Teeranachaideekul, V.; Iempridee, T. Longevity cosmeceuticals as the next frontier in cosmetic innovation: A scientific framework for substantiating product claims. *Front. Aging* **2025**, *6*, 1586999. [CrossRef] [PubMed]

https://doi.org/10.3390/cosmetics13010002

cosmetics

MDPI

Article

Multifunctional Effects of N-Carbamylglutamate on Skin-Related Cells: Antioxidant, Anti-Aging, Anti-Melanogenic and Anti-Inflammatory Activities

Sa Rang Choi [1,†], Nu Ri Song [2,†], Seo Yeon Shin [2], Ki Min Kim [2], Jae Hee Byun [2], Seon Ju Kim [2], Dai Hyun Jung [3], Su Jung Kim [3] and Kyung Mok Park [2,*]

[1] Department of Otolaryngology–Head and Neck Surgery, Chonnam National University Medical School and Chonnam National University Hospital, 42 Jaebong-ro, Dong-gu, Gwangju 61469, Republic of Korea; choisalang_1234@naver.com

[2] Department of Biocosmetics, Dongshin University, 185, Gunjae-ro, Naju 58245, Jeonnam, Republic of Korea; nuri980424@naver.com (N.R.S.); ssy33144@naver.com (S.Y.S.); kimin4643@naver.com (K.M.K.); sad1055@naver.com (J.H.B.); kimsunjoo147@naver.com (S.J.K.)

[3] BIO-FD&C Co., Ltd., 106, Sandan-gil, Hwasun 58141, Jeonnam, Republic of Korea; dhjung@biofdnc.com (D.H.J.); sjkim@biofdnc.com (S.J.K.)

* Correspondence: parkkm@dsu.ac.kr; Tel.: +82-61-330-3273

† These authors contributed equally to this work.

Abstract: Skin aging is accelerated by both environmental factors—including ultraviolet (UV) radiation and pollution—and intrinsic processes such as chronic inflammaging. N-carbamylglutamate (NCG), an arginine precursor known for its benefits for gut and reproductive health, has not been extensively studied in dermatological applications. To explore its suitability as a multifunctional cosmetic ingredient, this study examines the protective role of NCG in counteracting UV-stimulated oxidative and inflammatory responses in skin cells. NCG significantly reduced UV-induced reactive oxygen species (ROS), indicating strong antioxidant properties. It also inhibited matrix metalloproteinase (MMP) activity, preserving collagen integrity and reducing wrinkle formation. In addition, NCG suppressed nitric oxide (NO) production and downregulated key inflammatory mediators—including cyclooxygenase-2 (COX-2), inducible nitric oxide synthase (iNOS), tumor necrosis factor-alpha (TNF-α), and interleukin-6 (IL-6)—highlighting its anti-inflammatory potential. Furthermore, NCG reduced melanin production and the expression of melanogenesis-related factors such as the microphthalmia-associated transcription factor (MITF), tyrosinase-related protein 1 (TRP-1), and TRP-2. These findings support the role of NCG as a promising multifunctional cosmetic ingredient with antioxidant, anti-inflammatory, anti-wrinkle, and skin-brightening properties.

Keywords: N-carbamylglutamate; skin aging; inflammation; antioxidant; hyperpigmentation

1. Introduction

Skin aging is a multifactorial biological process influenced by both intrinsic physiological factors and extrinsic environmental stressors, ultimately leading to structural and functional deterioration of the skin [1,2]. Clinically, skin aging presents as wrinkle formation, reduced elasticity, and hyperpigmentation—conditions that compromise both skin health and appearance, negatively impacting quality of life. Emerging evidence has identified inflammaging, a state of chronic, low-grade inflammation, as a pivotal contributor to skin aging. This chronic inflammatory state is exacerbated by environmental insults such as ultraviolet (UV) radiation, pollution, and lipopolysaccharide (LPS), a bacterial endotoxin [3–5]. Among these, UV radiation is considered the most potent contributor, inducing DNA damage, oxidative stress, and inflammatory cytokine production, which degrade extracellular matrix (ECM) components like collagen and elastin [6–11]. Although UVA penetrates more deeply into the dermis, UVB exerts stronger biological effects by directly inducing oxidative stress, inflammation, and photoaging-related responses such as MMP activation and collagen degradation. Accordingly, UVB irradiation has been widely employed as a representative in vitro model for studying photoaging in human dermal fibroblasts (HDFs) [12,13]. Similarly, LPS exposure activates innate immunity, stimulating the release of inflammatory mediators that not only accelerate skin aging but also contribute to hyperpigmentation [14–16]. Additionally, α-melanocyte-stimulating hormone (α-MSH) upregulation promotes melanogenesis, thereby exacerbating pigmentary disorders like melasma and lentigines [17,18].

Given this mechanistic understanding, there is a growing demand for functional cosmetic ingredients capable of preventing or ameliorating inflammaging-related skin aging. Although several anti-aging agents such as retinoids, peptides, and hydroxy acids (AHA, BHA) are currently in use, these compounds often present limitations including skin irritation, suboptimal safety profiles, and diminished long-term efficacy [19–21]. Therefore, the development of novel functional materials with enhanced safety and efficacy remains an urgent priority.

N-carbamylglutamate (NCG), an analog of N-acetylglutamate, serves as an effective precursor for endogenous arginine synthesis and has demonstrated anti-inflammatory and antioxidant effects in systemic studies. Mechanistically, NCG is known to suppress the ERK1/2–mTOR–S6K1 signaling pathway, downregulating pro-inflammatory cytokines such as interleukin-1β (IL-1β), interleukin-6 (IL-6), and interleukin-8 (IL-8) [22–24]. Despite these benefits, NCG's potential in skin health has not been systematically evaluated.

The present study investigates the biological effects and underlying mechanisms of NCG in skin cells. We focused on its antioxidant, anti-inflammatory, anti-wrinkle, and anti-melanogenic activities using in vitro models exposed to UVB and LPS. Our aim was to validate NCG as a multifunctional cosmetic agent capable of alleviating inflammaging and enhancing skin health.

2. Materials and Methods

2.1. Chemicals and Reagents

The following reagents were used in this study, sourced from the respective manufacturers: NCG (C4375) from Sigma-Aldrich Corporation (St. Louis, MO, USA); Dulbecco's Modified Eagle Medium (DMEM) and phosphate-buffered saline (PBS) from Lonza Group Ltd. (Walkersville, MD, USA). For primary fibroblast culture, fibroblast medium (FM), fibroblast growth supplement (FGS), and penicillin-streptomycin (P/S) were purchased from ScienCell (Carlsbad, CA, USA). 3-(4,5-dimethylthiazol-2-yl)-2,5-diphenyltetrazolium bromide (MTT), dimethyl sulfoxide (DMSO), α-MSH, arbutin, and lipopolysaccharide (LPS) were also obtained from Sigma-Aldrich. ELISA kits for human Pro-MMP-1, Pro-MMP-3, and Pro-Collagen I alpha 1 were purchased from R&D Systems (Minneapolis, MN, USA). Antibodies used in Western blotting were sourced as follows: GAPDH from Enogen Biotechnology (New York, NY, USA); TRP-1, TRP-2, MMP-1, JNK, p-JNK, TNF-α, NF-κB, and p38 from Santa Cruz Biotechnology (Dallas, TX, USA); MITF, p-p38 MAPK, and ERK from Cell Signaling Technology (Beverly, MA, USA); COX-2, IL-6, and iNOS from Invitrogen (Carlsbad, CA, USA).

2.2. Cell Culture

B16F10 melanoma cells, human epidermal keratinocytes (HaCaTs), and RAW 264.7 macrophages were obtained from the American Type Culture Collection (ATCC, Manassas, VA, USA). Cells were maintained in DMEM supplemented with 10% fetal bovine serum (FBS) and 1% penicillin-streptomycin in a humidified incubator at 37 °C with 5% CO_2. HDFs were obtained from ScienCell (Carlsbad, CA, USA).

2.3. Cell Viability Assay

The MTT assay was conducted to determine cell viability. Following seeding into 96-well plates, cells were treated with varying doses of NCG. After incubation (24 h for HaCaT and RAW 264.7; 72 h for B16F10 and HDF), MTT solution (0.5 mg/mL) was added and incubated for 3 h. DMSO was then added to dissolve the resulting formazan crystals, and absorbance was measured at 570 nm using a Multiskan Sky spectrophotometer (Thermo Fisher Scientific, Waltham, MA, USA).

2.4. Analysis of Intracellular Reactive Oxygen Species (ROS) Reduction Activity

HaCaT cells were cultured for 24 h in 24-well plates prior to further treatment. Based on our preliminary experiments, UVB irradiation at an intensity of 15 mJ/cm^2 was selected because this level effectively induced oxidative stress without causing cytotoxicity in HaCaTs, consistent with previous studies employing similar UVB doses in keratinocyte-based oxidative stress models [25,26]. After UVB exposure, cells were treated with various NCG concentrations and incubated for an additional 24 h. ROS levels were assessed using CellROX™ Orange Reagent (Thermo Fisher Scientific). Cells were stained with 5 µM CellROX™ for 30 min in the dark, and fluorescence was visualized using the EVOS™ M5000 imaging system with appropriate filters (excitation/emission: 545/565 nm). Fluorescence intensity was quantified from three randomly selected fields per condition.

2.5. Measurement of Melanin Content

B16F10 melanoma cells were seeded in 6-well plates and incubated overnight. Cells were subsequently exposed to varying concentrations of NCG for 72 h. After treatment, the cells were washed with PBS, harvested by trypsinization, and centrifuged at 13,000 rpm for 5 min. Following centrifugation, the collected cell pellets were treated with 1 N NaOH containing 10% DMSO and maintained at 80 °C for 1 h to ensure complete solubilization. Readings at 475 nm were used to determine the level of absorbance.

2.6. Determination of Cellular Tyrosinase Activity

The tyrosinase activity within B16F10 melanoma cells was assessed according to a previously described method with slight modifications [27]. Cells were plated in 6-well plates and treated with different concentrations of NCG for a duration of 72 h. Supernatants obtained after centrifugation (13,000 rpm, 5 min, 4 °C) were diluted with 0.1 M phosphate buffer (pH 6.8) and aliquoted into 96-well plates. L-DOPA (1 mg/mL) was added, and samples were incubated for 1 h prior to reading absorbance at 475 nm.

2.7. Type 1 Procollagen Synthesis

Type I procollagen was quantified using an enzyme immunoassay specific to the Procollagen Type I C-peptide (PIP), following the manufacturer's protocol. HDFs were plated and incubated for 24 h, exposed to UVB (10 mJ/cm^2), and treated with different NCG concentrations. Collected culture supernatants were centrifuged (12,000 rpm, 10 min, 4 °C). Antibody-POD and supernatant were added, incubated, washed, and reacted with substrate. Reaction was stopped using 1 N H_2SO_4, and absorbance was read at 450 nm.

2.8. Inhibitory Activity of MMP-1 and MMP-3

The inhibitory effects of NCG on MMP-1 and MMP-3 were evaluated using ELISA kits specific to human pro-MMP-1 and pro-MMP-3, respectively. HDFs were seeded into 6-well plates and cultured. Following UVB irradiation, cells were treated with NCG for 72 h. The culture media were collected and centrifuged to obtain the supernatant. Thereafter, 100 μL of the collected supernatant along with the RD1-52 reagent was dispensed into each well and allowed to react for 2 h. The wells were then emptied, followed by the addition of either anti-human pro-MMP-1 or pro-MMP-3 conjugate antibodies, which were incubated for another 2 h. Following this, the wells were washed, and 100 μL of substrate solution was added and incubated for 20 min. The reaction was terminated using stop solution. Absorbance was measured at 450 nm with a reference at 540 nm, and the final value was calculated as the difference between the two readings.

2.9. Measurement Nitric Oxide (NO) Inhibitory Activity

The inhibitory effect of NCG on NO production was assessed by measuring nitrite accumulation via the Griess reaction. RAW 264.7 macrophage cells were seeded into 96-well plates, stimulated with lipopolysaccharide (LPS, 1 μg/mL), and subsequently treated with various concentrations of NCG. Following incubation, equal volumes of the culture supernatant and Griess reagent were mixed in a new 96-well plate and allowed to react at room temperature for 20 min. The absorbance was measured at 540 nm to quantify nitrite levels, which reflect NO production.

2.10. Western Blot Analysis

Western blotting was conducted with minor modifications to established protocols [28]. Cells were seeded and maintained under standard culture conditions, then treated with NCG. Following incubation, cells were lysed, and the resulting lysates were centrifuged to isolate the supernatant. Following quantification using a protein assay method, quantities of each protein sample were loaded onto SDS−PAGE gels and subsequently transferred to polyvinylidene difluoride (PVDF) membranes for detection. Following 1 h of blocking in 5% skim milk diluted in TBST, the membranes were incubated with specific primary antibodies at 4 °C for approximately 24 h. (as detailed in Table 1). Membranes were washed four times with TBST and subsequently incubated at room temperature for 1 h with secondary antibodies. Protein bands were visualized using an enhanced chemiluminescence (ECL) detection system.

Table 1. Primary antibodies used in Western blot analysis.

Target Protein	Catalog No.	Host Species	Dilution Ratio
GAPDH	E12-057	Mouse	1:5000
TRP-1	ab178676	Rabbit	1:10,000
TRP-2	ab2211144	Rabbit	1:1000
MITF	#12590	Rabbit	1:1000
MMP-1	sc-58377	Mouse	1:400
MMP-3	AF513	Goat	1:5000
TNF-α	#12744	Mouse	1:200
COX-2	#35-8200	Mouse	1:500
IL-6	#P620	Rabbit	1:1000
iNOS	#PA1-036	Rabbit	1:1000

2.11. Statistical Analysis

All experiments were independently repeated three times, and the data are expressed as mean ± standard deviation (SD). To evaluate statistical differences, analyses were conducted in SPSS (v27.0; IBM, Chicago, IL, USA). Group differences were examined using one-way ANOVA or Student's t-test where appropriate. Statistical significance was set at * $p < 0.05$, ** $p < 0.01$, and *** $p < 0.001$.

3. Results

3.1. Effect of NCG on Cell Viability

The cytotoxicity of NCG was evaluated in HaCaT, HDF, B16F10, and RAW 264.7 cells using the MTT assay. NCG exhibited no cytotoxic effects at concentrations ranging from 100–1000 μM in HaCaT and RAW 264.7 cells, 100–500 μM in B16F10 cells, and 25–200 μM in HDFs. These results confirm that NCG is biocompatible within the tested ranges for each cell type (Figure 1a–d).

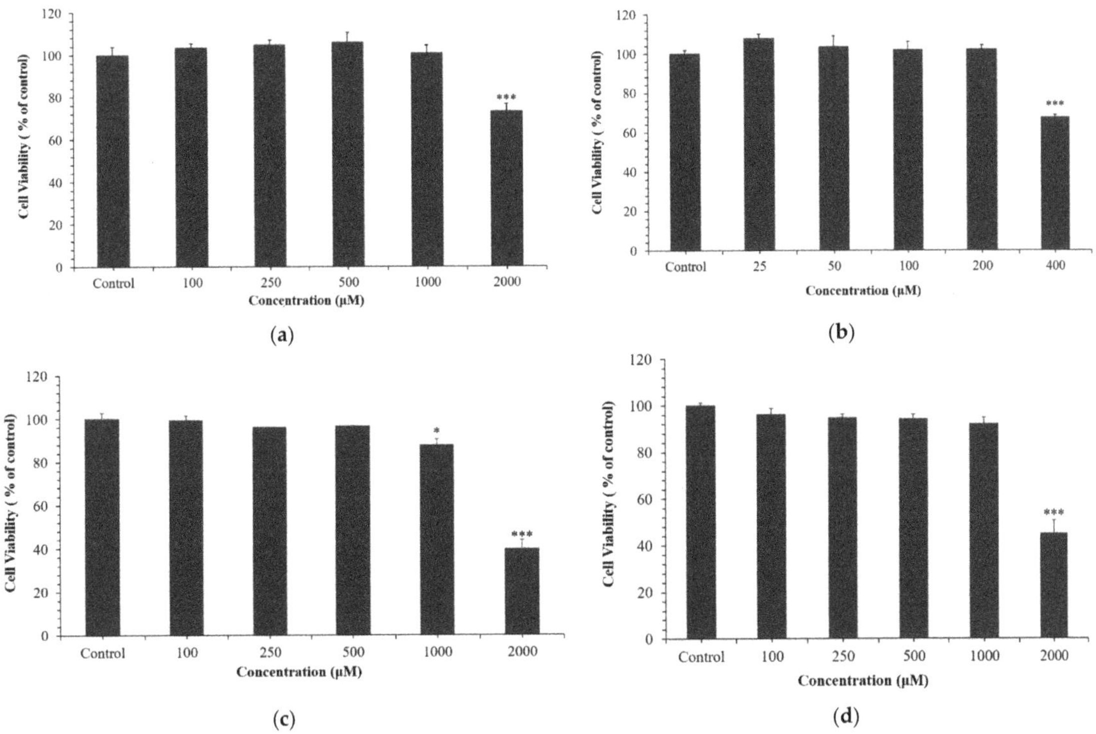

Figure 1. Effects of NCG on cell viability. (**a**) HaCaTs, (**b**) HDFs, (**c**) B16F10 cells, and (**d**) RAW 264.7 macrophages. Cells were treated with NCG and incubated for 24 h (HaCaTs, RAW 264.7) or 72 h (B16F10, HDF). Data are presented as mean ± SD based on triplicate experiments. * $p < 0.05$, *** $p < 0.001$ vs. control.

3.2. Effects of NCG on Intracellular and Mitochondrial ROS Levels

To assess the antioxidant properties of NCG, ROS production was induced in HaCaT cells using UVB irradiation (15 mJ/cm^2). Post-irradiation treatment with NCG significantly suppressed intracellular ROS levels in a dose-dependent manner. Notably, concentrations between 500–1000 μM reduced ROS generation by over 90%. These findings demonstrate the potent ROS-scavenging capacity of NCG. Importantly, 500 μM NCG exhibited a ROS scavenging effect comparable to that of 50 mM NAC, a well-established antioxidant used as a positive control (Figure 2). These findings demonstrate the potent ROS-scavenging capacity of NCG, even at much lower concentrations than conventional antioxidants.

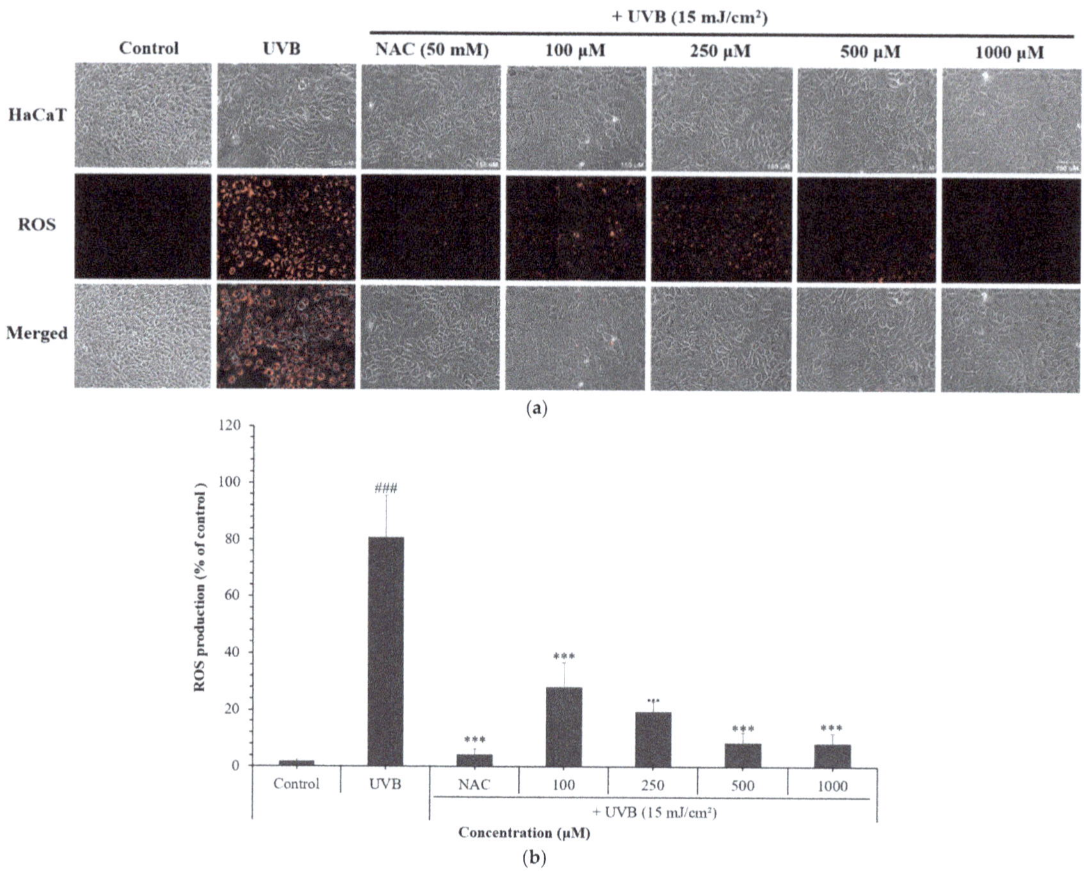

Figure 2. Effects of NCG on intracellular and mitochondrial ROS levels. (**a**) Representative fluorescence images of ROS production in HaCaTs exposed to UVB (15 mJ/cm^2) and treated with NCG (100–1000 µM) for 24 h. NAC (N-acetylcysteine, 50 mM) was used as a positive control. Images were captured under identical acquisition settings at 20×; scale bar = 150 µm (applies to all images in each row). (**b**) Quantification of intracellular ROS production expressed as a percentage of control. Data are presented as mean ± SD from at least three independent experiments. $^{###}$ $p < 0.001$ vs. control; *** $p < 0.001$ vs. UVB.

3.3. *Effects of NCG on UVB-Induced Skin Aging in HDFs*

To evaluate NCG's protective effects against UVB-induced dermal aging, HDFs were exposed to UVB and subsequently treated with NCG. NCG significantly restored type I procollagen synthesis, achieving up to a 30% increase compared to UVB-treated controls (Figure 3a). Additionally, NCG inhibited enzymatic activities of MMP-1 and MMP-3 by more than 53% (Figure 3b,c). Western blot analysis confirmed reduced protein expression of both MMPs following NCG treatment, suggesting a robust protective effect on ECM integrity (Figure 3d,e).

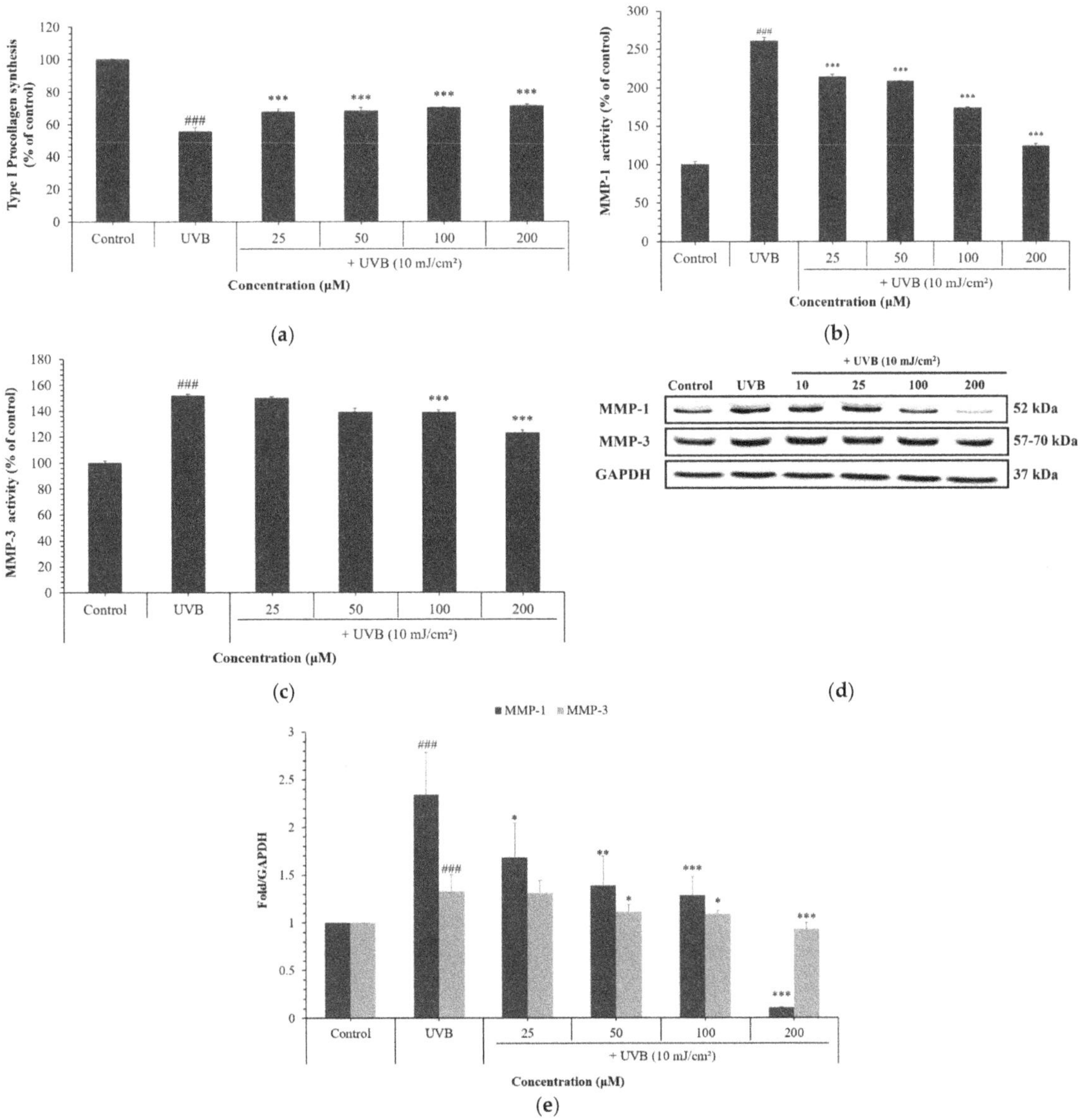

Figure 3. Effects of NCG on UVB-induced skin aging in HDFs. (**a**) Type I procollagen synthesis measured by ELISA. (**b**) MMP-1 enzymatic activity measured by ELISA. (**c**) MMP-3 enzymatic activity measured by ELISA. (**d**) Protein expression levels of MMP-1 and MMP-3 analyzed by Western blot. (**e**) Densitometric quantification of Western blot bands from panel (**d**), showing relative expression of MMP-1 and MMP-3. The results are expressed as the mean ± standard deviation from three independent experiments. ### $p < 0.001$ vs. control; * $p < 0.05$, ** $p < 0.01$, *** $p < 0.001$ vs. UVB-treated group.

3.4. *Effects of NCG on Melanogenesis in B16F10 Cells*

To determine the impact of NCG on melanogenesis, B16F10 cells were stimulated with α-MSH (100 nM), and arbutin (400 μM) was used as a positive control. NCG treatment significantly inhibited melanin synthesis and reduced intracellular tyrosinase activity by 18.3%. Western blot analysis revealed that NCG downregulated melanogenic proteins including microphthalmia-associated transcription factor (MITF), tyrosinase-related protein (TRP-1), and 2 (TRP-2) by over 50%. Notably, 500 μM NCG exhibited anti-melanogenic effects comparable to those of the positive control arbutin, both in melanin content reduction and tyrosinase inhibition assays, indicating that NCG has a similar level of efficacy at a comparable concentration (Figure 4). These results highlight the ability of NCG to attenuate melanin production by modulation of key melanogenic pathways.

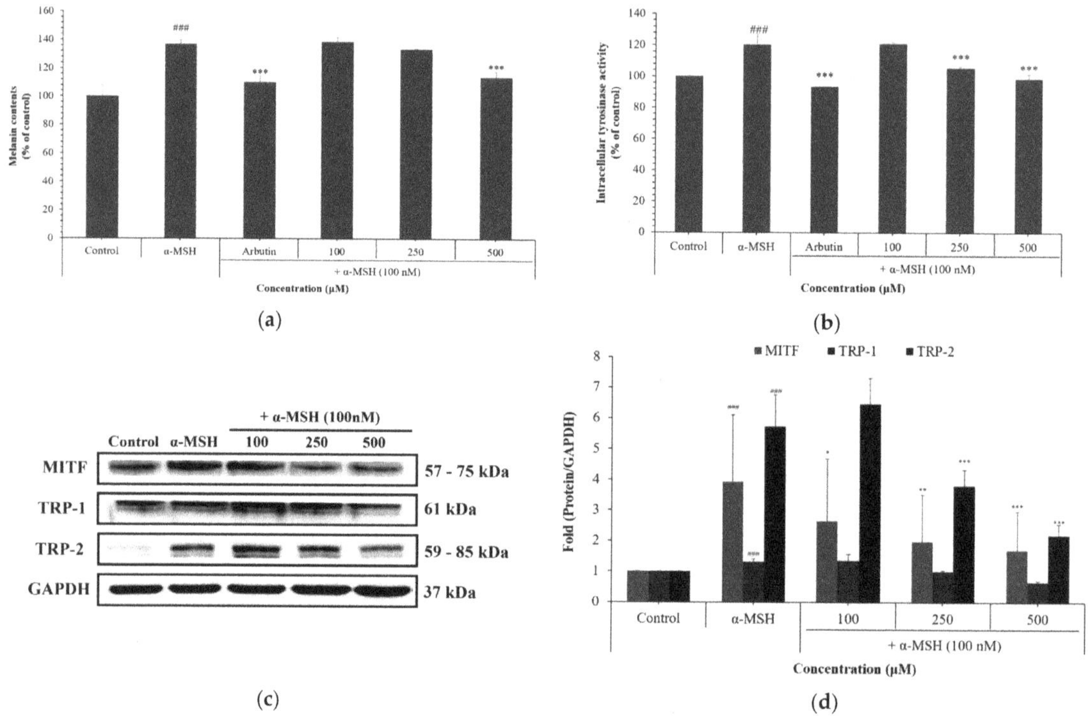

Figure 4. Effects of NCG on melanogenesis in B16F10 melanoma cells. (**a**) Melanin content after treatment with NCG (100–500 μM) for 72 h. (**b**) Intracellular tyrosinase activity following NCG treatment. (**c**) Protein expression of MITF, TRP-1, and TRP-2 determined by Western blot; normalized to GAPDH. (**d**) Densitometric analysis of protein levels. As a standard for comparison, arbutin was included as the positive control. Experimental data are displayed as mean ± standard deviation, based on three independently repeated experiments. ### $p < 0.001$ vs. control; * $p < 0.05$, ** $p < 0.01$, *** $p < 0.001$ vs. α-MSH group.

3.5. Anti-Inflammatory Effects of NCG

The anti-inflammatory potential of NCG was assessed in RAW 264.7 macrophages stimulated with LPS. NCG significantly decreased NO production in a dose-dependent manner, with a maximum reduction of 47.5% at 1000 μM. Importantly, this reduction in NO levels did not affect cell viability. NCG effectively inhibited the protein expression of key inflammatory mediators—cyclooxygenase-2 (COX-2), inducible Nitric Oxide Synthase (iNOS), tumor necrosis factor-alpha (TNF-α), and IL-6—as confirmed by Western blotting, indicating a strong anti-inflammatory response at the molecular level (Figure 5).

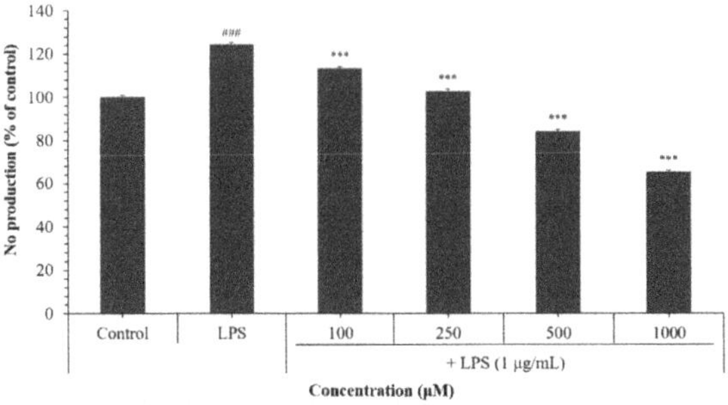

(a)

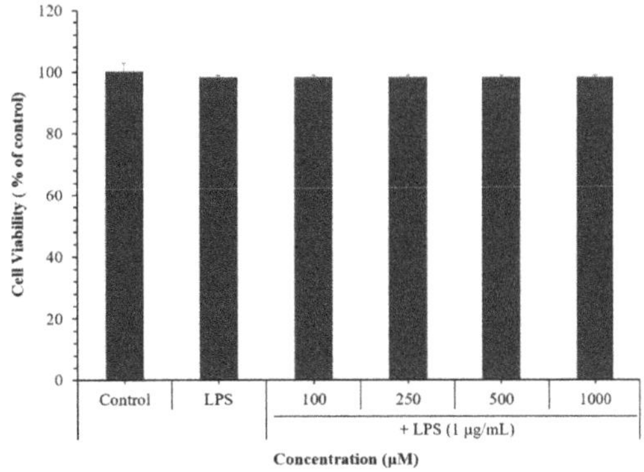

(b)

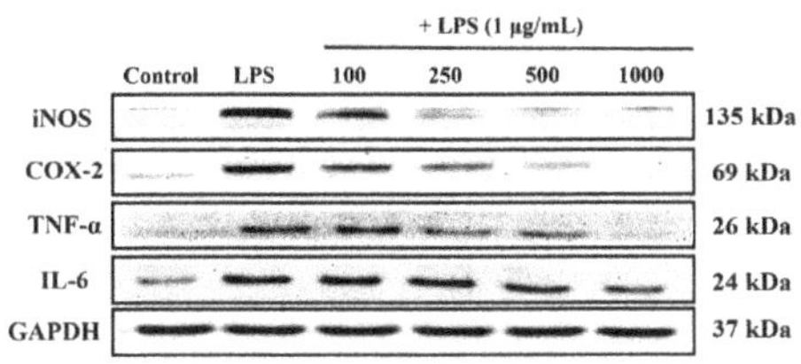

(c)

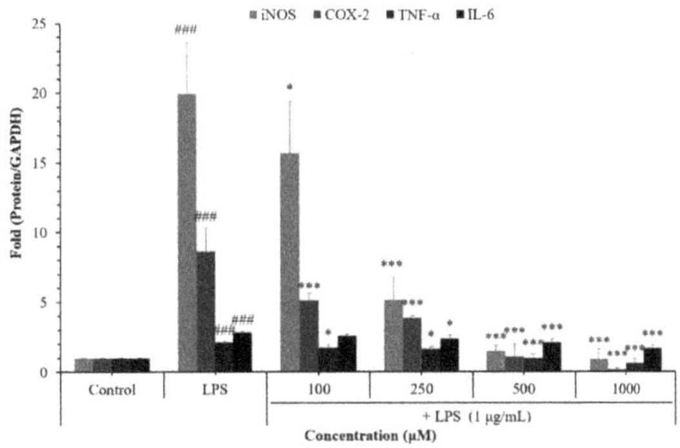

(d)

Figure 5. Anti-inflammatory effects of NCG in RAW 264.7 macrophages. (**a**) Cell viability after NCG treatment (100–1000 μM) for 24 h. (**b**) NO production measured by the Griess assay. (**c**) Protein expression levels of iNOS, COX-2, IL-6, and TNF-α analyzed by Western blot; normalized to GAPDH. (**d**) Quantification of Western blot bands. The results are presented as mean ± SD obtained from three independent experiments. Statistical significance was indicated as follows: ### $p < 0.001$ vs. control; * $p < 0.05$, *** $p < 0.001$ vs. LPS group.

4. Discussion

Skin aging is a complex biological process influenced by intrinsic and extrinsic factors, including genetic background, oxidative stress, and chronic inflammation [29]. Among these, pro-inflammatory cytokines have emerged as central mediators of structural degradation and functional decline in the skin. Notably, the concept of "inflammaging"—chronic, low-grade inflammation associated with aging has garnered increasing attention as a critical contributor of skin deterioration [30,31].

Prolonged inflammatory stimuli also influence melanogenesis [32,33]. Inflammatory cytokines such as IL-6 and TNF-α stimulate melanocyte activity, leading to increased melanin synthesis [34–37]. While melanin serves as a protective pigment against UV radiation, its excessive accumulation may exacerbate oxidative stress by interacting with reactive species, further damaging skin cells and contributing to pigmentation disorders such as melasma and freckles [38,39].

Oxidative stress induced by UV radiation and environmental pollutants promotes ROS generation, leading to inflammation, hyperpigmentation, and degradation of extracellular matrix components [40–43]. Specifically, ROS trigger the expression of MMPs, such as

MMP-1 and MMP-3, which degrade collagen and elastin fibers, reducing dermal elasticity and accelerating wrinkle formation [44,45].

MITF is a key transcriptional regulator of melanogenesis [46]. It governs the expression of tyrosinase and related proteins (TRP-1 and TRP-2), which are essential enzymes in melanin biosynthesis [47]. Overactivation of MITF contributes to melanin overproduction and the development of hyperpigmentation disorders [48].

Inflammaging is further perpetuated by transcriptional regulators including iNOS, COX-2, TNF-α, and IL-6. iNOS generates nitric oxide (NO), COX-2 synthesizes pro-inflammatory prostaglandins, and cytokines such as TNF-α and IL-6 amplify inflammatory cascades [49,50]. Chronic overexpression of these mediators disrupts skin homeostasis, impairs the skin barrier, and accelerates aging.

Our study demonstrates that NCG modulates these interconnected biological pathways: First, NCG attenuated UVB-induced ROS generation, reducing oxidative stress and preventing dermal damage. Although the precise mechanism remains to be fully elucidated, this antioxidant effect may result from both direct ROS scavenging activity and the modulation of endogenous antioxidant systems, such as glutathione metabolism or redox-sensitive signaling pathways. Based on these findings, we believe that NCG exerts various biological effects—such as anti-inflammatory, anti-melanogenic, and anti-aging activities—by suppressing UVB-induced ROS accumulation. This hypothesis is supported by previous studies reporting that NCG enhances antioxidant capacity through glutathione metabolism and redox-sensitive pathways [51–53]. Second, it suppressed MMP-1 and MMP-3 activity while enhancing type I procollagen synthesis, thereby preserving ECM structure and skin elasticity. Third, NCG modulated melanogenesis by downregulating MITF, tyrosinase, and TRP-1 expression, leading to reduced melanin production. Finally, NCG significantly downregulated iNOS, COX-2, TNF-α, and IL-6, thereby alleviating inflammation and maintaining skin homeostasis. In addition, although the intracellular uptake and skin permeability of NCG were not directly assessed in this study, the observed cellular effects suggest that NCG may be bioavailable and capable of penetrating the skin barrier to exert its effects in dermal cells. Future studies employing uptake assays, skin diffusion models, or reconstructed human skin systems will be essential to validate its delivery and efficacy in realistic dermatological applications. To gain a more comprehensive understanding of these biological effects, we plan to explore the involvement of key regulatory signaling pathways—such as NF-κB, MAPKs (p38, JNK, ERK), and CREB—in future studies. Identifying these upstream modulators will help elucidate the precise mechanism of action of NCG and further validate its potential as a targeted bioactive compound.

Taken together, these results position NCG as a promising multifunctional compound capable of targeting multiple mechanisms of skin aging. Its simultaneous antioxidant, anti-inflammatory, anti-wrinkle, and anti-melanogenic effects offer considerable potential for use in dermocosmetic applications. Further formulation studies and clinical evaluations will be instrumental in commercializing NCG-based skin health products.

5. Conclusions

Collectively, our findings demonstrate that NCG possesses antioxidant, anti-inflammatory, and anti-melanogenic properties that contribute to the attenuation of inflammaging and skin aging (Figure 6). The current in vitro data highlight NCG's potential as a safe and multifunctional active compound for use in cosmetic and dermatological products. Still, in vivo validation and clinical assessments remain necessary before it can be considered for practical application in commercial formulations.

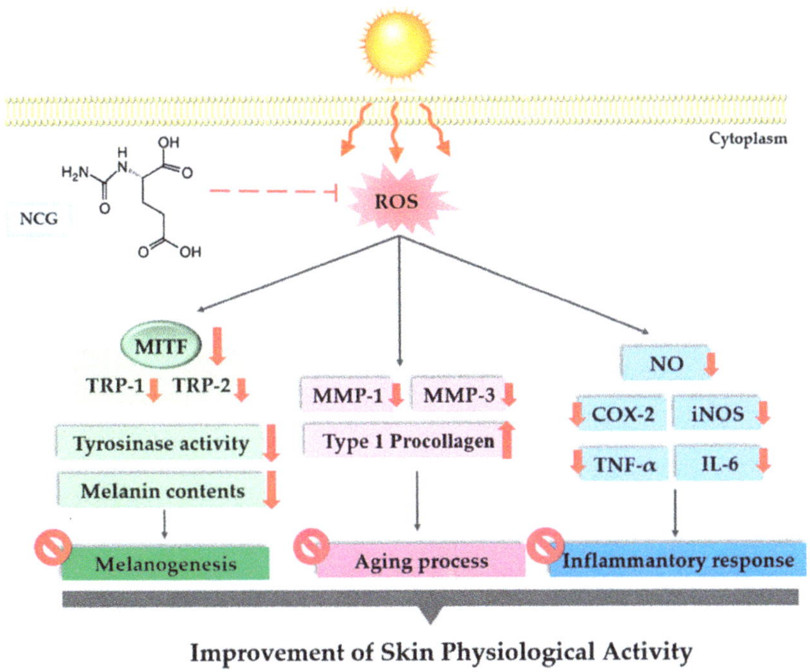

Figure 6. Summary of the molecular mechanism by which NCG enhances skin physiological activity through anti-melanogenic, anti-aging, and anti-inflammatory pathways.

Author Contributions: Conceptualization, methodology, and writing—original draft preparation: S.R.C.; investigation and writing—review and editing: N.R.S.; investigation and methodology: S.Y.S., K.M.K., J.H.B. and S.J.K. (Seon Ju Kim); methodology: D.H.J. and S.J.K. (Su Jung Kim); project administration and writing—review and editing: K.M.P. All authors have read and agreed to the published version of the manuscript.

Funding: This research was supported by a Korea Innovation Foundation (INNOPOLIS) grant funded by the Korean government (Ministry of Science and ICT) through the "Science and Technology Project that Opens the Future of the Region" (grant number: 2021-DD-UP-0380). This research was supported by the Regional Innovation System & Education (RISE) program through the Jeollanamdo RISE center, funded by the Ministry of Education (MOE) and the Jeollanamdo, Republic of Korea (2025-RISE-14-001).

Institutional Review Board Statement: Not applicable.

Informed Consent Statement: Not applicable.

Data Availability Statement: Data available on request due to restrictions.

Acknowledgments: During the preparation of this manuscript, the authors used ChatGPT 4o for the purposes of preparing the manuscript for general editing of this paper. The authors have reviewed and edited the output and take full responsibility for the content of this publication.

Conflicts of Interest: S.J.K. and D.H.J. are employees of BIO-FD&C Co., Ltd. BIO-FD&C provided no funding for this study. The company had no role in the study design; data collection, analysis, or interpretation; manuscript preparation; or the decision to submit for publication.

References

1. Shin, S.H.; Lee, Y.H.; Rho, N.K.; Park, K.Y. Skin aging from mechanisms to interventions: Focusing on dermal aging. *Front. Physiol.* **2023**, *14*, 1195272. [CrossRef]
2. Park, S. Biochemical, structural and physical changes in aging human skin, and their relationship. *Biogerontology* **2022**, *23*, 275–288. [CrossRef]
3. Ding, Y.; Jiratchayamaethasakul, C.; Lee, S.H. Protocatechuic aldehyde attenuates UVA-induced photoaging in human dermal fibroblast cells by suppressing MAPKs/AP-1 and NF-κB signaling pathways. *Int. J. Mol. Sci.* **2020**, *21*, 4619. [CrossRef]
4. Pilkington, S.M.; Bulfone-Paus, S.; Griffiths, C.E.; Watson, R.E. Inflammaging and the Skin. *J. Investig. Dermatol.* **2021**, *141*, 1087–1095. [CrossRef]
5. Yadav, E. Inflammation and Aging: The Skin Inflammasome in the Context of Longevity Science. *J. Cell. Immunol.* **2025**, *7*, 37–42. [CrossRef]
6. Salminen, A.; Kaarniranta, K.; Kauppinen, A. Photoaging: UV radiation-induced inflammation and immunosuppression accelerate the aging process in the skin. *Inflamm. Res.* **2022**, *71*, 817–831. [CrossRef] [PubMed]
7. Singh, D.; Rai, V.; Agrawal, D.K. Regulation of collagen I and collagen III in tissue injury and regeneration. *Cardiol. Cardiovasc. Med.* **2023**, *7*, 5. [CrossRef] [PubMed]
8. Ha, A.T.; Cho, J.Y.; Kim, D. MLK3 regulates inflammatory response via activation of AP-1 pathway in HEK293 and RAW264. 7 cells. *Int. J. Mol. Sci.* **2022**, *23*, 10874. [CrossRef] [PubMed]
9. Gorman, A.; Golovanov, A.P. Lipopolysaccharide structure and the phenomenon of low endotoxin recovery. *Eur. J. Pharm. Biopharm.* **2022**, *180*, 289–307. [CrossRef]
10. Tang, X.; Yang, T.; Yu, D.; Xiong, H.; Zhang, S. Current insights and future perspectives of ultraviolet radiation (UV) exposure: Friends and foes to the skin and beyond the skin. *Environ. Int.* **2024**, *185*, 108535. [CrossRef]
11. Sharma, M.R.; Mitrani, R.; Werth, V.P. Effect of TNFα blockade on UVB-induced inflammatory cell migration and collagen loss in mice. *J. Photochem. Photobiol. B Biol.* **2020**, *213*, 112072. [CrossRef] [PubMed]
12. Cavinato, M.; Jansen-Dürr, P. Molecular mechanisms of UVB-induced senescence of dermal fibroblasts and its relevance for photoaging of the human skin. *Exp. Gerontol.* **2017**, *94*, 78–82. [CrossRef]
13. Wang, P.W.; Hung, Y.C.; Lin, T.Y.; Fang, J.Y.; Yang, P.M.; Chen, M.H.; Pan, T.L. Comparison of the biological impact of UVA and UVB upon the skin with functional proteomics and immunohistochemistry. *Antioxidants* **2019**, *8*, 569. [CrossRef]
14. Rybnikář, M.; Malaník, M.; Šmejkal, K.; Švajdlenka, E.; Shpet, P.; Babica, P.; Dall'Acqua, S.; Ondřej, J.; Treml, J. Dibenzocyclooctadiene Lignans from Schisandra chinensis with Anti-Inflammatory Effects. *Int. J. Mol. Sci.* **2024**, *25*, 3465. [CrossRef]
15. Tamagawa-Mineoka, R.; Ueta, M.; Arakawa, Y.; Nakanishi, M.; Nishigaki, H.; Katoh, N. Roles of interferon regulatory factor 3 in skin inflammation: Possible involvement of regulatory mechanisms. *Exp. Dermatol.* **2023**, *32*, 715–717. [CrossRef] [PubMed]
16. Tong, W.; Chen, X.; Song, X.; Chen, Y.; Jia, R.; Zou, Y.; Li, L.; Yin, Z.; He, C.; Liang, X.; et al. Resveratrol inhibits LPS-induced inflammation through suppressing the signaling cascades of TLR4-NF-κB/MAPKs/IRF3. *Exp. Ther. Med.* **2020**, *19*, 1824–1834. [CrossRef] [PubMed]
17. Bayazid, A.B.; Jang, Y.A.; Jeong, S.A.; Lim, B.O. Cypress tree (*Chamaecyparis obtusa*) Bark extract inhibits melanogenesis through repressing CREB and MITF signalling pathways in α-MSH-stimulated B16F10 cells. *Food Agric. Immunol.* **2022**, *33*, 498–510. [CrossRef]
18. Shin, S.Y.; Sun, S.O.; Ko, J.Y.; Oh, Y.S.; Cho, S.S.; Park, D.H.; Park, K.M. New synthesized Galloyl-RGD inhibits Melanogenesis by regulating the CREB and ERK signaling pathway in B16F10 melanoma cells. *Photochem. Photobiol.* **2020**, *96*, 1321–1331. [CrossRef]
19. Cha, S.Y.; Woo, I.K.; Cha, Y.J.; Lee, N.K.; Jang, H.J.; Paik, H.D. Anti-melanogenic and Antioxidant Activities of Lactiplantibacillus plantarum Strains in Skin Cells via the CREB/MITF and Nrf2/HO-1 Pathways. *Probiotics Antimicrob. Proteins* **2025**, 17. [CrossRef]
20. Prakoeswa, F.R.S.; Maharani, F.; Satria, Y.A.A.; Awanis, G.S.; Febrianty, A.F. Topical anti-aging agents: State-of-the-art review. *J. Med. Sci.* **2023**, *55*, 372–383. [CrossRef]
21. Milani, M.; Colombo, F. Skin anti-aging effect of oral vitamin a supplementation in combination with topical retinoic acid treatment in comparison with topical treatment alone: A randomized, prospective, Assessor-blinded, parallel trial. *Cosmetics* **2023**, *10*, 144. [CrossRef]
22. Mew, N.A.; Payan, I.; Daikhin, Y.; Nissim, I.; Nissim, I.; Tuchman, M.; Yudkoff, M. Effects of a single dose of N-carbamylglutamate on the rate of ureagenesis. *Mol. Genet. Metab.* **2009**, *98*, 325–330. [CrossRef]
23. Hu, Y.; Shao, D.; Wang, Q.; Xiao, Y.; Zhao, X.; Shen, Y.; Shan, Z.; Haibing, T.; Shi, S. Effects of dietary N-carbamylglutamate supplementation on growth performance, tissue development and blood parameters of yellow-feather broilers. *Poult. Sci.* **2019**, *98*, 2241–2249. [CrossRef]
24. Hu, N.; Mao, P.; Xiong, X.; Ma, Z.; Xie, Z.; Gao, M.; Wu, Q.; Ma, W. Effect of N-Carbamylglutamate Supplementation on Growth Performance, Jejunal Morphology, Amino Acid Transporters, and Antioxidant Ability of Weaned Pigs. *Animals* **2023**, *13*, 3183. [CrossRef]

25. Zhu, X.; Li, N.; Wang, Y.; Ding, L.; Chen, H.; Yu, Y.; Shi, X. Protective effects of quercetin on UVB irradiation-induced cytotoxicity through ROS clearance in keratinocyte cells. *Oncol. Rep.* **2017**, *37*, 209–218. [CrossRef]
26. Kong, Y.H.; Xu, S.P. Juglanin administration protects skin against UVB-induced injury by reducing Nrf2-dependent ROS generation. *Int. J. Mol. Med.* **2020**, *46*, 67–82. [CrossRef]
27. Kavinda, M.H.D.; Park, J.; Kim, N.; Choi, Y.H.; Kim, G.Y. Anti–melanogenic properties of FBCC–EP850 derived from Carex pumila Thunb. *Asian Pac. J. Trop. Biomed.* **2024**, *14*, 477–485. [CrossRef]
28. Pillai-Kastoori, L.; Schutz-Geschwender, A.R.; Harford, J.A. A systematic approach to quantitative Western blot analysis. *Anal. Biochem.* **2020**, *593*, 113608. [CrossRef] [PubMed]
29. Krutmann, J.; Schikowski, T.; Morita, A.; Berneburg, M. Environmentally-induced (extrinsic) skin aging: Exposomal factors and underlying mechanisms. *J. Investig. Dermatol.* **2021**, *141*, 1096–1103. [CrossRef]
30. Naharro-Rodriguez, J.; Bacci, S.; Hernandez-Bule, M.L.; Perez-Gonzalez, A.; Fernandez-Guarino, M. Decoding Skin Aging: A Review of Mechanisms, Markers, and Modern Therapies. *Cosmetics* **2025**, *12*, 144. [CrossRef]
31. He, X.; Gao, X.; Xie, W. Research progress in skin aging and immunity. *Int. J. Mol. Sci.* **2024**, *25*, 4101. [CrossRef]
32. Fu, C.; Chen, J.; Lu, J.; Yi, L.; Tong, X.; Kang, L.; Pei, S.; Ouyang, Y.; Jiang, L.; Ding, Y.; et al. Roles of inflammation factors in melanogenesis. *Mol. Med. Rep. Mol.* **2020**, *21*, 1421–1430. [CrossRef]
33. Thawabteh, A.M.; Jibreen, A.; Karaman, D.; Thawabteh, A.; Karaman, R. Skin pigmentation types, causes and treatment—A review. *Molecules* **2023**, *28*, 4839. [CrossRef] [PubMed]
34. Emanuelli, M.; Sartini, D.; Molinelli, E.; Campagna, R.; Pozzi, V.; Salvolini, E.; Simonetti, O.; Campanati, A.; Offidani, A. The double-edged sword of oxidative stress in skin damage and melanoma: From physiopathology to therapeutical approaches. *Antioxidants* **2022**, *11*, 612. [CrossRef] [PubMed]
35. Hossain, M.R.; Ansary, T.M.; Komine, M.; Ohtsuki, M. Diversified stimuli-induced inflammatory pathways cause skin pigmentation. *Int. J. Mol. Sci.* **2021**, *22*, 3970. [CrossRef]
36. Di Costanzo, L.; Scala, E.; Caiazzo, G.; Lembo, S.; Marino, R.; Megna, M.; Patri, A.; Di Caprio, R.; Balato, A. Possible role of BMP-4 in the hyper-pigmentation of psoriatic plaques after anti-TNF-α treatment. *Exp. Ther. Med.* **2019**, *18*, 4120–4124. [CrossRef]
37. Markiewicz, E.; Karaman-Jurukovska, N.; Mammone, T.; Idowu, O.C. Post-inflammatory hyperpigmentation in dark skin: Molecular mechanism and skincare implications. *Clin. Cosmet. Investig. Dermatol.* **2022**, *15*, 2555–2565. [CrossRef] [PubMed]
38. Bocheva, G.; Slominski, R.M.; Janjetovic, Z.; Kim, T.K.; Böhm, M.; Steinbrink, K.; Reiter, R.J.; Kleszczyński, K.; Slominski, A.T. Protective role of melatonin and its metabolites in skin aging. *Int. J. Mol. Sci.* **2022**, *23*, 1238. [CrossRef]
39. Fahad, D.; Mohammed, M.T. Oxidative stress: Implications on skin diseases. *Plant Arch.* **2020**, *20*, 4150–4157.
40. Papaccio, F.; Caputo, S.; Bellei, B. Focus on the contribution of oxidative stress in skin aging. *Antioxidants* **2022**, *11*, 1121. [CrossRef]
41. Sparavigna, A. Role of the extracellular matrix in skin aging and dedicated treatment-state of the art. *Plast. Aesthetic Res.* **2020**, *7*, 14. [CrossRef]
42. Xing, X.; Dan, Y.; Xu, Z.; Xiang, L. Implications of oxidative stress in the pathogenesis and treatment of hyperpigmentation disorders. *Oxid. Med. Cell. Longev.* **2022**, *2022*, 7881717. [CrossRef]
43. Ansary, T.M.; Hossain, M.R.; Kamiya, K.; Komine, M.; Ohtsuki, M. Inflammatory molecules associated with ultraviolet radiation-mediated skin aging. *Int. J. Mol. Sci.* **2021**, *22*, 3974. [CrossRef]
44. Loo, Y.C.; Hu, H.C.; Yu, S.Y.; Tsai, Y.H.; Korinek, M.; Wu, Y.C.; Chang, F.R.; Chen, Y.J. Development on potential skin anti-aging agents of Cosmos caudatus Kunth via inhibition of collagenase, MMP-1 and MMP-3 activities. *Phytomedicine* **2023**, *110*, 154643. [CrossRef] [PubMed]
45. Wan, J.; Zhang, G.; Li, X.; Qiu, X.; Ouyang, J.; Dai, J.; Min, S. Matrix metalloproteinase 3: A promoting and destabilizing factor in the pathogenesis of disease and cell differentiation. *Front. Physiol.* **2021**, *12*, 663978. [CrossRef] [PubMed]
46. Chung, Y.C.; Kim, M.J.; Kang, E.Y.; Kim, Y.B.; Kim, B.S.; Park, S.M.; Hyun, C.G. Anti-melanogenic effects of hydroxyectoine via mitf inhibition by jnk, p38, and akt pathways in b16f10 melanoma cells. *Nat. Prod. Commun.* **2019**, *14*, 1934578X19858523. [CrossRef]
47. Lim, J.W.; Ha, J.H.; Jeong, Y.J.; Park, S.N. Anti-melanogenesis effect of dehydroglyasperin C through the downregulation of MITF via the reduction of intracellular cAMP and acceleration of ERK activation in B16F1 melanoma cells. *Pharmacol. Rep.* **2018**, *70*, 930–935. [CrossRef]
48. Nguyen, N.T.; Fisher, D.E. MITF and UV responses in skin: From pigmentation to addiction. *Pigment. Cell Melanoma Res.* **2019**, *32*, 224–236. [CrossRef]
49. Zhang, Z.; Li, L.; Huang, G.; Zhou, T.; Zhang, X.; Leng, X.; Chen, Z.; Lin, J. Embelia Laeta aqueous extract suppresses acute inflammation via decreasing COX-2/iNOS expression and inhibiting NF-κB pathway. *J. Ethnopharmacol.* **2021**, *281*, 114575. [CrossRef]
50. Heo, Y.J.; Lee, N.; Choi, S.E.; Jeon, J.Y.; Han, S.J.; Kim, D.J.; Kang, Y.; Lee, K.W.; Kim, H.J. Amphiregulin Induces iNOS and COX-2 Expression through NF-κB and MAPK Signaling in Hepatic Inflammation. *Mediat. Inflamm.* **2023**, *2023*, 2364121. [CrossRef]

51. Xiao, L.; Cao, W.; Liu, G.; Fang, T.; Wu, X.; Jia, G.; Chen, X.; Zhao, H.; Wang, J.; Wu, C.; et al. Arginine, N-carbamylglutamate, and glutamine exert protective effects against oxidative stress in rat intestine. *Anim. Nutr.* **2016**, *2*, 242–248. [CrossRef] [PubMed]
52. Liu, G.; Xiao, L.; Cao, W.; Fang, T.; Jia, G.; Chen, X.; Zhao, H.; Wu, C.; Wang, J. Changes in the metabolome of rats after exposure to arginine and N-carbamylglutamate in combination with diquat, a compound that causes oxidative stress, assessed by 1 H NMR spectroscopy. *Food Funct.* **2016**, *7*, 964–974. [CrossRef] [PubMed]
53. Gu, F.F.; Jiang, L.Y.; Wang, D.M.; Zhao, F.Q.; Liu, J.X. Supplementation with N-carbamoylglutamate during the transition period improves the function of neutrophils and reduces inflammation and oxidative stress in dairy cows. *J. Dairy Sci.* **2022**, *105*, 5786–5795. [CrossRef] [PubMed]

cosmetics

MDPI

Article

Peony Root Extract Controls AGE–RAGE Interaction, Suppresses AGE Formation, and Reduces Skin Dullness

Kyoko Kanai, Kazal Boron Biswas, Asuka Hirasawa, Misaki Futamura, Kiyotaka Tanaka and Kotaro Sakamoto *

Ichimaru Pharcos Co., Ltd., 318-1 Asagi, Motosu 501-0475, Gifu, Japan; kanai-kyoko@ichimaru.co.jp (K.K.); kazal@ichimaru.co.jp (K.B.B.); hirasawa-asuka@ichimaru.co.jp (A.H.); futamura-misaki@ichimaru.co.jp (M.F.); ktanaka@ichimaru.co.jp (K.T.)

* Correspondence: sakamoto-kotaro@ichimaru.co.jp

Abstract: Skin dullness contributes to a fatigued and aged appearance, often exceeding one's biological age. It is a common dermatological concern influenced by aging and poor lifestyle habits, regardless of ethnicity or age. This study aimed to examine advanced glycation end products (AGEs) and their receptor (receptor for AGEs [RAGE]) as contributing factors to skin dullness. AGEs themselves have a yellowish hue, contributing to "yellow dullness." Additionally, AGE–RAGE signaling promotes melanin production in melanocytes and impairs keratinocyte differentiation as a result of inflammation. Therefore, regulating the AGE–RAGE interaction may help reduce skin dullness. Through screening various natural ingredients, we found that peony root extract (PRE) inhibits AGE formation and blocks AGE–RAGE binding. Furthermore, the presence of PRE leads to the suppression of AGE-induced melanin production in melanocytes and the restoration of impaired keratinocyte differentiation in glycated basement membrane components. In a human clinical study, topical application of a 1% PRE-containing lotion for 2 weeks significantly reduced melanin content, with a trend toward decreased AGE accumulation and visible spots on the cheeks. These findings support the potential of PRE as a multifunctional cosmetic ingredient that comprehensively addresses skin dullness by modulating the AGE–RAGE interaction.

Keywords: peony root extract; advanced glycation end products; receptor for AGEs; melanocyte; keratinocyte; skin dullness

1. Introduction

"Dullness" is a common dermatological concern, often characterized by an uneven skin tone, which negatively affects one's overall appearance [1]. Despite the development of cosmetic ingredients and final products aimed at reducing skin dullness and enhancing brightness, many individuals continue to struggle with dull skin. This persistence may be attributable to the multifactorial nature of skin dullness, thereby not depending on a single underlying mechanism [2]. Moreover, the contributing factors vary in balance among individuals, making it difficult to accurately identify the cause of their dullness and address it appropriately.

In this study, we focused on the multifaceted role of advanced glycation end products (AGEs) in skin dullness. AGEs are a collective term for products formed through nonenzymatic glycation reactions wherein reducing sugars bind to proteins or lipids [3]. Given that glycation is a nonenzymatic process, factors such as heat, excess sugar, and oxidative stress can accelerate its reaction rate. AGEs that accumulate in the body are broadly classified

https://doi.org/10.3390/cosmetics12040163

into those derived from external sources (e.g., food) and those produced endogenously. Approximately 10–30% of AGEs ingested through food are absorbed into the body, although a portion is subsequently excreted [4]. Endogenous AGEs are primarily formed in response to blood glucose levels and can accumulate in vascular endothelium, as well as in the eyes, kidneys, and skin [5]. Oxidative processes occur in the later stages of glycation, and age-related declines in antioxidant capacity contribute to AGE accumulation [6]. Glycated proteins undergo structural alterations that can cause function loss and increased resistance to degradation. In the dermis, long-lived proteins such as collagen and elastin are prone to glycation, leading to wrinkle formation and sagging. In the epidermis, AGEs themselves have a yellowish hue, which causes "yellow dullness" in the skin [7].

Recent studies have confirmed that AGEs contribute not only to yellow dullness but also to various other skin issues. AGEs activate their receptor (receptor for AGEs [RAGE]), by binding to it, thereby triggering various signaling pathways [8]. RAGE is a transmembrane protein expressed on the surface of epidermal keratinocytes, epidermal melanocytes, and dermal fibroblasts. RAGE-mediated responses are diverse. For example, it stimulates melanocytes to produce melanin [9]. Additionally, RAGE signaling induces inflammation [10]. Moreover, glycation in the cellular microenvironment suppresses keratinocyte differentiation [11], leading to impaired skin turnover. Given these backgrounds, both AGEs themselves and the RAGE-mediated signaling can be considered major contributing factors to skin dullness development.

We hypothesized that controlling AGE formation and RAGE-mediated signaling can mitigate skin dullness and promote a more radiant complexion. To this end, we screened natural ingredients capable of inhibiting the AGE–RAGE interaction and successfully identified the peony root extract (PRE) derived from the root of *Paeonia albiflora*. Although peonies are widely appreciated for their ornamental value because of their striking flowers and numerous cultivars, the root has long been used in traditional medicine for its anti-spasmodic and anti-inflammatory properties [12]. *Paeonia albiflora* roots are also rich in bioactive phytochemicals such as paeoniflorin and albiflorin, which have been reported to possess anti-inflammatory, antioxidant, and skin-conditioning activities. However, its potential to inhibit AGE formation or affect RAGE signaling has not yet been reported. In this study, we demonstrate the multifaceted antiglycation activity of PRE, its efficacy in reducing skin dullness, and its potential as a high-value cosmetic ingredient.

2. Materials and Methods

2.1. Preparation of the PRE

The roots of Japanese-grown *Paeonia albiflora* were purchased from Tochimoto Tenkaido Co., Ltd. (Osaka, Japan), and the roots of Chinese-grown *Paeonia albiflora* were obtained from Koshiro Co., Ltd. (Kyoto, Japan). Dried peony root was immersed in 15 times its weight of 50% 1,3-butylene glycol (BG) and extracted at room temperature for 1 week. The mixture was then filtered and allowed to stand at 30 °C for another 1 week. After a second filtration, the resulting extract was prepared so that the solid concentration was 2% *w*/*w*, and this was used as PRE. BG was selected as the extraction solvent due to its high skin compatibility, moisturizing properties, and widespread use in cosmetic formulations. Its ability to effectively extract glycosidic and polar constituents from peony roots made it a suitable choice for both laboratory extraction and subsequent clinical application.

2.2. Evaluation of the Inhibitory Activity of PRE Against the AGE–RAGE Interaction

Recombinant RAGE protein (BT4127; R&D Systems, Minneapolis, MN, USA) was dissolved in Dulbecco's phosphate-buffered saline (D-PBS), added to 96-well plates (Nunc, 439454, Thermo Fisher Scientific, Waltham, MA, USA) at 100 ng/50 μL/well concentration, and incubated overnight at 4 °C for coating. After blocking the wells with D-PBS containing 0.1% BSA (Jackson ImmunoResearch, West Grove, PA, USA, 001-000-162), we washed them with D-PBS containing 0.1% Tween 20 (washing buffer). PRE or a solvent control was prepared in blocking buffer and added to the wells, followed by incubation at room temperature for 1 h. After washing the wells with the washing buffer, we added biotin-labeled AGEs (BT4127; R&D Systems) prepared in blocking buffer and incubated them at room temperature for 1 h. Next, the wells were washed again, followed by color development using HRP-conjugated streptavidin (ab7403; Abcam, Cambridge, UK) and 1-Step™ Ultra TMB ELISA Substrate Solution (34028; Thermo Fisher Scientific, Waltham, MA, USA). To detect biotin-labeled AGEs bound to RAGE, we measured the absorbance at 450 nm. The AGE–RAGE binding rate was calculated by defining the absorbance of wells treated with solvent with RAGE coating as 100% binding and that of wells treated with solvent alone (without RAGE coating) as 0% binding. This study used paeoniflorin (CAS: 23180-57-6) (164-21731) and albifolorin (CAS: 39011-90-0) (016-22201), which were purchased from FUJIFILM WAKO (Osaka, Japan).

2.3. Evaluation of the Inhibitory Effect of PRE on Melanin Production by AGEs

Normal human epidermal melanocytes (FC-0030, P5; LIFELINE Cell Technology, Frederick, MD, USA) were seeded into 12-well plates and cultured in an MGM™-4 Melanocyte Growth Medium-4 BulletKit™ (MGM) (CC-3249; Lonza, Walkersville, MD, USA) until reaching 70% confluency. After replacing the culture medium with MGM that lacked fetal bovine serum and incubating it overnight, this medium was replaced with fresh medium containing PRE at 0%, 0.2%, or 0.4%. After 1 h of treatment, AGEs (22968; Cayman, Ann Arbor, MI, USA) were added to a final concentration of 400 μg/mL, and the cells were cultured for 6 days. Cell proliferation was assessed using the water-soluble tetrazolium salt 1 method with Cell Counting Kit-8 (DOJINDO, Kumamoto, Japan, 343-07623). Thereafter, the wells were washed with D-PBS, and the cells were detached and collected using trypsin. Ethanol/ether (3:1 mixture) was added to the collected samples, followed by centrifugation at 2000 rpm to obtain melanin pellets. The pellets were then dried overnight, resuspended in 25 μL of 1 N NaOH/10% dimethyl sulfoxide solution, and heated at 95 °C to disperse the melanin. Melanin content was quantified by measuring the absorbance at 420 nm. Based on preliminary cytotoxicity screening, PRE concentrations were limited to 0.4% or lower in cell-based assays to avoid non-specific effects.

2.4. Restorative Effects of Pretreatment on Keratinocyte Differentiation Defects in the Glycosylated Pseudobasal Layer of PRE

Laminin (iMatrix-332) (892031; Nippi, Tokyo, Japan) was dissolved in D-PBS to a final concentration of 4.2 $\mu g/cm^2$ and added to 12-well plates. The plates were incubated at 37 °C for 1 h to allow coating. After washing the wells with D-PBS, we added either D-PBS alone or D-PBS containing glucose (100 mg/mL) and incubated the plates at 37 °C for 2 weeks to induce laminin glycation. Normal human epidermal keratinocytes (NHEKs) (KK-4009, P3; KURABO, Osaka, Japan) were precultured in Keratinocyte Growth Medium (00195130; Lonza) until reaching 80% confluence. The cells were then detached and seeded onto glycated or nonglycated laminin-coated plates. Once confluence was achieved, the culture medium was replaced with a differentiation medium containing PRE at final concentrations of 0%, 0.2%, or 0.4% and 0.15 mM calcium and lacking transferrin, hydrocortisone, and epinephrine. After culturing at 37 °C for 6 h, we extracted total RNA by using an RNeasy Kit (74104; QIAGEN, Venlo, The Netherlands). According to the measured total RNA concentrations, cDNA was synthesized using RevaTra Ace qPCR RT Master Mix (FSQ201; TOYOBO, Shiga, Japan). Gene expression was then quantitatively analyzed using THUNDERBIRD Next SYBR qPCR Mix (TOYOBO, QPX201).

Target genes included the differentiation markers transglutaminase 1 (TGM1) and keratin 10 (K10). We used the housekeeping gene RPS18 as an internal control. Primers for TGM1 and K10 were obtained from Takara Bio (TGM1: HA171645, K10: HA347462). The primers RPS18 Forward (TTTGCGAGTACTCAACACCAACATC) and RPS18 Reverse (GAGCATATCTTCGGCCCACAC) were used. Based on preliminary cytotoxicity screening, PRE concentrations were limited to 0.4% or lower in cell-based assays to avoid non-specific effects.

2.5. Evaluation of the Inhibitory Effect of PRE on AGE Formation

Albumin and glucose were dissolved in 0.4 M phosphate buffer (pH 7.4) at final concentrations of 1 and 10 mg/mL, respectively. As a control, PRE or 50% BG was added to a final concentration of 1%, and the solution was then filter-sterilized. After incubation at 37 °C for 4 weeks, the autofluorescence derived from AGEs formed by the albumin–glucose reaction was measured (Ex: 370 nm, Em: 440 nm).

2.6. Clinical Studies in Humans

A randomized, single-blinded, split-face study was conducted to evaluate the clinical efficacy of PRE. Seventeen healthy male and female participants (9 males and 8 females; age range: 40–62 years; mean age: 50.5 ± 6.8 years) were enrolled after providing written informed consent. Inclusion criteria included individuals with visible facial dullness or pigmentation but otherwise healthy skin. Exclusion criteria included known allergies to cosmetics, ongoing dermatological treatments, or any acute or chronic skin diseases. They applied a lotion containing 1% PRE to one-half of the face and a placebo lotion to the other half, twice daily (morning and evening) for 2 weeks. At baseline and after 2 weeks, skin glycation levels on the cheeks were measured using the AGE Scanner (DiagnOptics Technologies B.V., Groningen, The Netherlands). Next, melanin content and facial spots were assessed using the Mexameter (Integral, Tokyo, Japan) and the VISIA imaging system (Integral), respectively.

The study was performed at Ichimaru Pharcos Co., Ltd. (Motosu, Japan), and ethical approval was granted by the company's Institutional Review Board (Approval Code: No. H24-001K, Approval Date: 25 October 2024). The study was conducted in accordance with the principles outlined in the Declaration of Helsinki (1975, revised in 2013).

2.7. Statistical Analysis

The melanocyte cell number, melanin content, and differentiation marker expression were statistically analyzed using Dunnett's test. For the human clinical trial, comparisons between baseline and two weeks posttreatment in each group were analyzed using the Wilcoxon signed-rank test. A p-value below 0.05 was considered statistically significant.

3. Results

3.1. PRE Inhibits the AGE–RAGE Interaction

Naturally derived ingredients capable of inhibiting the binding between AGEs and RAGE were screened using an ELISA-based binding assay. Among the 380 natural ingredients [13,14] evaluated, PRE showed to be a potent inhibitor. Figure 1 shows that PRE inhibits the binding of AGEs in a concentration-dependent manner. At 1% concentration, PRE derived from Japanese-grown *Paeonia albiflora* reduced AGE–RAGE binding by approximately 90% compared with the untreated group (Figure 1A). Similarly, PRE derived from Chinese-grown *Paeonia albiflora* exhibited approximately 70% inhibition at the same concentration (Figure 1B). Furthermore, similar inhibitory effects were confirmed using PREs obtained from different production regions within Japan and China, suggesting that the observed activity is reproducible and not limited to a single source. Since the wells were washed after PRE treatment and AGEs were added afterward in the binding assay, PRE acts on RAGE to inhibit its interaction with AGEs. The PRE contains paeoniflorin and albiflorin as constituent compounds [15], both of which are well-known phytochemicals in *Paeonia* roots and were thus selected as candidate active components for further evaluation. Therefore, the binding between AGEs and RAGE was evaluated in the presence of paeoniflorin (0.26, 1.3, and 6.5 μg/mL) and albiflorin (0.1, 0.5, 2.5 μg/mL). As a result, neither compound exhibited inhibitory activity even at the highest concentration, which was significantly higher than the levels present in PRE [15]. Thus, paeoniflorin and albiflorin are unlikely to be the active components responsible for the effects of PRE observed in this study.

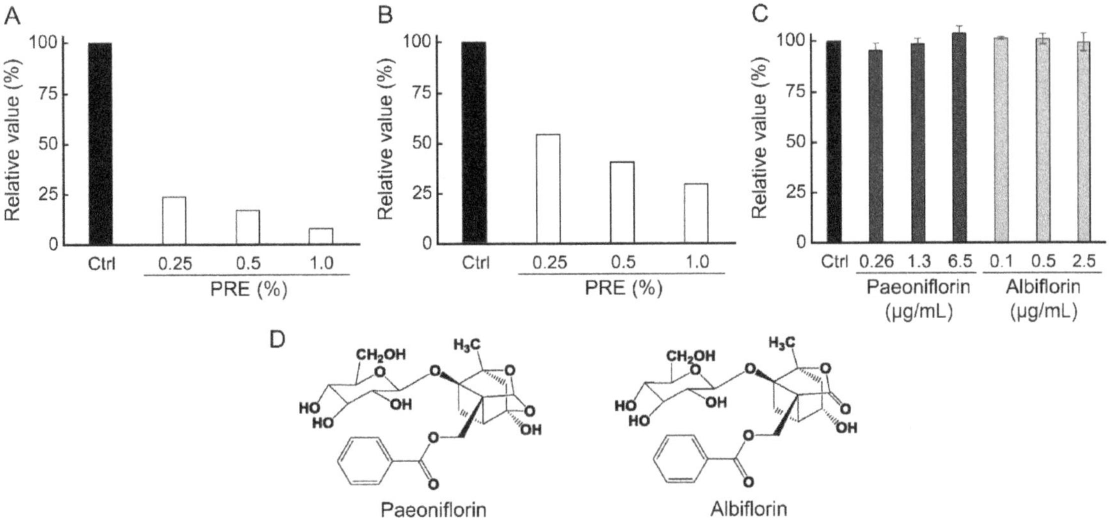

Figure 1. Inhibitory effect of PRE on advanced glycation end products–Receptor for AGEs interaction. AGE–RAGE interaction rate was evaluated using recombinant protein. The data are presented as percentages (*n* = 1), with the control (Ctrl) representing 100%. (**A**) Inhibition effect of PRE, which extracted from Japanese-grown *Paeonia albiflora*. (**B**) Inhibition effect of PRE, which extracted from Chinese-grown *Paeonia albiflora*. (**C**) Inhibition effect of the compounds (*n* = 3, mean ± SEM). According to Dunnett's test, none of the compounds showed statistically significant differences from the control at any tested concentration. (**D**) Chemical structures of paeoniflorin and albiflorin.

3.2. PRE Suppresses AGE-Induced Melanin Production

The effect of PRE, which inhibits the binding of AGEs to RAGE, on AGE-induced melanin production was evaluated (Figure 2). In normal human epidermal melanocytes (NHEM), the addition of AGEs resulted in a significant 37% increase in melanin production compared with the untreated control group. In the presence of PRE, melanin production of NHEM by AGEs inhibited without cytotoxicity. The effect was concentration-dependent, with 0.4% PRE reducing melanin production by 12%. These results indicate that AGEs promote melanin production in NHEMs, and that the presence of PRE suppresses AGE-induced melanin synthesis.

3.3. Pretreatment Restores the Differentiation Ability of Keratinocytes on Glycosylated Laminin 332

Laminin 332 is a key component of the epidermal basement membrane that aids in maintaining NHEK homeostasis in the basal layer. When basal layer components become glycated, their protein functions are impaired, negatively affecting keratinocyte differentiation. Suppressed keratinocyte differentiation can lead to abnormal keratinization and uneven skin turnover, ultimately resulting in uneven skin tone. As shown in Figure 3, the effects of glycated laminin 332 and PRE treatment on NHEK differentiation were evaluated using laminin 332–coated wells as a model of basal layer differentiation. In the glycated laminin group, the mRNA expression level of *K10*, a marker of early to midstage keratinocyte differentiation, was significantly reduced, while that of *TGM1*, a marker of terminal differentiation, also showed a decreasing trend. Once PRE was added, both the *K10* and *TGM1* mRNA expression levels showed a dose-dependent recovery, with *K10* expression being significantly increased in the 0.4% PRE-treated group compared with that in the glycated group. Therefore, laminin 332 glycation may suppress NHEK differentiation, and PRE treatment may restore it.

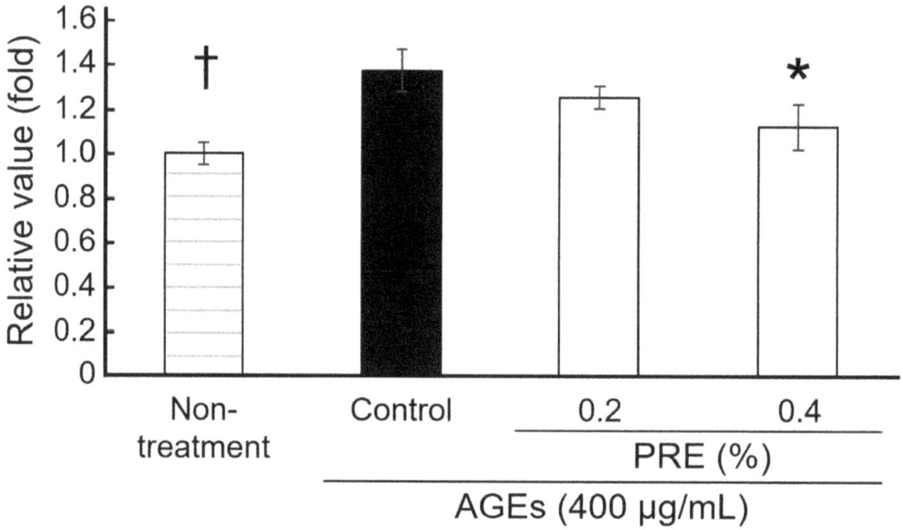

Figure 2. Effects of PRE on melanin production. Melanin contents of melanocytes were evaluated in the presence or absence of PRE ($n = 3$, mean ± SEM, * $p < 0.05$ by Dunnett's test vs. control, † $p < 0.05$ by Student t-test vs. control).

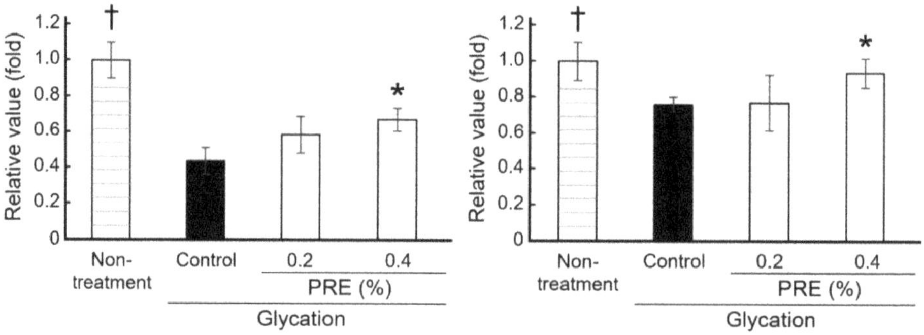

Figure 3. Effects of PRE on keratinocyte differentiation. The mRNA expression levels of *K10* (**left**) and *TGM1* (**right**) were evaluated in keratinocytes cultured on glycated laminin–coated plates (Ctrl; control, $n = 3$, mean ± SEM, * $p < 0.05$ by Dunnett's test vs. control, † $p < 0.05$ by Student t-test vs. control).

3.4. PRE Suppresses AGE Formation

The ability of PRE to suppress AGE formation was evaluated. After 4 weeks, the level of AGE-derived autofluorescence in the PRE-treated group was less than half that observed in the 0.5% BG control group (Figure 4). Therefore, PRE may not only suppress melanin production from melanocytes by modulating the AGE–RAGE interaction and restore keratinocyte differentiation (Figures 2 and 3) but also directly inhibit AGE formation.

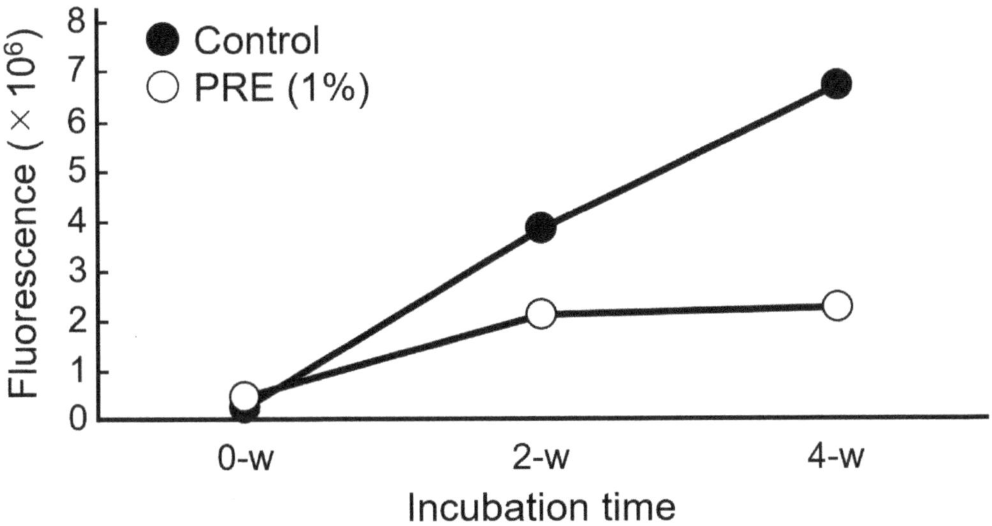

Figure 4. Anti-glycation effect of PRE in vitro. The AGE production level was measured by the autofluorescence of AGEs (n = 1).

3.5. PRE Exhibits Cosmetic Effects in Human Clinical Trials

In the human clinical trial, facial parameters such as skin glycation levels on the cheeks, melanin content, and skin pigmentation were measured using the AGE Scanner (DiagnOptics Technologies B.V.), Mexameter (Integral), and VISIA (Integral), respectively. The AGE Scanner measures the number of AGEs deposited approximately 1 mm beneath the skin surface by detecting their inherent autofluorescence. The cheek AGE levels in the group treated with the PRE-containing lotion showed a decreasing trend (Figure 5A). In the same group, melanin levels were significantly reduced (Figure 5B). Conversely, these two parameters demonstrated no significant changes in the placebo group. Regarding skin pigmentation, the number of dark spots tended to increase in the placebo group but decrease in the PRE-treated group (Figure 5C). Therefore, PRE may suppress the production of AGEs and melanin in the human skin, thereby reducing pigmentation spots and dullness and subsequently contributing to a more even and radiant complexion. No local irritation or adverse reactions were observed.

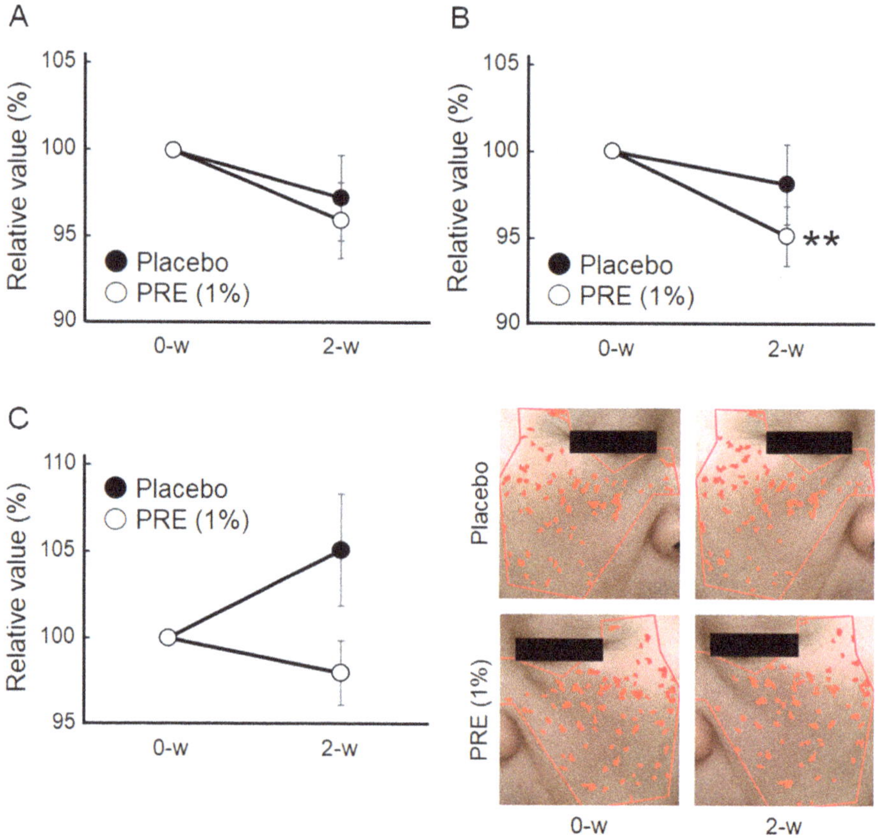

Figure 5. Cosmetic effects of PRE in vivo. (**A**) Impact of PRE on AGEs in the human skin. The amount of AGEs was measured using the AGE Scanner. (**B**) Impact of PRE on melanin production in the human skin. Melanin content was measured using the Mexameter. (**C**) Impact of PRE on pigmentation spots. The number of pigmentation spots was evaluated using VISIA. Representative pictures are shown. Areas identified as visible spots are indicated in red. Data are expressed as mean ± SEM (n = 17, ** $p < 0.01$ by Wilcoxon signed-rank test).

4. Discussion

This study identifies PRE, a naturally derived ingredient that blocks the interaction between AGEs and RAGE, as a novel cosmetic ingredient for mitigating skin dullness. PRE is known to exhibit various effects, including wound healing [16], anti-inflammatory activity [17], melanin production inhibition [18], and blood circulation promotion [19]. However, its effect on the AGE–RAGE interaction has not yet been documented. AGE accumulation is accelerated in UV-exposed skin, resulting in a negative loop that further activates RAGE signaling by AGEs and boosts melanin production [20], making it an increasingly important target in cosmetic product development. Currently, PRE's specific compounds that are responsible for inhibiting the AGE–RAGE interaction remain unknown. We fractionated PRE using a column packed with the synthetic adsorbent HP20 and evaluated the inhibitory activity of the resulting fractions; however, none of the fractions inhibited AGE–RAGE binding. These findings suggest that the active compound(s) may require synergistic interactions among multiple constituents, or that they may have been lost or inactivated during the fractionation process. Although HPLC-based fractionation or mass spectrometry analysis was not performed in the present study, further phytochemical investigation is warranted. Detailed chemical profiling, such as activity-guided HPLC fractionation followed by LC-MS/MS analysis, is considered an important direction for future research. Furthermore, paeoniflorin and albiflorin, which are representative compounds known to be present in PRE, showed no inhibitory activity against AGE–RAGE binding (Figure 1). Although paeoniflorin suppresses RAGE expression [21], our data suggest that neither of these compounds directly acts on RAGE. We also demonstrated that PRE suppresses AGE formation (Figure 4). Natural substances such as quercetin and epigallocatechin gallate reportedly inhibit AGE formation [22,23], but no studies demonstrated that these compounds are present in PRE. Polyphenols have also been reported to inhibit AGE formation and the AGE–RAGE interaction [24]. Given that PRE is rich in polyphenols [25], these compounds might be responsible for the observed inhibitory activities of PRE.

Moreover, we found that PRE suppresses melanin production induced by AGEs (Figure 2). Paeoniflorin inhibits melanin production triggered by α-MSH stimulation [26]. In this mechanism, cAMP response element–binding protein (CREB) activation is suppressed, and tyrosinase and the tyrosinase-related proteins 1 and 2 are downregulated. Conversely, AGEs promote melanin production through a different signaling pathway, triggering downstream ERK phosphorylation via RAGE and subsequently activating CREB [9]. Thus, the entry points of the signaling cascades differ between α-MSH and AGEs. PRE suppresses both the α-MSH and RAGE pathways; thus, it may provide a more robust inhibition of melanin production.

Although RAGE signaling suppression through the application of a lotion containing 1% PRE in human clinical trials has not yet been confirmed, the observed reductions in skin AGE levels and melanin content are nonetheless significant (Figure 5). The reduction in AGE levels might be caused by the inhibitory effect of PRE on AGE formation, whereas the reduction in melanin content might be attributed to the suppression of RAGE signaling by PRE. Given that AGEs are metabolized slowly, their removal involves phagocytosis by anti-inflammatory M2 macrophages [27]. Furthermore, RAGE knockout leads to the polarization of macrophages to the M2 type [28]. Based on these reports and our results, we hypothesize that PRE possibly suppresses RAGE signaling, thereby influencing macrophage polarization and contributing to reduced AGE accumulation. However, it should be emphasized that this remains speculative in the absence of direct experimental evidence, such as analysis of RAGE expression or downstream signaling (e.g., ERK/CREB phosphorylation). Future studies, including in vitro assays, are necessary to directly clarify the effects of PRE on RAGE modulation. This observation represents an important direction for future investigation.

5. Conclusions

This study is the first to identify that PRE possesses the abilities of inhibiting the AGE–RAGE interaction and suppressing AGE formation. Additionally, human clinical trials have demonstrated that PRE has a cosmetic effect of reducing skin dullness. Given that AGEs are waste products associated with lifestyle factors and aging, PRE can potentially improve AGE-induced skin dullness across various ethnicities. Therefore, PRE is a comprehensive cosmetic ingredient that mitigates skin dullness, offering a promising novel approach that addresses the unmet needs of dullness care.

Author Contributions: K.K. conducted data curation, investigation, methodology, writing—original draft; K.B.B. conducted investigation and writing—original draft; A.H. and M.F. conducted data curation; K.T. conducted investigation and supervision; and K.S. conducted investigation, writing—original draft, writing—review and editing. All authors have read and agreed to the published version of the manuscript.

Funding: This research received no external funding.

Institutional Review Board Statement: The study was conducted in accordance with the Declaration of Helsinki, and approved by the Institutional Review Board of Ichimaru Pharcos Co., Ltd. (Approval Code: No. H24-001K, Approval Date: 25 October 2024).

Informed Consent Statement: Informed consent was obtained from all subjects involved in the study.

Data Availability Statement: All data generated for this study are contained in the manuscript. Raw data are available from the corresponding author upon reasonable request.

Conflicts of Interest: Kyoko Kanai, Kazal Boron Biswas, Asuka Hirasawa, Misaki Futamura, Kiyotaka Tanaka, and Kotaro Sakamoto are employees of Ichimaru Pharcos Co., Ltd. The authors declare that the research was conducted in the absence of any commercial or financial relationships that could be construed as a potential conflict of interest. The authors affirm that their affiliations did not influence the scientific integrity of the article.

Abbreviations

The following abbreviations are used in this manuscript:

AGE	advanced glycation end product
BG	1,3-butylene glycol
K10	keratin 10
MGM	MGM™-4 Melanocyte Growth Medium-4 BulletKit™
NHEK	normal human epidermal keratinocytes
PRE	peony root extract
RAGE	receptor of AGE
TGM1	transglutaminase 1

References

1. De Rigal, J.; Des Mazis, I.; Diridollou, S.; Querleux, B.; Yang, G.; Leroy, F.; Barbosa, V.H. The effect of age on skin color and color heterogeneity in four ethnic groups. *Ski. Res. Technol.* **2010**, *16*, 168–178. [CrossRef] [PubMed]
2. Nurani, A.M.; Kikuchi, K.; Iino, M.; Shirasugi, Y.; Sonoki, A.; Fujimura, T.; Hasegawa, K.; Shibata, T. Development of a method for evaluating skin dullness: A mathematical model explaining dullness by the color, optical properties, and microtopography of the skin. *Ski. Res. Technol.* **2023**, *29*, e13407. [CrossRef] [PubMed]
3. Chen, C.Y.; Zhang, J.Q.; Li, L.; Guo, M.M.; He, Y.F.; Dong, Y.M.; Meng, H.; Yi, F. Advanced glycation end products in the skin: Molecular mechanisms, methods of measurement, and inhibitory pathways. *Front. Med.* **2022**, *9*, 837222. [CrossRef]
4. Koschinsky, T.; He, C.J.; Mitsuhashi, T.; Bucala, R.; Liu, C.; Buenting, C.; Heitmann, K.; Vlassara, H. Orally absorbed reactive glycation products (glycotoxins): An environmental risk factor in diabetic nephropathy. *Proc. Natl. Acad. Sci. USA* **1997**, *94*, 6474–6479. [CrossRef]
5. Zheng, D.L.; Wu, Q.R.; Zeng, P.; Li, S.M.; Cai, Y.J.; Chen, S.Z.; Luo, X.S.; Kuang, S.J.; Rao, F.; Lai, Y.Y.; et al. Advanced glycation end products induce senescence of atrial myocytes and increase susceptibility of atrial fibrillation in diabetic mice. *Aging Cell* **2022**, *21*, e13734. [CrossRef]
6. Dyer, D.G.; Dunn, J.A.; Thorpe, S.R.; Bailie, K.E.; Lyons, T.J.; McCance, D.R.; Baynes, J.W. Accumulation of Maillard reaction products in skin collagen in diabetes and aging. *J. Clin. Investig.* **1993**, *91*, 2463–2469. [CrossRef] [PubMed]
7. Fang, B.; Li, L.; Winget, J.; Laughlin, T.; Hakozaki, T. Identification of yellow advanced glycation end products in human skin. *Int. J. Mol. Sci.* **2024**, *25*, 5596. [CrossRef]
8. Stern, D.; Du Yan, S.; Yan, S.F.; Schmidt, A.M. Receptor for advanced glycation endproducts: A multiligand receptor magnifying cell stress in diverse pathologic settings. *Adv. Drug Deliv. Rev.* **2002**, *54*, 1615–1625. [CrossRef]
9. Lee, E.J.; Kim, J.Y.; Oh, S.H. Advanced glycation end products (AGEs) promote melanogenesis through receptor for AGEs. *Sci. Rep.* **2016**, *6*, 27848. [CrossRef]
10. Weinhage, T.; Wirth, T.; Schütz, P.; Becker, P.; Lueken, A.; Skryabin, B.V.; Wittkowski, H.; Foell, D. The receptor for advanced glycation endproducts (rage) contributes to severe inflammatory liver injury in mice. *Front. Immunol.* **2020**, *11*, 1157. [CrossRef]
11. Radziszewski, M.; Galus, R.; Łuszczyński, K.; Winiarski, S.; Wąsowski, D.; Malejczyk, J.; Włodarski, P.; Ścieżyńska, A. The RAGE pathway in skin pathology development: A comprehensive review of its role and therapeutic potential. *Int. J. Mol. Sci.* **2024**, *25*, 13570. [CrossRef]

12. He, D.Y.; Dai, S.M. Anti-inflammatory and immunomodulatory effects of *Paeonia lactiflora* pall., a traditional Chinese herbal medicine. *Front. Pharmacol.* **2011**, *2*, 10. [CrossRef]
13. Sakamoto, K.; Watanabe, C.; Masutani, T.; Hirasawa, A.; Wakamatsu, K.; Iddamalgoda, A.; Kakumu, Y.; Yamauchi, K.; Mitsunaga, T. *Arnica montana* L. extract containing 6-O-methacryloylhelenalin and 6-O-isobutyrylhelenalin accelerates growth and differentiation of human subcutaneous preadipocytes and leads volumizing of skin. *Int. J. Cosmet. Sci.* **2023**, *45*, 1–13. [CrossRef]
14. Sakamoto, K.; Fujimoto, R.; Nakagawa, S.; Kamiyama, E.; Kanai, K.; Kawai, Y.; Kojima, H.; Hirasawa, A.; Wakamatsu, K.; Masutani, T. Juniper berry extract containing anthricin and yatein suppresses lipofuscin accumulation in human epidermal keratinocytes through proteasome activation, increases brightness and decreases spots in human skin. *Int. J. Cosmet. Sci.* **2023**, *45*, 655–671. [CrossRef]
15. He, C.N.; Peng, Y.; Zhang, Y.C.; Xu, L.J.; Gu, J.; Xiao, P.G. Phytochemical and biological studies of paeoniaceae. *Chem. Biodivers.* **2010**, *7*, 805. [CrossRef] [PubMed]
16. Malviya, N.; Jain, S. Wound healing activity of aqueous extract of Radix paeoniae root. *Acta Pol. Pharm.* **2009**, *66*, 543–547.
17. Lee, B.; Shin, Y.W.; Bae, E.A.; Han, S.J.; Kim, J.S.; Kang, S.S.; Kim, D.H. Antiallergic effect of the root of *Paeonia lactiflora* and its constituents paeoniflorin and paeonol. *Arch. Pharm. Res.* **2008**, *31*, 445–450. [CrossRef]
18. Lin, D.; Wang, S.H.; Song, T.Y.; Hsieh, C.W.; Tsai, M.S. Safety and efficacy of tyrosinase inhibition of *Paeonia suffruticosa* Andrews extracts on human melanoma cells. *J. Cosmet. Dermatol.* **2019**, *18*, 1921–1929. [CrossRef]
19. Goto, H.; Shimada, Y.; Akechi, Y.; Kohta, K.; Hattori, M.; Terasawa, K. Endothelium-dependent vasodilator effect of extract prepared from the roots of *Paeonia lactiflora* on isolated rat aorta. *Planta Med.* **1996**, *62*, 436–439. [CrossRef] [PubMed]
20. Zgutka, K.; Tkacz, M.; Tomasiak, P.; Tarnowski, M. A role for advanced glycation end products in molecular ageing. *Int. J. Mol. Sci.* **2023**, *24*, 9881. [CrossRef] [PubMed]
21. Song, S.; Xiao, X.; Guo, D.; Mo, L.; Bu, C.; Ye, W.; Den, Q.; Liu, S.; Yang, X. Protective effects of paeoniflorin against AOPP-induced oxidative injury in HUVECS by blocking the ROS-HIF-1α/VEGF pathway. *Phytomedicine* **2017**, *34*, 115–126. [CrossRef]
22. Li, X.; Zheng, T.; Sang, S.; Lv, L. Quercetin inhibits advanced glycation end product formation by trapping methylglyoxal and glyoxal. *J. Agric. Food Chem.* **2014**, *62*, 12152–12158. [CrossRef]
23. Poojary, M.M.; Zhang, W.; Olesen, S.B.; Rauh, V.; Lund, M.N. Green tea extract decreases Arg-derived advanced glycation endproducts but not Lys-derived AGEs in UHT milk during 1-year storage. *J. Agric. Food Chem.* **2020**, *68*, 14261–14273. [CrossRef]
24. González, I.; Morales, M.A.; Rojas, A. Polyphenols and AGEs/RAGE axis. Trends and challenges. *Food Res. Int.* **2020**, *129*, 108843. [CrossRef] [PubMed]
25. Huang, Z.; Yang, J.; Shen, L.; Wu, L.; Wang, C.; Liu, Y. The innovative extraction and purification process of insoluble polyphenols from *Paeonia ostii* roots: Optimum study and in vitro activities. *Process Biochem.* **2024**, *142*, 13–23. [CrossRef]
26. Wen, S.Y.; Wu, Y.S.; Liu, H.; Ng, S.C.; Padma, V.V.; Huang, C.Y.; Kuo, W.W. Paeoniflorin found in *Paeonia lactiflora* root extract inhibits melanogenesis by regulating melanin-related signal transduction in B16F10 cells. *J. Cosmet. Dermatol.* **2023**, *22*, 2824–2830. [CrossRef] [PubMed]
27. Madsen, D.H.; Leonard, D.; Masedunskas, A.; Moyer, A.; Jürgensen, H.J.; Peters, D.E.; Amornphimoltham, P.; Selvaraj, A.; Yamada, S.S.; Brenner, D.A.; et al. M2-like macrophages are responsible for collagen degradation through a mannose receptor–mediated pathway. *J. Cell Biol.* **2013**, *202*, 951–966. [CrossRef]
28. Osonoi, S.; Mizukami, H.; Takeuchi, Y.; Sugawa, H.; Ogasawara, S.; Takaku, S.; Sasaki, T.; Kudoh, K.; Ito, K.; Sango, K.; et al. RAGE activation in macrophages and development of experimental diabetic polyneuropathy. *JCI Insight* **2022**, *7*, e160555. [CrossRef]

cosmetics

MDPI

Article

The Effectiveness of a Topical Rosehip Oil Treatment on Facial Skin Characteristics: A Pilot Study on Wrinkles, UV Spots Reduction, Erythema Mitigation, and Age-Related Signs

Diana Patricia Oargă (Porumb) [1,†], Mihaiela Cornea-Cipcigan [2,*,†], Silvia Amalia Nemeș [3] and Mirela Irina Cordea [1,*]

[1] Faculty of Horticulture and Business in Rural Development, University of Agricultural Sciences and Veterinary Medicine Cluj-Napoca, Manastur 3-5, 400372 Cluj-Napoca, Romania; diana-patricia.oarga@usamvcluj.ro

[2] Faculty of Veterinary Medicine, University of Agricultural Sciences and Veterinary Medicine Cluj-Napoca, Manastur 3-5, 400372 Cluj-Napoca, Romania

[3] Faculty of Food Science and Technology, University of Agricultural Sciences and Veterinary Medicine Cluj-Napoca, Manastur 3-5, 400372 Cluj-Napoca, Romania; amalia.nemes@usamvcluj.ro

* Correspondence: mihaiela.cornea@usamvcluj.ro (M.C.-C.); mcordea@usamvcluj.ro (M.I.C.)

† These authors contributed equally to this work.

Abstract: Skin aging is a complex process influenced by several factors, including UV exposure, environmental stressors, and lifestyle choices. The demand for effective, natural skincare products has driven research into plant-based oils rich in bioactive compounds. Rosehip oil has garnered attention for its high content of carotenoids, phenolics, and antioxidants, which are known for their anti-aging, photoprotective, and skin-rejuvenating properties. Despite the growing interest in rosehip oil, limited studies have investigated its efficacy on human skin using advanced imaging technologies. This study aims to fill this gap by evaluating the efficacy of cold-pressed *Rosa canina* seed oil on facial skin characteristics, specifically wrinkles, ultraviolet (UV) spot reduction, and erythema mitigation, using imaging technologies (the VISIA analysis system). Seed oil pressed from *R. canina* collected from the Băișoara area of Cluj County has been selected for this study due to its high carotenoid, phenolic, and antioxidant contents. The oil has also been analyzed for the content of individual carotenoids (i.e., lutein, lycopene, β Carotene, and zeaxanthin) using HPLC-DAD (High-Performance Liquid Chromatography—Diode Array Detector), along with lutein and zeaxanthin esters and diesters. After the preliminary screening of multiple Rosa species for carotenoid, phenolic, and antioxidant contents, the *R. canina* sample with the highest therapeutic potential was selected. A cohort of 27 volunteers (aged 30–65) underwent a five-week treatment protocol, wherein three drops of the selected rosehip oil were topically applied to the face daily. The VISIA imaging was conducted before and after the treatment to evaluate changes in skin parameters, including the wrinkle depth, UV-induced spots, porphyrins, and texture. Regarding the bioactivities, rosehip oil showed a significant total carotenoids content (28.398 μg/mL), with the highest levels in the case of the β-carotene (4.49 μg/mL), lutein (4.33 μg/mL), and zexanthin (10.88 μg/mL) contents. Results indicated a significant reduction in mean wrinkle scores across several age groups, with notable improvements in individuals with deeper baseline wrinkles. UV spots also showed visible declines, suggesting ideal photoprotective and anti-pigmentary effects attributable to the oil's high vitamin A and carotenoid content. Porphyrin levels, often correlated with bacterial activity, decreased in most subjects, hinting at an additional antimicrobial or microbiome-modulatory property. However, skin responses varied, possibly due to individual differences in skin sensitivity, environmental factors, or compliance with sun protection. Overall, the topical application of *R. canina* oil appeared to improve the facial skin quality, reduce the appearance of age-related markers, and support skin health. These findings reinforce the potential use of rosehip oil in anti-aging skincare formulations. Further long-term, large-scale studies are warranted to refine dosing regimens, investigate mechanisms of action, and explore synergistic effects with other bioactive compounds.

Keywords: *Rosa canina*; skin complexion; hyperpigmentation; skin aging; topical treatment

1. Introduction

In recent years, the field of aging research has evolved significantly, bringing to the forefront advanced technologies that allow a detailed and accurate assessment of skin quality [1]. The VISIA system's analysis of the skin provides insights into various parameters such as texture, wrinkles, blemishes, and pores, thus contributing to the understanding of the mechanisms of aging and the effectiveness of treatments. The VISIA analysis system, developed by Canfield Scientific Inc., provides assessments of skin conditions through multi-spectral and 3D imaging and allows the efficacy of applied treatments to be evaluated.

Bioactive compounds present in *Rosa canina* oil are widely recognized for their role in skin protection and regeneration, mitigating oxidative damage and preventing premature aging [2–4]. *Rosa canina* L., commonly known as rosehip, is a widespread species in Europe and Asia, recognized for its nutrient-rich pseudofruits. The oil extracted from rosehip seeds has gained significant cosmetic and therapeutic attention due to its high concentration of bioactive compounds. These compounds include vitamin A, carotenoids, polyphenols, and polyunsaturated fatty acids, known for their remarkable antioxidant properties [2,3]. *R. canina* oil's most notable characteristic is its extremely high vitamin A content. Studies have shown that rosehip (fruits) contain vitamin C concentrations of up to 1252.3 mg/100 g fresh weight, thus surpassing other vitamin C-rich fruits, such as sea buckthorn (*Hippophae rhamnoides*) and black elder (*Sambucus nigra*) [5,6]. Subsequently, the abundance in vitamin A contributes significantly to the antioxidant capacity of the oil, protecting the skin from free radicals and oxidative damage caused by UV exposure [4]. In addition to vitamin C, *Rosa canina* is a natural source of phenolic compounds and carotenoids. Phenolic compounds, including quercetin and its derivatives, provide antioxidant and anti-inflammatory benefits, contributing to skin health and protections against premature aging [7]. Carotenoids, a prominent class of plant pigments, are responsible for the characteristic colors of various fruits and vegetables and have been extensively documented to exhibit antioxidant activity [8]. The antioxidant properties of rosehip oils are a crucial factor in determining their stability and suitability for medical applications. Studies have reported varying total carotenoid contents ranging from 36.4 to 107.7 mg/kg in oil samples from Poland [9] and 86.3 mg.kg in samples from Turkey [10]. Notably, the extraction technique employed has been shown to significantly influence the carotenoid content of rosehip oils [11], highlighting the importance of optimized extraction methods in preserving the bioactive compounds. A comprehensive study by Fromm et al. (2012) reported a total carotenoid content of 39.15 mg/kg, with β-carotene being the predominant carotenoid (9.28 mg/kg), accounting for approximately 24% of the total carotenoid content [12]. Notably, the study also detected appreciable amounts of all-trans isomers of lutein, zeaxanthin, rubixanthin, and lycopene [12]. In terms of a dermatological assessment, carotenoids, including lycopene and β-carotene, not only protect against oxidative stress, but are also involved in collagen synthesis, which is essential for maintaining skin elasticity and firmness [13].

The present study evaluated the topical application of rosehip oil for its beneficial effects on skin conditions, specifically in terms of wrinkle reduction, the management of UV spots, and the elimination of porphyrins. Each of these outcomes can be explained through plausible mechanistic pathways linked to the bioactive components of rosehip oil, which include enhancing collagen synthesis, inhibiting melanin formation, and exhibiting antibacterial effects [3,14,15]. One of the primary mechanisms by which rosehip oil reduces wrinkles is through the promotion of collagen synthesis. Collagen is an essential protein in the skin that maintains its structure and elasticity. The rich composition of rosehip oil, particularly its high concentration of polyunsaturated fatty acids (PUFAs) like linoleic and alpha-linolenic acids, supports the dermal structure and promotes fibroblast activity, which is crucial for collagen production [16,17]. Rosehip oil can activate type III collagen and accelerate collagen synthesis, leading to improved wound contraction and repair. This process is likely facilitated by the antioxidant properties of rosehip oil, attributed to its vitamin A content and phenolic compounds, which help protect skin cells from oxidative stress that can damage collagen fibers, supporting overall skin integrity and elasticity. Furthermore, studies have shown that rosehip oil has demonstrated wound-healing properties due to its richness in fatty acids, particularly linoleic, linolenic, and oleic acids [18]. However, the absorption of these compounds may be limited due to their relatively high molecular weight (>500 Da). An FTIR analysis of skin treated with *R. rubiginosa* oil did not detect vibration transitions characteristic of the oil, suggesting that it remains on the skin surface, which is consistent with the proposition that high molecular weight molecules may not readily penetrate the skin. Nevertheless, the topical application of *R. rubiginosa* fixed oil may still provide benefits through an occlusive effect, creating a protective barrier on the skin surface that prevents moisture loss and reduces transepidermal water loss [19,20]. In this regard, the study conducted by der Walt (2017) involved the formulation and preparation of two oil-in-water emulsions (*o*/*w*) [15]. The results revealed significant physical and chemical changes in both formulations, exceeding the 5% threshold in key parameters, such as active ingredients, viscosity, and conductivity. Clinical studies showed that both formulations (20% and 100% rosehip seed oil) effectively improved skin hydration levels. Additionally, both products demonstrated beneficial effects for reducing wrinkles and enhancing skin viscoelasticity [15]. In a different study, the use of chitosan films with rosehip oil-loaded nanocapsules as a future prospective application in skincare regimes has been evaluated. It was revealed that nanoencapsulation protects the oil from UV-induced oxidation, and the nanocapsules incorporated into a chitosan gel and film show promise for topical applications in dermatology and cosmetics [19].

The effectiveness of rosehip oil in managing UV spots is primarily linked to its ability to inhibit melanin formation. Melanin is produced by melanocytes in response to UV radiation, and an excess can result in pigmentation irregularities, including spots. Studies have demonstrated that rosehip oil can reduce the activity of key enzymes involved in melanin synthesis, such as tyrosinase, thereby decreasing melanin production [17,20]. This effect is further enhanced by the oil's antioxidant properties, which counteract oxidative stress-induced melanin production, providing a dual mechanism in addressing UV-induced pigmentation [21]. Additionally, its anti-inflammatory properties may help alleviate conditions, such as post-inflammatory hyperpigmentation, that arise from UV damage [22].

The presence of porphyrins, often related to acne and other dermal issues, has been shown to be effectively managed with rosehip oil due to its antibacterial properties. The oil contains various phytochemicals that have demonstrated antimicrobial efficacy, including the ability to inhibit bacterial growth, which is critical in reducing skin lesions associated with porphyrins [23]. The antimicrobial action helps clear bacteria that might otherwise exacerbate inflammatory responses or contribute to acne development, promoting healthier skin overall. By controlling bacterial populations on the skin surface, rosehip oil may also indirectly support the prevention of porphyrin accumulation, leading to clearer skin and reduced lesion formation [24].

Therefore, the topical application of rosehip oil proves to be associated with multiple mechanistic pathways that support skin health. Its ability to enhance collagen synthesis contributes to anti-wrinkle effects, while its capacity to inhibit melanin formation addresses UV spots. Finally, its antibacterial properties play a significant role in managing porphyrins. The integration of these actions makes rosehip oil a versatile therapeutic option for various skin conditions.

The present pilot study focuses on the analysis using the VISIA system to assess the outcome of using rosehip oil as topical treatment to alleviate several skin characteristics, including color-related (i.e., red areas, spots, brown, and UV spots) and perception-related characteristics (i.e., texture, wrinkles, pores, and porphyrins). Furthermore, this study aims to assess the potential of the studied *Rosa canina* oil to improve the appearance and health of the skin, supporting its use in skin care products and anti-aging treatments. To the best of our knowledge, this is the first study to comprehensively investigate the effects of a topical *R. canina* seed oil treatment on various skin characteristics. By utilizing a systematic and quantitative approach, this pilot study aims to provide novel insights into the benefits of rosehip oil for skin health, shedding light on its potential as a natural and effective skincare solution. The findings contribute to the growing body of evidence supporting the use of natural products in dermatology and esthetics and pave the way for future research into the therapeutic applications of rosehip oil.

2. Materials and Methods

2.1. Seed Material

The fruits of *Rosa canina* (Municipality Baisoara, Cluj county, Romania) were carefully harvested to avoid damage. After harvest, the seeds were extracted, washed, and air-dried to ensure complete removal of impurities, an essential step in order to obtain an oil free from contaminants that could affect its properties. The seeds were subjected to a controlled drying process to reduce internal moisture, an important step for efficient oil pressing. Drying was carried out in an oven at temperature below 40 °C, to prevent degradation of heat-sensitive bioactive components, according to [25,26]. Under-drying the seeds can lead to the growth of molds and bacteria, while over-drying can lead to the loss of valuable nutrients. After complete drying, the seeds were ground to a fine powder using a grinding mill (Grindomix®, GM 200—Retsch Gmbh, Haan, Germany). The particle size obtained by grinding increases the contact surface area of the plant material, thus facilitating complete oil extraction during pressing. The powder obtained must be uniform to ensure efficient pressing and maximum oil yield.

2.2. Cold Seed Pressing and Oil Filtration

The seed powder was placed in a cold press where mechanical pressure was applied to extract the oil. The cold-pressing process was carried out at temperatures between 24 and 28 °C to reduce as much as possible the loss of bioactive components, such as essential fatty acids and antioxidants. Cold-pressing prevents the denaturation of these heat-sensitive compounds, ensuring high quality oil with therapeutic properties. The crude oil, extracted by pressing, underwent a filtration process to remove remaining solids and impurities. Filtration was carried out using nylon filters which do not react with the oil, ensuring its clarity and purity. This step is essential to obtain a high-quality end product suitable for therapeutic and cosmetic uses [27]. The filtered oil was stored in opaque brown glass containers to prevent oxidation caused by exposure to light. The containers were kept in a cool and dark place at temperatures between 18 and 22 °C and away from humidity.

2.3. Determination of Total Phenolic Compounds (TPC) Content

To determine the content of phenolic compounds in rosehip oil, a multi-step structured extraction protocol was followed. Initially, 3 mL of oil was mixed with 3 mL of hexane and vortexed to ensure uniform dispersion of the substances. Next, 5 mL of methanol/water solution was added in a 3:2 ratio to facilitate solubilization of the phenolic compounds. The resulting mixture was subjected to treatment in an ultrasonic bath for 15 min, followed by centrifugation at 25 °C for 10 min to separate the phases [28]. The sonication and centrifugation process was repeated 23 times, each time adding 3 mL of fresh hexane to maximize the extraction of the compounds. Finally, the oil samples were subjected to an evaporation step to remove residual solvents and concentrate the phenolic compounds. The concentrated residue was dissolved in 1 mL of methanol and filtered through a nylon filter with a porosity of 0.45 μm to obtain a clear extract, optimal for further analysis.

Total phenolic content (TPC) was assessed using the Folin–Ciocalteu method [29]. For this purpose, 25 μL of the sample was taken and mixed with 1.515 mL of distilled water in a 24-well microplate. Each extract was mixed with 120 μL of Folin–Ciocalteu reagent (0.2 N) and left at room temperature for 5 min. Subsequently, 340 μL of 7.5% (*w*/*v*) Na2CO3 solution was added to create primary conditions (pH~10) for the redox reaction between the phenolic compounds and Folin–Ciocalteu reagent. The resulting solution was incubated in the dark at 25 °C for 30 min. Methanol was used as a control, and absorbances were measured at 750 nm using a microplate reader (BioTek Instruments, Winooski, VT, USA). Gallic acid (0.013–1.00 mg/mL) was used to create the standard curve, and TPC in the samples was expressed as gallic acid equivalents (GAEs) in mg/100 g of plant material. The analysis was performed in triplicate.

2.4. Antioxidant Activity

The antioxidant activity of the samples was determined using the DPPH (1,1-diphenyl-2-picrylhydrazyl) free radical neutralizing capacity technique developed by [30]. To determine the antioxidant response of the samples studied, 35 μL of oil samples previously extracted in methanol was prepared in triplicates, mixed with 250 μL methanolic DPPH solution. The reaction solution was incubated for 30 min at room temperature in the dark before measuring the absorbance at 515 nm using a multi-mode plate reader (BioTek, Winuschi, Winuschi, VT, USA). The results were presented as Trolox equivalents, (μmol TE)/100 g sample.

2.5. Determination of Individual Carotenoids via HPLC-DAD Analysis

In order to determine the individual carotenoids, the standard chromatogram was first performed. For this purpose, the oil sample was dissolved in hexane in 1/1 (v/v) ratio, filtered through a Chromafil Xtra nylon 0.45 μm and 20 μL Chromafil Xtra nylon 0.45 μm filter, then injected into the HPLC system (High-Performance Liquid Chromatography). Agilent 1200 series HPLC system was equipped with degasser for solvents, quaternary pumps, DAD (Diode Array Detector). and automatic injector (Agilent Technologies, Santa Clara, CA, USA). Separation was performed on EC 250/4.6 Nucleodur 300-5 C-18 ec. column (250 × 4.6 mm, 5 μm) at a temperature of 25 °C (Macherey-Nagel, Düren, Germany). Mobile phases of acetonitrile/app/triethylamine 90/10/10/0.25 (A) and ethyl acetate/triethylamine 100/0.25 (B) were used in the following gradient: at min. 0, 90% A; at min. 10, 50% A; at min. 20, 10% A, and at min. 26 returns to 90% A. The flow rate was 1 mL/min, and chromatograms were recorded at the wavelength λ = 450 nm. Data recording and interpretation of results was achieved using Agilent ChemStation software, version Rev. B.02.01-SR2. Acetonitrile and ethyl acetate (HPLC purity) were purchased from Merck (Darmstadt, Germany) and triethylamine (99.5% purity) from Fluka (Buchs, Switzerland). Ultrapure water was purified with the Direct-Q UV system from Millipore Corporation (Burlington, VT, USA). For the identification and quantification of carotenoids in the samples, lutein, lycopene and β-carotene standards from Sigma, Virginia Beach, VA, USA, were used. For the quantification of carotenoids, calibration curves were performed by injecting five different concentrations of lutein, lycopene, and β-carotene dissolved in ethyl acetate.

2.6. Assay and Assessment of Therapeutic Effects on Human Skin Conditions

VISIA® facial scanning (Canfield Scientific Inc., BV Proostwetering, Utrecht, The Netherlands) used 3D images to identify and quantify all aspects of skin esthetics, even before there are visible signs of deterioration or aging. The system integrates imaging technologies to support clinical trials in evaluating skin health and the efficacy of various treatments (software version 8.5). The 3D imaging allows multi-angle images of the face, facilitating a complete examination of skin topography, whereas RBX enhances visualization of subcutaneous conditions, such as pigmentation and vascular problems, providing information beyond the superficial layer of skin. The TruSkin age simulates the effect of aging and predicts skin health, particularly for long-term studies. The system quantifies the depth and distribution of wrinkles and accurately identifies and measures hyperpigmented spots and pore size, helping to assess skin texture and clarity. Furthermore, the system detects sun damage not visible to the naked eye, which is important for photoprotection and skin repair studies.

2.7. Subjects and Selection Criteria

The open-label, non-blinded study was conducted on a cohort of 86 volunteers, from which 27 participants (23 females and 4 males) were selected based on skin type (i.e., normal) and stringent inclusion and exclusion criteria as part of the PhD thesis study protocol. The exclusion of 59 volunteers was due to the presence of skin conditions that could potentially confound the study outcomes or compromise the comprehensiveness of the results. This study commenced on 14 October 2024 and concluded on 23 November 2024, allowing sufficient time for the assessment of therapeutic effects. The sample size was determined based on the minimum customary criterion for pilot studies, aiming to involve at least 20 individuals with baseline and 5-week scores (ranging from 0 to 100). The study protocol was approved by an Ethics Committee with the reference ID 194/3 October 2024. Participation was contingent upon volunteers providing informed consent, ensuring their conscious and voluntary involvement. This study adopted an inclusive approach, with no gender restrictions, and enrolled volunteers aged 25–65 years, encompassing a broad adult demographic. Participants were stratified into four age groups: AG1 (25–35 years, $n = 8$), AG2 (36–45 years, $n = 8$), AG3 (46–55 years, $n = 5$), and AG4 (56–65 years, $n = 6$). Further details on participant characteristics are provided in Supplementary Table S1.

2.8. Study Design

To assess the therapeutic effects of cold-pressed *R. canina* seed oil on the complexion, each volunteer applied three drops of oil daily using a pipette. Application was carried out over the entire face, avoiding the eye area. This protocol was followed consistently throughout the study to ensure uniformity of treatment. Skincare routines were standardized (e.g., participants refrained from new or potentially confounding cosmetic treatments). The individuals refrained from using anything other than rosehip oil in their skincare routine. They were allowed to use regular make-up during the day and to cleanse in the evening with micellar water. Furthermore, clear instructions were given to participants regarding sunscreen use, sun exposure, or other daily skincare products. Given that the study took place in late fall, in a country where the average temperature is around 10 °C, the participants did not use SPF in their skincare routine, so as to not influence the results' outcome on the evaluated skin characteristics, particularly UV spots and redness.

With this method of application and evaluation, the aim was to determine the improvements brought about by rosehip oil on the skin.

2.9. The Allocation of Volunteers According to the Treated Area and the Selection of the Skin Assessment Parameter Set

To be eligible for the study, volunteers had to meet certain inclusion criteria. Thus, men and women without pre-existing skin or other health conditions that could affect the integrity of the results were included. They also had not undergone any minimally invasive or invasive skin resurfacing procedures, such as fractional laser treatments, ablative laser, chemical peels, or microneedling, in the previous 12 months.

On the other hand, exclusion criteria were clearly defined to ensure the integrity and safety of the assay. Volunteers with keloid scars or a history of eczema in the treatment area, psoriasis, or other chronic skin conditions considered by the investigator as disqualifying were not considered eligible. Those with a history of actinic (solar) keratoses in the treatment area, hemophilia, or diabetes were also excluded. Volunteers with raised moles or warts on the treatment area were also excluded. Furthermore, other exclusion criteria included conditions such as scleroderma, collagen vascular disease, or cardiac abnormalities, blood clotting problems, active bacterial or fungal infections, facial melanosis, malignant tumors, or immunosuppressant. Also, the use of anticoagulants or prednisone, pregnant or breastfeeding women, corticosteroid administration in the last two weeks before the procedure, chronic liver disease, porphyria, or other skin diseases were reasons for exclusion.

2.10. Statistical Analysis

The statistical analyses were assessed using SPSS software (version 19). As a pilot study, the primary focus of the statistical analyses was descriptive statistics reflecting acceptability and outcome scores of the topical treatment. Dermatology parameters measures were tested for normality using the Shapiro–Wilk test. Mann–Whitney pairwise comparison with Bonferroni corrected p values was used for comparing groups with normal distributions (Supplementary File). Cohen's d was employed to assess effect sizes; values in the range between 0.20 and 0.49 represent minor effect sizes, between 0.50 and 0.79 denote medium effect sizes, values between 0.80 and 1.19 indicate high effect sizes, and values exceeding 1.20 reveal very large impact sizes [31]. Statistical power and sample size using goal seek have been generated to provide insights into the implications and necessary sample size, which can be used in future investigations (Supplementary Table S5).

The data are represented as means ± standard deviation. Box plots were generated using the ggplot2 package. Principal component analysis (PCA) was employed to visualize trends associated with the age groups and evaluated skin parameters according to each side of facial characteristics. Pearson's correlation coefficients were constructed to evaluate the associations before and after 5 weeks of topical treatment with rosehip with the assessed skin parameters. Heatmaps and hierarchical cluster analysis (HCA) were created, employing the Euclidean distance to point out the resemblances and distinctions in the improvements or detrimental effects imposed by the topical application of rosehip oil. The packages used included cluster R, dendextend, and ggplot from the R database (version 2024.12.1).

3. Results

3.1. R. canina Oil TPC and Antioxidant Activity

Prior to the initiation of the clinical study, a detailed laboratory analysis was conducted to determine the chemical composition and therapeutic potential of *R. canina* oil. The results revealed a high total carotenoid content, reaching values of up to 32.687 μg/mL in samples collected from the Băișoara region, indicating a significant concentration of natural antioxidants. Furthermore, the assessment of phenolic compounds showed a total content of 2.637 mg gallic acid equivalents (GAEs)/100 g, and the antioxidant potential measured by the DPPH assay was of 5.572 μmol Trolox/100 g. These findings suggest that *R. canina* oil not only exhibits a high antioxidant capacity but also possesses skin-protective and regenerative properties, thereby justifying its inclusion in the proposed clinical study to investigate its therapeutic effects on human skin.

3.2. Determination of Carotenoids by HPLC-DAD Method

Carotenoids (lutein, lycopene, and β-carotene) were identified by comparing retention times and UV-VIS absorption spectra with those of standard substances (Supplementary Table S2), and lutein and zeaxanthin esters and diesters were assessed by a comparison with literature data [13].

The chromatogram for the oil extracted from *R. canina* seeds from the Băișoara area, recorded at a wavelength of 450 nm, is displayed in Supplementary Figure S1. The chromatogram was obtained using HPLC, an advanced method for the separation and identification of chemical compounds, with a specific detection for carotenoids and other compounds absorbing in the 450 nm spectral region. The major peaks at Rt = 12.306 and Rt = 19.542 are mostly representative for lutein, β-carotene, and other carotenoids present in R. canina oil, which are essential for antioxidant properties and various human health benefits. A detailed overview of the chemical composition of the *R. canina* oil observed by the diversity of peaks indicates chemical complexity, which may contribute to its therapeutic and antioxidant potential. The dominant peak at Rt = 19.542 is presumably associated with β-carotene, known for its beneficial effects as a potent antioxidant. This implies that *R. canina* oil from the Băișoara area is rich in bioactive compounds, which is relevant for therapeutic and cosmetic applications. The chromatogram confirms the therapeutic and antioxidant value of the oil, demonstrating the importance of optimizing extraction methods to maximize the yield of bioactive compounds. Table 1 shows the content of free and esterified carotenoids detected in the *Rosa canina* oil from the Băișoara area. The rosehip oil showed a significant total carotenoids content (28.398 μg/mL), with the highest levels in the case of β-carotene (4.49 μg/mL), lutein (4.33 μg/mL), and zexanthin (10.88 μg/mL) contents.

Table 1. Individual carotenoids identified in *R. canina* oil.

No.	Rt (min.)	Compound	*R. canina* Oil Collected from the Băișoara Area (μg/mL)
1	5.81	Lutein	1.291
2	12.30	Lycopene	2.835
3	14.06	β Carotene	4.495
4	17.45	Lutein-C12:0-C12:0	2.513
5	17.91	Zeaxantin-C12:0-C12:0	2.052
6	18.75	Lutein-C12:0-C14:0	4.333
7	19.15	Zeaxantin-C12:0-C14:0	5.901
8	19.54	Zeaxantin-C14:0-C14:0	4.979
	Total carotenoids		**28.398**

3.3. Overall Skin Improvement Results Evaluated with VISIA System on Different Parameters

The VISIA system provided an overview of various parameters, including texture, wrinkles, blemishes, and pores, according to the subject's skin type, age, and gender.

Several skin parameters were evaluated, including color-related (i.e., red areas, spots, brown, and UV spots) and perception-related characteristics (i.e., texture, wrinkles, pores, and porphyrins). Regarding color-related parameters, the red areas score registered an overall slight increase after the topical application of rosehip oil from 12.79 to 13.37, which might denote different skin types and sensitivities. The score of spots was revealed to have a comparable tendency before (31.17) and after (31.42) the end of the treatment, as also noticed in the case of UV spots, which registered a trivial intensification from 16.53 to 18.85. Surprisingly, the score in brown spots diminished post-treatment, with a score from 18.22 to 17.47, highlighting the effectiveness of rosehip oil in reducing skin pigmentation and signs of aging and promoting skin brightness. Subsequently, the texture score presented a reduction from 6.90 to 5.03, highlighting the efficiency of the treatment in reducing roughness and promoting glossiness. The porphyrins registered a decrease in score from 11.13 to a final score of 10.37 post-treatment. The porphyrin content is known to be highly related to proinflammatory outcomes owing to the presence of bacterial activity, which may also promote the formation of acne predominantly in individuals with an oily type skin. A diminished score in facial pores has also been recorded from 12.51 to 11.23. A positive outcome has been observed in the case of wrinkle visibility with values from 17.20 to 15.75, which underlines the effectiveness of rosehip oil in promoting skin elasticity.

3.3.1. True Skin Age

The true skin age parameter evaluated the changes and improvement in the skin complexion and can be visualized in Figure 1. Significant differences have been observed before and after the topical treatment, particularly in AG1 on the right side of the face before (33.5 $\pm$ 4.87) and after (29.625 $\pm$ 6.07); however, there was a small effect size before and after treatment (95% conf. interval = −4.9679, 8.8938; Cohen's d = 0.1547). Insignificant differences have been recorded in the other age groups with similar values before and after the treatment. The same outcome has been recorded on the left facial side (95% conf. interval = −6.294, 7.9237; Cohen's d = 0.0626), with an improvement in skin complexion in AG1 as observed before (32.571 $\pm$ 5.782) and after (30 $\pm$ 6.141) 5 weeks. The following age groups recorded a slight improvement in their complexion, such as in AG2 from 37.714 $\pm$ 4.627 to 36.625 $\pm$ 5.423, followed by AG3 with similar positive effects from 54.75 $\pm$ 8.62 to 52 $\pm$ 7.906. Lastly, individuals in AG4 presented a trivial enhancement with values starting from 60.40 $\pm$ 5.762 to 59.01 $\pm$ 6.633.

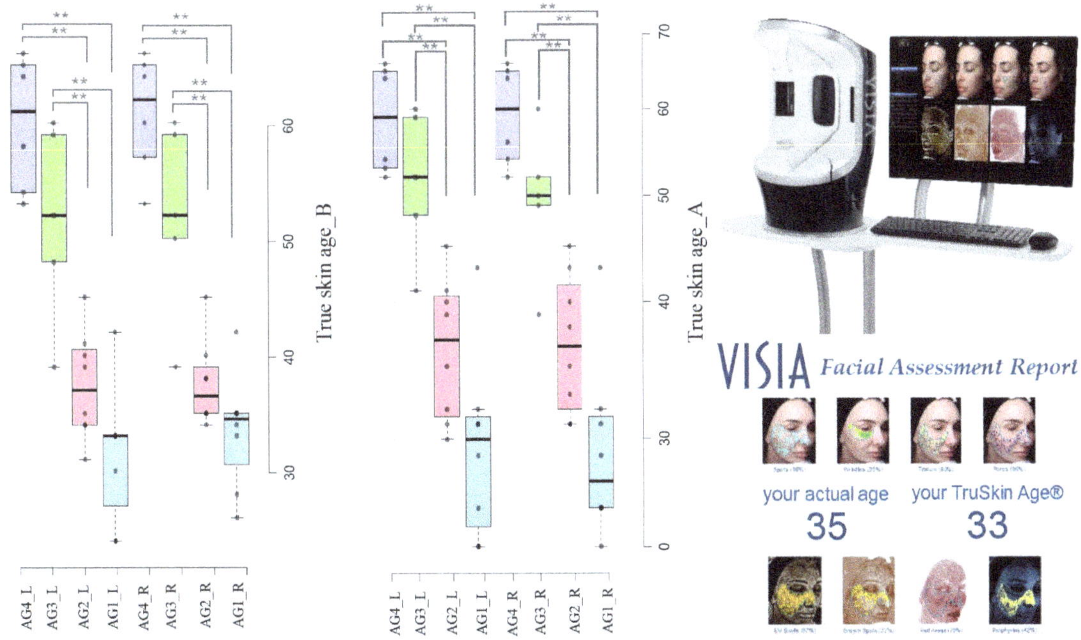

Figure 1. The assessment of the TruSkin age evaluated with the VISIA analysis system for the right and left side of the face, before (**left chart**) and after (**right chart**) 5 weeks of the rosehip oil treatment; AG—age group; L—left side of the face; R—right side of the face; 1–4, individuals divided by age, starting from younger to older. Significant differences are represented by asterisks and *p* values, where * represents a significance level less than 0.05, ** represents a significance level less than 0.01, and *** represents a significance level less than 0.001.

3.3.2. Assessment of Spots

The evaluated score of spots highlighted the effectiveness of the treatment in terms of alleviating their appearance based on the age group and the scanned side of the face (Figure 2). Consequently, the differences between the age groups proved to be significant, mainly regarding AG1 and the others ($p < 0.01$). Regarding the front side of the face, the score of spots presented a slight reduction before (26.963 ± 4.014) and after (26.026 ± 2.385) the treatment in AG1 (Table 2).

The most notable change has been recorded in RC01, with a spots score of 32.234 before and a score of 23.681 after 5 weeks. Nonetheless, a higher effectiveness of the treatment has been observed in the following group, in which the VISIA system recorded a substantial decrease in their score as evidenced before (32.209 ± 5.427) and after 5 weeks (33.871 ± 6.264). In the following age group, the spots score presented similar values before (32.303 ± 2.361) and after the treatment (33.792 ± 3.843), which might be due to the skin type (i.e., oily due to lack of moisture). Lastly, in AG4 a considerable diminished score has been visualized with the VISIA system, as seen with the values before (36.915 ± 5.799) and after the end of the treatment period (33.403 ± 5.389). Overall, significant differences between the age groups have been observed mainly after the topical oil application ($p < 0.01$); however, there was a small effect size (95% conf. interval = −2.8396, 3.4191; Cohen's $d = 0.05057$).

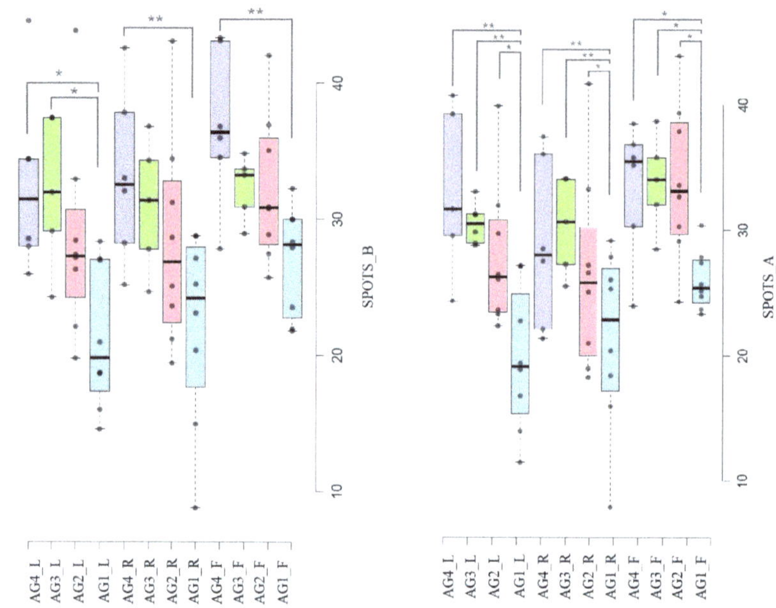

Figure 2. Assessment of spots evaluated with VISIA analysis system before (**left chart**) and after (**right chart**) 5 weeks of rosehip oil treatment. Significant differences are represented by asterisks and p values, where * represents a significance level less than 0.05, ** represents a significance level less than 0.01.

An improvement in the spots' appearance and visibility has been regarded on the right side of the face, with significant differences in all age groups (Table 3) but an overall small effect size (95% conf. interval = −3.0946, 5.0082; Cohen's $d = 0.129$). The score in spots recorded a slight decrease in AG1 after 5 weeks (21.388 ± 7.173), with similarities also observed in the front side of the face. Positive alterations in the spot appearance have been recorded in RC16, with a reduction from 20.354 to 18.39, followed by RC21 with comparable outcomes from 23.086 to 20.393 at the end of the treatment. Although insignificant, equivalent positive responses have been attained in both AG2 and AG3. Interestingly, in AG4 the spots score presented a moderate improvement using the rosehip oil, with a starting score of 33.129 ± 6.313 and a final score of 28.823 ± 6.777. Relatively dissimilar changes occurred on the left side of the face compared to the right side (Table 4). The most notable changes in terms of spots occurred in AG1, which recorded a score of 22.388 ± 5.378 compared with the final score of 19.685 ± 5.725, suggesting patients' tendency of sleeping on the right side of the face. Similar outcomes have been observed in AG2 and AG3 at the end of the treatment with a score beginning from 28.67 ± 7.36 and 32.76 ± 5.653 to a slightly lower score of 27.912 ± 5.844 and 30.498 ± 1.64. The final age group presented insignificant changes. In general, significant differences in the right and left sides of the face have been recorded between AG1 and AG2 ($p < 0.05$) and between AG1 and both AG3 and AG4 ($p < 0.01$). The results suggest that topical oil applications may have a more pronounced effect on the spot appearance and visibility in younger age groups, indicating that age may play a role in the efficacy of this treatment method for improving skin spots.

Table 2. Differential characteristics of the evaluated front side of the face, the correlation between age groups, and the evaluation with the VISIA analysis before (B) and after (A) the topical treatment with rosehip oil.

Skin Characteristics	AG1	AG2	AG3	AG4	*p* Value	Significance	AG 1-AG2	AG 1-AG 3	AG 1-AG 4	AG 2-AG 3	AG 2-AG 4	AG 3-AG 4
Skin Characteristics for the Front Side of the Face												
Spots_B	26.963 ± 4.014	32.209 ± 5.427	32.303 ± 2.361	36.915 ± 5.799	0.007	** $p < 0.01$	ns	ns	0.004	ns	ns	ns
Spots_A	26.026 ± 2.385	33.871 ± 6.264	33.792 ± 3.843	33.403 ± 5.389	0.009	* $p < 0.05$	0.015	0.040	0.040	ns	ns	ns
Texture_B	4.542 ± 3.062	5.138 ± 3.325	8.008 ± 6.149	11.492 ± 9.038	0.115	ns	ns	ns	ns	ns	ns	ns
Texture_A	2.906 ± 3.031	4.758 ± 2.92	6.496 ± 3.797	7.029 ± 3.764	0.119	ns	ns	ns	ns	ns	ns	ns
Pores_B	10.716 ± 7.613	13.999 ± 4.224	12.858 ± 6.851	12.648 ± 5.238	0.757	ns	ns	ns	ns	ns	ns	ns
Pores_A	9.58 ± 5.479	12.891 ± 4.432	10.98 ± 6.881	11.431 ± 3.912	0.650	ns	ns	ns	ns	ns	ns	ns
Uv Spots_B	11.735 ± 7.664	20.277 ± 5.795	16.496 ± 9.693	17.991 ± 9.773	0.221	ns	ns	ns	ns	ns	ns	ns
Uv Spots_A	15.56 ± 6.575	20.749 ± 5.165	18.728 ± 4.454	20.838 ± 7.645	0.318	ns	ns	ns	ns	ns	ns	ns
Brown Spots_B	14.79 ± 4.503	21.044 ± 6.657	17.599 ± 5.105	19.571 ± 5.434	0.165	ns	ns	ns	ns	ns	ns	ns
Brown Spots_A	14.062 ± 2.696	19.627 ± 6.569	17.003 ± 4.418	19.57 ± 4.702	0.112	ns	ns	ns	ns	ns	ns	ns
Porphyrins_B	14.372 ± 9.908	12.101 ± 10.241	11.761 ± 11.974	9.785 ± 5.474	0.851	ns	ns	ns	ns	ns	ns	ns
Porphyrins_A	12.924 ± 7.661	12.148 ± 12.42	11.036 ± 9.098	10.471 ± 4.199	0.960	ns	ns	ns	ns	ns	ns	ns
Wrinkles_B	5.199 ± 4.089	8.77 ± 6.15	26.102 ± 30.801	37.059 ± 28.041	0.016	* $p < 0.05$	ns	ns	0.022	ns	0.048	ns
Wrinkles_A	5.634 ± 5.781	11.69 ± 11.456	21.17 ± 18.811	30.176 ± 13.872	0.008	** $p < 0.01$	ns	ns	0.007	ns	ns	ns
Red Areas_B	10.401 ± 1.672	12.603 ± 3.94	12.392 ± 2.442	16.568 ± 2.895	0.007	** $p < 0.01$	ns	ns	0.003	ns	ns	ns
Red Areas_A	10.459 ± 1.332	12.264 ± 2.291	12.381 ± 2.512	19.582 ± 7.308	0.002	** $p < 0.01$	ns	ns	0.001	ns	0.009	0.02496

Note: Significant differences are represented by asterisks and *p* values, where * represents a significance level less than 0.05, ** represents a significance level less that 0.01., ns, not significant.

Table 3. Differential characteristics of the evaluated right side of the face, the correlation between age groups, and the evaluation with the VISIA analysis before and after the topical treatment with rosehip oil.

Skin Characteristics	AG1	AG 2	AG 3	AG 4	*p* Value	Significance	AG 1-AG2	AG 1-AG 3	AG 1-AG 4	AG 2-AG 3	AG 2-AG 4	AG 3-AG 4
Skin Characteristics for the Right Side of the Face												
True Skin Age_B	33.5 ± 4.87	37.5 ± 3.665	52 ± 8.456	60.833 ± 5.115	6.19×10^9	*** $p < 0.001$	ns	0.001	0.001	0.001	0.001	ns
True Skin Age_A	29.625 ± 6.07	37 ± 5.372	50 ± 7.517	59.5 ± 6.189	2.72×10^8	*** $p < 0.001$	ns	0.001	0.001	0.006	0.001	ns
Spots_B	22.131 ± 7.13	28.325 ± 7.797	30.982 ± 4.886	33.129 ± 6.313	0.035	* $p < 0.05$	ns	ns	0.032	ns	ns	ns
Spots_A	21.388 ± 7.173	26.535 ± 7.948	30.32 ± 3.883	28.823 ± 6.777	0.001	** $p < 0.01$	0.027	0.006	0.001	ns	ns	ns
Texture_B	5.455 ± 3.304	8.575 ± 4.536	12.012 ± 7.954	19.144 ± 16.592	0.050	* $p < 0.05$	ns	ns	0.044	ns	ns	ns

Table 3. *Cont.*

Skin Characteristics	AG1	AG 2	AG 3	AG 4	*p* Value	Significance	AG 1-AG2	AG 1-AG 3	AG 1-AG 4	AG 2-AG 3	AG 2-AG 4	AG 3-AG 4
Skin Characteristics for the Right Side of the Face												
Texture_A	3.657 ± 3.383	7.211 ± 4.309	11.525 ± 8.082	13.451 ± 9.32	0.026	* $p < 0.05$	ns	ns	0.028	ns	ns	ns
Pores_B	10.703 ± 9.052	13.87 ± 4.811	9.894 ± 4.47	10.025 ± 3.485	0.586	ns	ns	ns	ns	ns	ns	ns
Pores_A	10.344 ± 8.132	11.516 ± 4.692	9.895 ± 5.246	9.722 ± 3.232	0.650	ns	ns	ns	ns	ns	ns	ns
Uv Spots_B	14.14 ± 8.28	24.996 ± 6.696	17.803 ± 8.088	23.048 ± 8.353	0.052	ns	ns	ns	ns	ns	ns	ns
Uv Spots_A	17.105 ± 6.413	25.344 ± 6.111	22.728 ± 5.914	24.419 ± 6.843	0.169	ns	ns	ns	ns	ns	ns	ns
Brown Spots_B	15.832 ± 5.402	24.4 ± 8.517	21.247 ± 3.642	24.518 ± 4.335	0.036	* $p < 0.05$	0.046	ns	ns	ns	ns	ns
Brown Spots_A	14.96 ± 3.528	23.385 ± 8.923	21.474 ± 6.027	23.093 ± 2.153	0.061	ns	ns	ns	ns	ns	ns	ns
Porphyrins_B	12.768 ± 9.438	12.903 ± 10.138	9.937 ± 7.781	12.216 ± 6.021	0.936	ns	ns	ns	ns	ns	ns	ns
Porphyrins_A	12.232 ± 5.575	11.851 ± 11.701	13.037 ± 10.628	12.561 ± 4.214	0.987	ns	ns	ns	ns	ns	ns	ns
Wrinkles_B	29.773 ± 7.433	28.781 ± 18.927	30.981 ± 5.259	46.619 ± 15.076	0.084	ns	ns	ns	ns	ns	ns	ns
Wrinkles_A	25.054 ± 11.802	27.891 ± 15.545	38.562 ± 8.947	47.056 ± 12.31	0.163	ns	ns	ns	ns	ns	ns	ns
Red Areas_B	10.648 ± 2.466	14.975 ± 7.291	13.353 ± 3.932	20.16 ± 5.864	0.024	* $p < 0.05$	ns	ns	0.015	ns	ns	ns
Red Areas_A	10.957 ± 2.183	14.354 ± 5.305	14.232 ± 4.319	24.42 ± 15.716	0.010	** $p < 0.01$	ns	ns	0.006	ns	0.048	ns

Significant differences are represented by asterisks and *p* values, where * represents a significance level less than 0.05, ** represents a significance level less than 0.01, and *** represents a significance level less than 0.001.

Table 4. Differential characteristics of the evaluated left side of thfe face, the correlation between age groups, and the evaluation with the VISIA analysis before and after the topical treatment with rosehip oil.

Skin Characteristics	AG1	AG 2	AG 3	AG 4	*p* Value	Significance	AG 1-AG2	AG 1-AG 3	AG 1-AG 4	AG 2-AG 3	AG 2-AG 4	AG 3-AG 4
Skin Characteristics for the Left Side of the Face												
True Skin Age_B	32.571 ± 5.782	37.714 ± 4.627	54.750 ± 8.62	60.400 ± 5.762	3.06×10^8	*** $p < 0.001$	ns	0.001	0.001	0.002	0.001	ns
True Skin Age_A	30 ± 6.141	36.625 ± 5.423	52.002± 7.906	59.001 ± 6.633	4.71×10^8	*** $p < 0.001$	ns	0.001	0.001	0.002	0.001	ns
Spots_B	22.388 ± 5.378	28.67 ± 7.36	32.760 ± 5.653	33.443 ± 6.794	0.012	* $p < 0.05$	ns	0.037	0.017	ns	ns	ns
Spots_A	19.685 ± 5.725	27.912 ± 5.844	30.498 ± 1.64	33.085 ± 6.843	0.001	** $p < 0.01$	0.027	0.006	0.001	ns	ns	ns
Uv Spots_B	15.459 ± 8.004	23.954 ± 7.309	20.078 ± 9.195	20.17 ± 7.925	0.163	ns	ns	ns	ns	ns	ns	ns
Uv Spots_A	17.928 ± 6.068	24.561 ± 6.869	22.996 ± 4.345	24.844 ± 7.992	0.169	ns	ns	ns	ns	ns	ns	ns
Brown Spots_B	17.166 ± 3.823	24.242 ± 8.704	20.549 ± 3.95	22.326 ± 5.606	0.113	ns	ns	ns	ns	ns	ns	ns
Brown Spots_A	15.301 ± 2.472	22.548 ± 9.016	20.84 ± 3.598	23.66 ± 5.887	0.061	ns	ns	ns	ns	ns	ns	ns
Wrinkles_B	31.649 ± 9.084	26.787 ± 19.709	29.967 ± 6.053	43.205 ± 8.283	0.166	ns	ns	ns	ns	ns	ns	ns

Table 4. *Cont.*

Skin Characteristics	AG1	AG 2	AG 3	AG 4	*p* Value	Significance	AG 1-AG2	AG 1-AG 3	AG 1-AG 4	AG 2-AG 3	AG 2-AG 4	AG 3-AG 4
Skin Characteristics for the Left Side of the Face												
Wrinkles_A	26.906 ± 10.186	32.026 ± 18.478	36.594 ± 9.098	42.602 ± 7.565	0.163	ns	ns	ns	ns	ns	ns	ns
Red Areas_B	11.21 ± 2.502	16.15 ± 4.945	13.676 ± 1.973	19.753 ± 7.787	0.030	* $p < 0.05$	ns	ns	0.020	ns	ns	ns
Red Areas_A	10.426 ± 2.01	13.382 ± 3.163	13.841 ± 2.459	22.618 ± 12.192	0.010	** $p < 0.01$	ns	ns	0.006	ns	0.048	ns
Pores_B	11.741 ± 8.471	13.243 ± 6.493	12.227 ± 5.089	9.957 ± 4.273	0.759	ns	ns	ns	ns	ns	ns	ns
Pores_A	9.842 ± 6.423	12.206 ± 5.766	9.465 ± 4.937	9.786 ± 3.626	0.759	ns	ns	ns	ns	ns	ns	ns
Texture_B	6.504 ± 3.09	8.917 ± 4.324	15.341 ± 8.585	17.404 ± 11.338	0.080	ns	ns	ns	ns	ns	ns	ns
Texture_A	3.952 ± 3.17	7.325 ± 4.168	12.698 ± 8.816	15.536 ± 11.338	0.026	* $p < 0.05$	ns	ns	0.028	ns	ns	ns
Porphyrins_B	12.897 ± 7.92	11.191 ± 8.366	15.887 ± 11.461	11.81 ± 5.149	0.980	ns	ns	ns	ns	ns	ns	ns
Porphyrins_A	12.00 ± 6.77	11.063 ± 10.976	12.819 ± 11.26	11.581 ± 4.108	0.987	ns	ns	ns	ns	ns	ns	ns

Significant differences are represented by asterisks and *p* values, where * represents a significance level less than 0.05, ** represents a significance level less than 0.01, and *** represents a significance level less than 0.001.

3.3.3. Assessment of Red Areas

A damaged skin barrier may augment the susceptibility to several external factors which eventually intensify inflammatory reactions, including burning sensations, itchiness, and a feeling of tightness. The topical treatment on the front side of the face revealed insignificant differences between the age groups at the beginning of the treatment, except AG1 and AG4 ($p < 0.01$) which is to be expected to be due to the age gap (Figure 3; Table 2), with a small effect size (95% CI = −1.7885, 2.9576; Cohen's d = 0.1345). Conversely, significant differences between AG4 and the other assemblages (AG1-AG3) have been regarded at the end of the treatment period. Thus, individuals in AG1 perceived discrepancies in the treatment (from 10.401 ± 1.672 to 10.459 ± 1.332), with several cases of diminished inflammatory sensations and redness visibility (e.g., RC02, and RC27); whereas others experienced a slight itchiness and an increased inflammatory response (e.g., RC01, RC 5, and RC 11). Improved outcomes were perceived from the second group, particularly in RC04, RC08, RC13, and RC20, with lessened unpleasant experiences and a visible improvement in red areas, as observed at the beginning (12.603 ± 3.94) and at the end (12.264 ± 2.291) of the treatment. Interestingly, the score in red areas presented similar values after 5 weeks in AG3, with no inflammatory responses reported by the individuals. Lastly, the majority in AG4 experienced a slight inflammatory response to the treatment and an increased visibility in redness with values from 16.568 ± 2.895 to 19.582 ± 7.308.

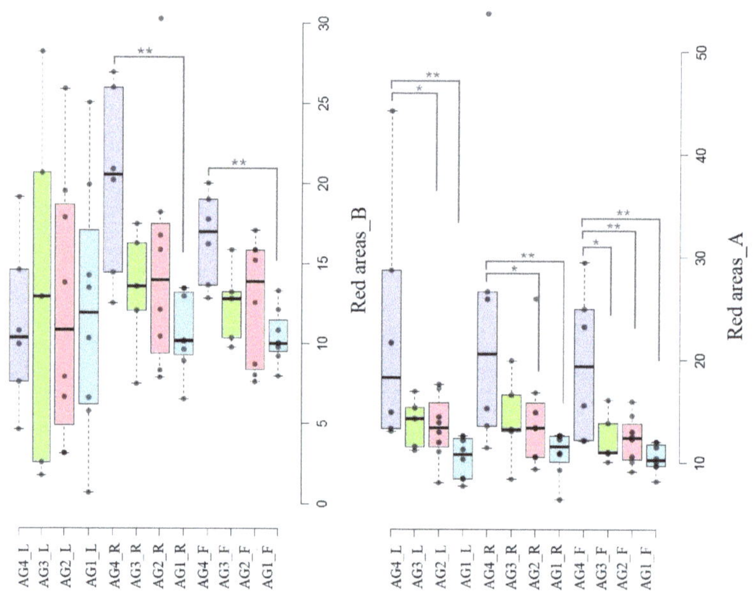

Figure 3. The assessment of redness or inflammatory responses evaluated with the VISIA analysis system before (**left chart**) and after (**right chart**) 5 weeks of the rosehip oil treatment. Significant differences are represented by asterisks and p values, where * represents a significance level less than 0.05, ** represents a significance level less than 0.01.

The changes before and after the 5 weeks of treatment proved to have a more significant effect on the right side of the face compared with the front (Table 3), although the effect size proved to be relatively small (95% CI = −3.2462, 5.2802; Cohen's d = 0.1303). Therefore, in AG1, a slight reduction in the appearance of redness has been observed in several individuals, including RC02 who experienced a positive outcome of the treatment with scores from 12.952 to 9.306; an effect was also observed in RC16 with scores from 13.44 to

12.672. Conversely, several individuals experienced a slight increase in the score of red areas, which may be due to the skin's inflammatory–sensitive type or a reaction to some existing components of the rosehip oil. The effect is observed in the overall appearance of redness in all evaluated patients, with a score of 10.648 ± 2.466 at the beginning of the treatment and finalizing with a value of 10.957 ± 2.183 at the end of the session. Significant differences in terms of red spots have been revealed between AG1 and AG4 ($p < 0.01$), which is to be expected due to the differential age category and changes in skin conditions and characteristics with aging (i.e., pronounced redness, inflammatory responses, and sensitivity). In AG2, responses to the treatment differed between the evaluated individuals, with an overall positive outcome with a decrease in the redness score from 14.975 ± 7.291 to 14.354 ± 5.305. The most notable change was shown in RC04 with a positive response and a diminished score from 18.204 to 9.368, followed by RC13 with a reduction in redness from 30.278 to 25.974. This group's results proved to have significant differences compared with AG4 ($p < 0.05$). Subsequently, AG3 individuals experienced a slight inflammatory response to the topical application, with an enhanced score from 13.353 ± 3.932 to 14.232 ± 4.319. Lastly, the individuals in AG4 experienced a positive outcome; however, a single patient (i.e., RC24) experienced a relatively high inflammatory response as evidenced from the change in the redness area from 26.938 to 53.691 (Table 4). These results suggest that AG3 and AG4 individuals had varying reactions to the topical application, with some experiencing improvements and others showing increased inflammation.

3.3.4. Wrinkle Assessment

Wrinkles are lines and folds that form on the skin, often as a result of natural aging, exposure to ultraviolet (UV) radiation, repetitive facial expressions, and the loss of skin elasticity. They are most commonly seen in areas frequently exposed to the sun, such as the face, neck, hands, and arms. A wrinkle assessment is essential in dermatologic and anti-aging studies to determine the effectiveness of treatments. The system allowed an accurate and objective analysis of the wrinkle depth and distribution, providing data for improving anti-aging products and procedures, as shown in Figure 4. Detailed pre- and post-treatment scans help monitor the progress and quantify the benefits of different therapies.

Figure 4 displays a comparison of facial scans performed with the VISIA® system, before (14 October 2024) and after 5 weeks of treatment (23 November 2024). Overall, the scans revealed a visible reduction in the number and depth of wrinkles, confirming the effectiveness of the treatment applied.

The reduction or increase in wrinkles has been confirmed by the relative fluctuations in their score based on the age group (AG), particularly between AG1 and AG4 ($p < 0.01$). Regarding the front side of the face from the AG1 group (25–35 years old), discrepancies in terms of wrinkle scores before (5.199 ± 4.089) and after the treatment (5.634 ± 5.781) have been observed among individuals, although a small effect size has been recorded (95% CI = −8.8424 11.742; Cohen's d = 0.07692). Interestingly, RC05 presented an intensification in the wrinkle score from 1.33 to 19.09. The same effect has been observed in RC11, which recorded an intensification in the wrinkle score from 1.95 to 4.56. These unpleasant outcomes may be due to some individuals' skin type and sensitivity to vitamin A and their living habits (i.e., environmental conditions, including region, temperature, humidity, etc.). Nonetheless, a significant diminution in wrinkle scores has been regarded in other individuals, such as in RC01, from 2.71 to 0.52, and in RC02, from 5.36 to 1.89. Additionally, in AG2 (36–42 years old) a similar pattern as observed in AG1 has been observed in the evaluated individuals, however, with significant results before (8.77 ± 6.15) and after treatment (11.69 ± 11.456). The highest variation in the wrinkle score has been recorded in RC07, from 17.58 to 36.13, with comparable outcomes in RC04, from 6.24

to 14.91. Conversely, a reduction in the wrinkle score has been regarded in RC10, from 10.80 to 4.82, and in RC20, from 8.56 to 5.51. Regarding AG3 (43–52 years old), in several individuals the wrinkle score was significantly reduced, particularly in RC25 from 75.34 to 54.24 and in RC17 from 37.71 to 16.79, demonstrating a major improvement of the skin in the periocular area. Overall, the wrinkle score recorded a considerable decrease from 26.102 ± 30.801 to 21.17 ± 18.811. Regarding the AG4 group (53–60 years old), discrepancies and noteworthy alterations have been recorded, particularly in RC23 in whom the wrinkle score decreased from 73.63 to 34.60, followed by RC24 with a gradual decrease in wrinkles from 51.34 to 37.38. In general, the wrinkle score recorded a substantial decrease from 37.059 ± 28.041 to 30.176 ± 13.872, showing a noticeable overall improvement in the skin condition. Therefore, the present results suggest the effectiveness of rosehip oil in improving the signs of aging and increasing skin quality in participants across different age groups. The specific improvements distinguished in specific sets of individuals indicate targeted benefits of topical applications of rosehip oil. Differential characteristics of the different sides of the face and the correlation between age groups and the evaluation with the VISIA analysis can be visualized in Tables 1–4. Concerning the right and left sides of the face, no significant variations in terms of wrinkle visibility have been recorded in each individually evaluated age groups, as also observed by the low effect size recorded in both the left (95% CI = −6.4869 8.1555; Cohen's d = 0.06224) and right side (95% CI = −7.4878 8.4385; Cohen's d = 0.0326).

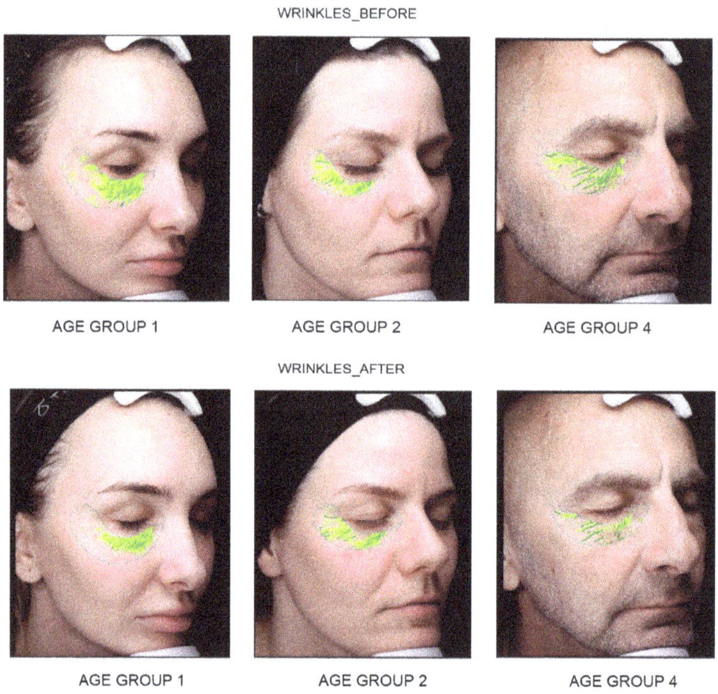

Figure 4. The assessment of wrinkles evaluated with the VISIA analysis system before and after 5 weeks of the rosehip oil treatment. The highlighted areas on the face represent the depth of the wrinkles.

3.3.5. Assessment of Spots Caused by UV Radiation

UV spots are skin manifestations resulting from the excessive exposure to the sun's ultraviolet (UV) radiation. These spots are often invisible to the naked eye and may appear on the face, hands, or other exposed parts of the body. They indicate cell damage caused by UV radiation, which can accelerate skin aging and contribute to dermatological conditions, including skin cancer. In Figure 5 a detailed assessment of skin health and the effectiveness of rosehip oil and anti-aging treatments can be visualized.

Figure 5. The assessment of wrinkles evaluated with the VISIA analysis system before and after 5 weeks of the rosehip oil treatment.

The VISIA system also identified and quantified UV spots, which indicate damage caused by ultraviolet (UV) radiation. The score of UV spots on the front side of the face recorded a slight increase in AG1 from 11.735 ± 7.664 to 15.56 ± 6.575 after 5 weeks of treatment, which may denote the skin's sensitivity to rosehip (Table 2). Furthermore, the AG2 group presented little to no modification in the UV spot visibility from 20.749 ± 5.165 to 20.277 ± 5.795. Nonetheless, although insignificant, a visible decrease in the UV spots score has been confirmed in several patients of this group, namely RC18 (from 23.50 to 21.06), RC10 (from 22.22 to 21.95), and RC22 (from 21.76 to 19.14). A similar pattern regarding the UV spots' visibility has also been observed in AG3 before, with a score of 17.803 ± 8.088, and after the end of the treatment, with a score of 22.728 ± 5.914. Lastly, the AG4 group also presented a slightly higher rate of UV spots from 23.048 ± 8.353 to 24.419 ± 6.843. Overall, while there were individual variations in UV spot scores among patients, the general trend showed a slight increase in the UV spot visibility post-treatment (Tables 3 and 4), as also denoted by the small effect size in the front (95% CI = −1.6947,

6.3331; Cohen's $d = 0.3156$), right (95% CI = −2.8422, 4.6293; Cohen's $d = 0.1306$), and left sides (95% CI = −6.4869, 8.1555; Cohen's $d = 0.06224$) of the face. This suggests that rosehip oil may not have been as effective in reducing UV spots for all participants, due to factors including skin types and the sensitivity to the oil components (i.e., vitamin A content).

3.3.6. Assessment of Brown Spots

Risk indicators for oily or sensitive skin types relate to elevated melanin contents and a high proportion of perioral lines. These features may indicate that the skin is inadequately shielded from UV rays, which can result in more noticeable pigmentation and aging symptoms. Since augmented melanocyte activity may affect the skin's defense towards external stimuli, these implications are linked to a degradation of the skin barrier's normal functioning. According to scientific investigations, UV rays can impair the function of the skin's barrier, cause inflammatory reactions in the skin, and trigger photoaging symptoms. Insignificant differences in terms of hyperpigmentation have been recorded in the front part of the face at the beginning and at the end of the treatment period, with a small effect size before and after the treatment (95% CI = −2.2581 3.7515; Cohen's $d = 0.1357$) (Figure 6; Table 2). Conversely, improved skin complexion has been observed on the right side of the face with significant differences between close-related age groups 1 and 2 (Table 3), although a small effect size has been recorded (95% CI = −2.8422 4.6293; Cohen's $d = 0.1306$).

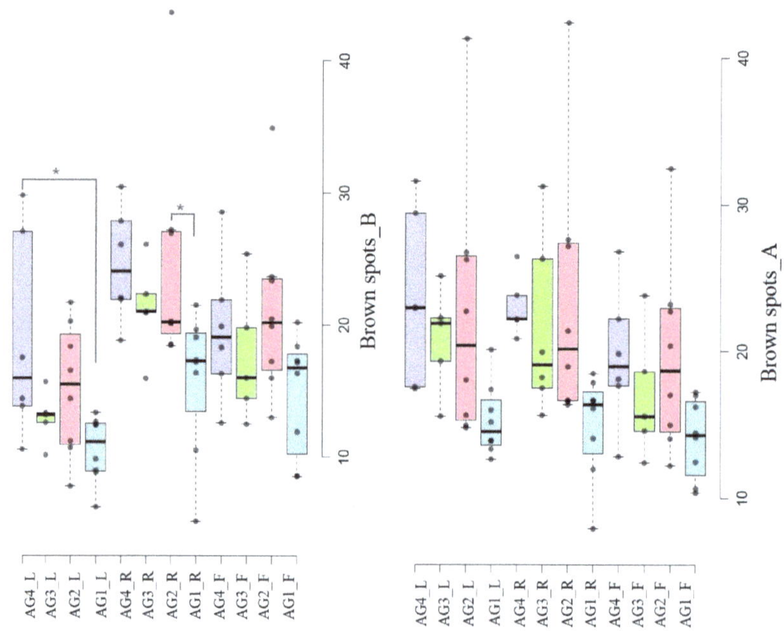

Figure 6. The assessment of brown spots evaluated with the VISIA analysis system before (**left chart**) and after (**right chart**) 5 weeks of the rosehip oil treatment. Significant differences are represented by asterisks and p values, where * represents a significance level less than 0.05.

With age, individuals with fair complexions are prone to freckle formation, particularly after a prolonged sun exposure. As highlighted in the previous sub-section, the use of rosehip oil may produce an inflammatory response or sensitivity due to several factors, including age and melanogenesis initiated by UV radiation exposure. Elevated scores in brown spots have been observed on the right side of the face irrespective of age groups.

In AG1, the score started from a value of 14.14 ± 8.28 to a final value of 17.105 ± 6.413. Although insignificant, a similar tendency has been recorded in AG2, with a score of 24.996 ± 6.696 to a final score of 25.344 ± 6.111. A higher inflammatory response has been recorded in AG3 after the topical treatment, with a relatively lower score of 17.803 ± 8.088 which gradually increased to 22.728 ± 5.914 3 after 5 weeks. In the last group, the rosehip did not have a detrimental effect, the score having a minimum change from 23.048 ± 8.353 to 24.419 ± 6.843. Relating to the left side, a similar tendency for elevated pigmentation scores on the right side has been recorded (Table 4), with a very small effect size (95% CI = −3.0548, 4.1229; Cohen's d = 0.08127). It is important to note that individuals with fair complexions should take extra precautions when exposed to the sun to prevent freckle formation. Additionally, proper skin care routines and the regular use of sunscreen can help minimize the effects of UV radiation on melanogenesis, which is not the case taking into account the implementation period of the present study (i.e., late fall).

3.3.7. Texture Level Assessment

Dermatologists demonstrate a particular emphasis for evaluating the surface skin texture, as these measurements are intended for skin diagnostics and evaluating cosmetic efficiency or therapeutic treatments, as shown in Figure 7. The front facial side presents a general improvement in skin texture, which has been positively perceived by most of the volunteers, particularly females who stated that the rosehip oil gave a similar "glass skin" effect as other cosmeceutical products (Table 1). Thus, AG1 recorded a decreased score in texture from 4.542 ± 3.062 to 2.906 ± 3.031. An equivalent achievement has been detected in the following AG2 group, with a starting value of 5.138 ± 3.325 and a final value of 4.758 ± 2.92. In AG3, the skin complexion improved as observed by the decrease in scores from 8.008 ± 6.149 to 6.496 ± 3.797. Lastly, a visible and considerable effect has been attained in AG4 with a score from 11.492 ± 9.038 to 7.029 ± 3.764. Regarding the right side of the face, a similar positive tendency for texture enhancements has been regarded particularly in AG1 and AG4 ($p < 0.05$), with a small effect size (95% CI = −2.3804, 6.9803; Cohen's d = 0.2684) (Table 2).

Individuals in AG1 recorded a development from 5.455 ± 3.304 to 3.657 ± 3.383, whereas volunteers in AG4 presented a noteworthy improvement from 19.144 ± 16.592 to 13.451 ± 9.32. Regarding the left side of the face, discrepancies with the right side have occurred which may be due to patients' sleeping habits, as observed by the very small effect size (95% CI = −2.8772 5.5836; Cohen's d = 0.1747) (Table 3). Thus, the texture score in AG1 presented comparable results with the other facial side before (6.504 ± 3.09) and after (3.952 ± 3.17) the treatment. Insignificant changes occurred in the following groups; namely AG2 and AG3, which recorded a texture score of 8.917 ± 4.324 and 15.341 ± 8.585 at the beginning of the treatment with a final score of 7.325 ± 4.168 and 12.698 ± 8.816. The last group recorded a general positive effect with a score of 15.341 ± 8.585 before the topical oil application and a final score of 12.698 ± 8.816, emphasizing the optimistic use of rosehip oil for complexion improvements based on both skin types (a younger and fair complexion and/or mature skin). It is important to note that a further investigation may be needed to determine the specific factors contributing to these differences in the texture enhancement between the right and left sides of the face. Additionally, considering individual lifestyle factors, such as sleep quality, could provide valuable insights into optimizing treatment outcomes for patients in future studies.

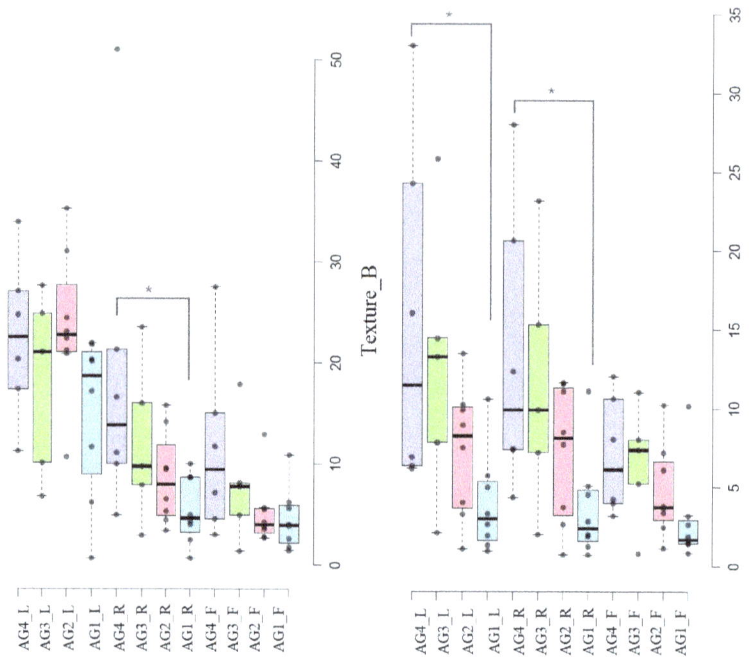

Figure 7. The assessment of texture evaluated with the VISIA analysis system before (**left chart**) and after (**right chart**) 5 weeks of the rosehip oil treatment. Significant differences are represented by asterisks and *p* values, where * represents a significance level less than 0.05.

3.3.8. Assessment of Pores and Porphyrins

Porphyrins are organic compounds that play an essential role in cell biology, being structural components of vital molecules such as hemoglobin and chlorophyll. In the context of dermatology and skin health, porphyrins are produced by bacteria living on the surface of the skin, including *Propionibacterium acnes*, which contribute to acne formation. The analysis of porphyrins on the skin can provide valuable information about the state of the skin microbiome and help identify problems such as acne and other bacterial infections. Porphyrins provide a detailed assessment of skin health and the effectiveness of dermatologic treatments, as can be seen in Figure 8.

The number of porphyrins on the front side of the face recorded a decrease from 1989 to 1894 after 5 weeks of treatment, demonstrating a reduction in bacterial activity and an improvement in the overall skin condition (Table 2), even though a small effect size has been observed (95% CI = −4.4836, 5.2776; Cohen's d = 0.04442). Regarding AG1, the porphyrin score decreased from 14.372 ± 9.908 to 12.924 ± 7.661, signifying a general enhancement in skin health compared to the reference group. The following age groups recorded similar outcomes after the topical treatment with rosehip oil: from 12.101 ± 10.241 to 12.148 ± 12.42 in AG2 and from 11.761 ± 11.974 to 11.036 ± 9.098 in AG3. Conversely, in AG4 a slight increase in the porphyrin score has been observed (from 9.785 ± 5.474 to 10.471 ± 4.199). Regarding the right side of the face, a positive outcome has been observed by the diminution in the porphyrins score in most of the evaluated individuals, mostly in AG1 and AG2 (Table 3), although a very small effect size has been recorded (95% CI = −4.3131, 4.6733; Cohen's d = 0.02189). In AG2 the porphyrins score slightly decreased after 5 weeks of the topical treatment from 12.903 ± 10.138 to 11.851 ± 11.701. Overall, the results suggest that rosehip oil may have a beneficial effect in reducing the porphyrins score in acne-affected or oily skin individuals. Regarding the score of pores, insignificant differences have been recorded both between age groups and the evaluated period (Supplementary Figure S2), which is supported by the very small effect sizes recorded on all sides of the face.

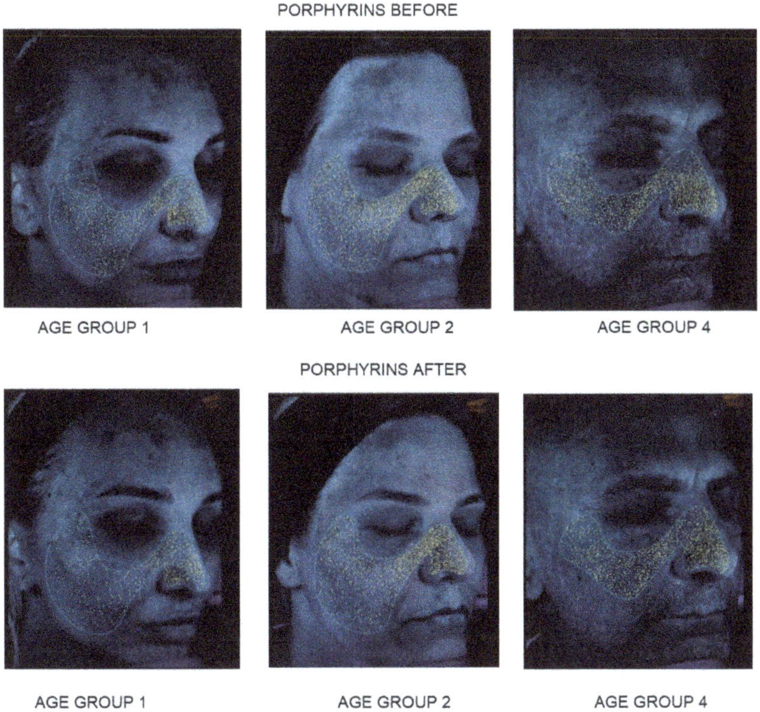

Figure 8. The assessment of porphyrins evaluated with the VISIA analysis system before and after 5 weeks of the rosehip oil treatment.

4. Discussion

The present study demonstrated the significant potential of *Rosa canina* oil in improving facial skin characteristics, particularly in reducing wrinkles, UV-induced spots, and porphyrin levels. These findings align with the previous research highlighting rosehip oil's antioxidant, anti-inflammatory, and skin-regenerative properties [2,3]. Several studies support the therapeutic potential of *R. canina* oil, attributing its bioactivity to its rich antioxidant composition. For instance, a study conducted by Roman et al. (2013) highlighted the variability of phenolic compounds and antioxidant activity in wild *R. canina* populations from Transylvania, reporting similar values for the polyphenol content and antioxidant potential [32]. Similarly, Koczka et al. (2018) confirmed the significant antioxidant activity of rosehip oil, emphasizing its protective effects against oxidative skin damage [33]. Moreover, studies reported total carotenoid contents ranging between 8.00 and 49.00 mg/100 g, which are consistent with the values obtained in the present study for cold-pressed oil [34]. In the study conducted by Szentmihályi et al. (2002) [11], it was reported that *R. canina* seed oil contains about 170 µg/g of total carotenoids, which is equivalent to 17 µg/mL. This result, although lower, is comparable to the value recorded for the rosehip oil under study [11]. Similarly, in the study conducted by Demir et al. (2014), carotenoid contents in the range of 25–35 µg/mL were reported, which demonstrates the methodological consistency and quality of the studied samples [35]. In comparison with other studies, the total carotenoid content in *R. canina* seed oil ranges from 20 to 86 µg/g, depending on the extraction method used [10]. The results obtained for the rosehip originated from Băișoara (32.687 μg/mL) fall within this range, being comparable with the values obtained by cold-pressing in the mentioned study. Furthermore, Barros et al. (2011) indicated a total carotenoid content of approximately 15–30 μg/mL in *R. canina* oil, values similar to those obtained in the present study for the rosehip from Băișoara [36]. Interestingly, ascorbic acid proves to be present in high amounts in seeds [36]; however, it proves to be absent in seed oil. Similarly, Koczka et al. (2018) confirmed the significant antioxidant activity of rosehip oil, emphasizing its protective effects against oxidative skin damage [33]. Recent studies have elucidated a diverse profile of individual carotenoids in rosehip seed oil. An analysis performed by Medveckienė et al. identified five carotenoids in the seed of *Rosa* spp., namely α-carotene, lutein, zeaxanthin, cis-lycopene, and trans-lycopene, underscoring the distinct carotenoid pattern compared to the flesh of the fruit [34,37]. Complementarily, rosehip seed oil also contains other major carotenoids, particularly lycopene, β-carotene, and rubixanthin, which substantially contribute to its bioactive potential [38]. In further support of this diversified carotenoid profile, it has been observed that while lycopene and β-carotene are predominant, only traces of lutein, zeaxanthin, and rubixanthin are present, emphasizing the complexity of the carotenoid distribution within different parts or extracts of rosehips [39].

Moreover, a comprehensive review expanded the known carotenoid spectrum in rosehip seed oil by detecting additional minor carotenoids, such as violaxanthin, γ-carotene, and neochrome, alongside rubixanthin and lycopene [40]. These findings are in agreement with reports that quantify the total carotenoid content in rosehip oil; for instance, the oil may contain carotene pigments at concentrations up to 107.7 mg/kg, further highlighting its richness in these bioactive compounds. The identification of both provitamin A carotenoids (e.g., α-carotene and β-carotene) and non-provitamin A carotenoids (such as lutein and zeaxanthin) suggests significant therapeutic implications, including antioxidant and photoprotective benefits [41]. Collectively, these studies provide robust evidence that rosehip seed oil possesses a complex and varied carotenoid composition. The presence of both major and minor carotenoids not only contributes to its color and potential health benefits, but also positions rosehip seed oil as a valuable source of natural bioactive compounds with applications in pharmaceutical and cosmetic fields. These findings, along with the laboratory analyses performed on the *R. canina* oil used in this study, reinforce the scientific foundation for its application in clinical research [22,42,43].

The application of topical rosehip oil has garnered interest in esthetic dermatology due to its potential benefits for skin rejuvenation and improvement. The VISIA skin analysis has been validated in several studies for its effectiveness in objectively quantifying changes in skin conditions, including the pore size, skin texture, wrinkles, and pigment spots [44–47]. Specific features, like the TruSkin Age score, help clinicians and researchers understand a patient's skin health relative to their chronological age, underlining skin conditions such as UV damage and discoloration [48]. This makes it an invaluable asset for both clinical studies and cosmetic practices involving topical treatments like rosehip oil. The research indicates that rosehip oil contains essential fatty acids and antioxidants, which may aid in hydrating and rejuvenating the skin [49,50]. Studies suggested that rosehip oil can improve the appearance of skin by reducing fine lines and enhancing overall texture [51,52]. Studies that monitor skin improvements using VISIA after topical applications demonstrate that it can measure enhancements in skin parameters, thus providing a quantitative basis for evaluating the efficacy of rosehip oil in skincare. The insights from the VISIA analysis not only quantify improvements but also assist practitioners in tailoring personalized skincare regimens based on individual skin responses. Given the clinical evidence supporting rosehip oil's efficacy in skin rejuvenation, its combined assessment with the VISIA system could help solidify its place in therapeutic and cosmetic dermatology. Thus the integration of topical treatments and this system is critical in advancing the field of esthetic medicine, where objective measurements can help inform strategies and evaluate outcomes more effectively [53,54]. Nonetheless, the VISIA system has several limitations, including a limited depth penetration, primarily assessing the skin surface and superficial features. Its accuracy may be affected by diverse skin types and tones, potentially leading to inconsistent results. Its operator-dependent variability and limited dynamic range may also impact results. Furthermore, the system primarily evaluates visual skin features, not functional aspects like the skin barrier function or hydration, highlighting the need for the careful interpretation of results and the consideration of potential biases.

The VISIA system may support the broader adoption of treatments that promote skin rejuvenation through naturally derived products like rosehip oil. These findings, along with the laboratory analyses performed on the *R. canina* oil used in this study, reinforce the scientific foundation for its application in clinical research [22,42,43]. This section discusses the results in the context of similar studies, emphasizing the consistency and discrepancies observed across different investigations.

4.1. Skin Improvements Evaluated with the VISIA System on the Front Side of the Face

The distinct positions made it possible to study the influence on the various skin parameters, including texture, wrinkles, blemishes, and pores, imposed by the topical treatment with rosehip oil (Figure 9A). According to the PCA, the first two components accounted for 55.4% of the total variation. The grouping pattern proved to be age-related, specifically in the case of AG1 and AG2. Conversely, individuals who were part of AG3 presented discrepancies in terms of evaluated skin parameters but also regarding the mature skin type, which influenced specific outcomes. Lastly, individuals in AG4 had specific outcomes according to their skin type, but also due to their sensitivity to the treatment (Figure 9B). Regarding the HCA, the primary cluster positioned the spots score appearance with slight changes after 5 weeks, as evidenced by the fluctuations in the importance score intensity (Figure 9C). Furthermore, changes in the overall score of wrinkles have been regarded, particularly in age groups 3 (43–52 years old) and 4 (53–60 years old) with a significant drop as evidenced by the changes in color intensity after the end of the treatment period. Regarding UV and brown spots, a slight improvement and reduction in their score has been observed, particularly in several individuals from AG2 and AG3. Concerning porphyrins, an overall improvement in their score has been noticed after topical applications of rosehip oil, as demonstrated by the fluctuations in the color intensity from light red to darker blue, which implies a potential antibacterial and anti-inflammatory effect and a lower risk of bacterial growth (i.e., *Propionibacterium acnes*) that may lead to the formation of acne and other skin conditions [55,56]. The slight increase in the porphyrins score, particularly in AG4, may indicate a natural variation in bacterial activity and diversity with age (i.e., hormonal imbalance in females) [57] or a temporary inflammatory skin response [58]. The statistical power of this pilot study has been evaluated as well as an appropriate sample size for the UV spots (power = 0.808) and porphyrins (power = 0.890), suggesting that for these skin characteristics, the number of allocated individuals proved to be sufficient to assess the outcome of the rosehip oil treatment (Supplementary Table S5).

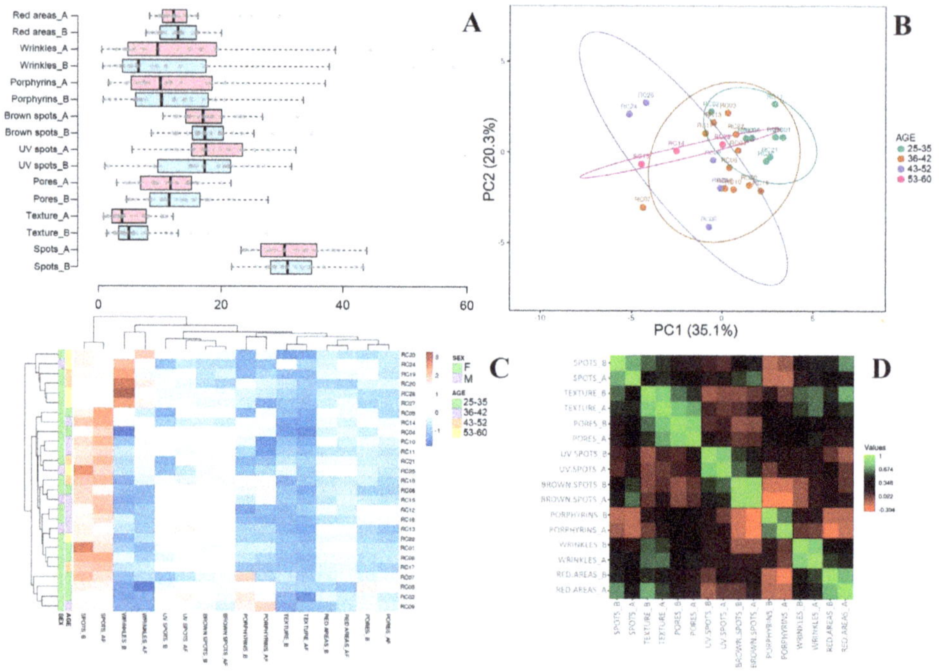

Figure 9. (**A**) A comparison of different skin parameters for the front side of the face, before and after topical applications with rosehip oil at the beginning and after 5 weeks of treatment. (**B**) The principal component analysis of the skin characteristics from the front side of the face and their respective cluster position divided by age groups. (**C**) The hierarchical clustering and heatmap visualization of the improved or detrimental outcomes following rosehip oil as a topical face treatment; the figure presents the skin parameters obtained from the front side of the face. (**D**) The Pearson correlation before and after 5 weeks for the front side evaluated skin parameters. Correlation coefficients between the evaluated characteristics are displayed in red and green.

The application of rosehip oil also has a positive effect on skin texture, which is among the most popular skin concerns as it increases the appearance and depth of pores [59]. Following the treatment, a reduction in the score of texture has been observed in all age groups, particularly in AG1 (23–35 years old). Regarding the red areas, a slight increase in their score has been observed in several individuals, which might be explained by the skin's sensitivity to higher concentrations of vitamin A. While the majority of participants experienced improvements, some individuals exhibited an increase in wrinkle scores and redness. These outcomes may be attributed to individual differences in skin sensitivity, environmental factors, or an inconsistent use of sun protection. Similar studies have noted that high concentrations of vitamin A, while beneficial [6], can sometimes lead to transient skin irritation (e.g., a burning sensation, dry skin, and exfoliation) in sensitive individuals [60]. Moreover, lifestyle factors such as UV exposure and sleep habits (i.e., sleeping on one side of the face) may have influenced the observed variations in skin improvements, as noted in previous dermatological studies [61,62]. According to the correlation plot, the spots showed a moderate and negative association with the porphyrins and a positive association with redness, both at the beginning and after 5 weeks of treatment. Following the texture and pores, they presented a significant and negative association with porphyrins and brown and UV spots before and after the topical treatment. Subsequently, UV spots showed a negative association with redness (Figure 9D; Supplementary Table S3).

The application of rosehip oil did not have a noteworthy effect on the appearance of pores in all the evaluated age group individuals. Overall, it can be discerned that rosehip oil can effectively improve the skin texture (i.e., pores, wrinkles, and porphyrins); however, it may show discrepancies in terms of redness, which is strongly influenced by each individual's skin type.

4.2. Skin Improvements Evaluated with the VISIA System on the Left Side of the Face

Regarding the left side of the face (Figure 10A), significant differences have been observed as compared with the front side. According to the PCA, the first two components accounted for 58% of the total variation (Figure 10B). The grouping pattern proved to highlight AG1, which presented relatively different outcomes compared with the other age groups. Conversely, individuals who were part of AG2 presented insignificant changes in terms of evaluated skin parameters, as seen by the grouping outline in the center of the plot. Lastly, individuals in AG3 and AG4 had specific and similar outcomes, mostly due to their skin type and age.

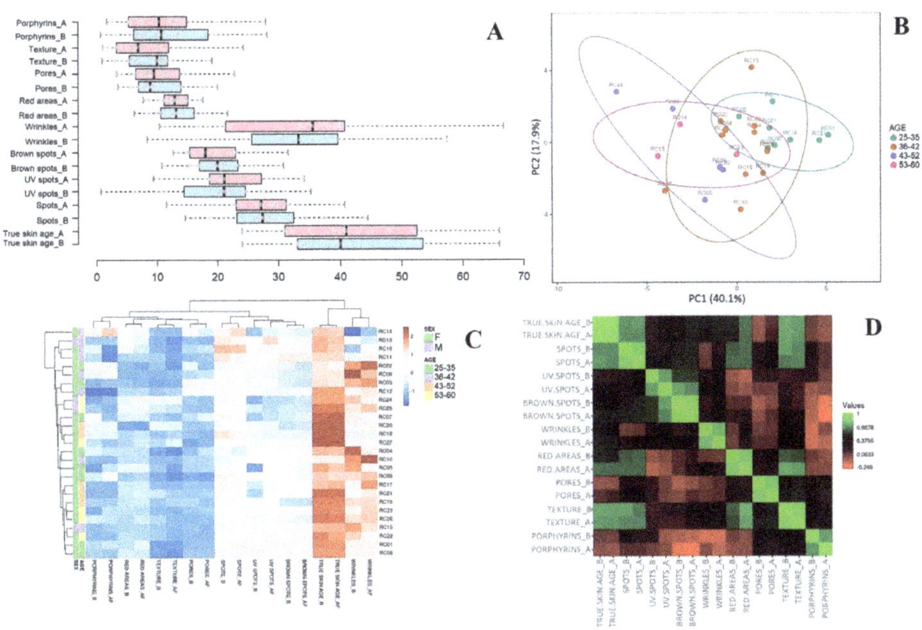

Figure 10. (**A**) A comparison of different skin parameters for the left side of the face, before and after topical applications with rosehip oil at the beginning and after 5 weeks of treatment. (**B**) The principal component analysis of the skin characteristics and their respective cluster position divided by age groups. (**C**) The hierarchical clustering and heatmap visualization of the improved or detrimental outcomes following rosehip oil as a topical face treatment; the figure presents the skin parameters obtained from the left side of the face. (**D**) The Pearson correlation before and after 5 weeks of treatment for the evaluated skin parameters. Correlation coefficients between the evaluated characteristics are displayed in red and green.

In the HCA, it can be seen that a considerable alteration in the wrinkles score has been regarded in all the evaluated individuals irrespective of their gender and age group (Figure 10C). Thus, although in several individuals the wrinkle score has been diminished, compared with others the score presented a significant upsurge. These fluctuations and discrepancies might be explained by the individual's skin type and their sleeping habits (i.e., the side of the face they usually sleep on).

The following skin characteristic (i.e., true skin age) underlines the positive outcome of the application of rosehip oil on the authentic appearance of the skin. Several individuals have noticed an improvement in their skin complexion and appearance after the conclusion of the treatment period. Regarding brown and UV spots, a general improvement of the skin and a reduction in pigmentation areas have been observed in most of the treatment groups, particularly in AG2 and AG3. The reduction in UV spots observed in this study corresponds with previous research indicating the photoprotective properties of rosehip oil. The present study observed a reduction in UV spots across all age groups, which further corroborates these findings.

The spots presented relatively trivial modifications in terms of the size reduction and appearance. Furthermore, the appearance of pores has been significantly diminished after the treatment period in all evaluated age groups. A similar pattern has been observed in the case of texture, with noteworthy outcomes similar to the popular skincare result of the "glass skin effect" in AG1 and AG2. Although rosehip oil has facilitated the reduction in red areas, a number of individuals experienced a slight increase in their score, which can be explained by the skin's sensitivity to vitamin C. Lastly, the score of porphyrins has been drastically reduced predominantly in AG2 and AG4, which demonstrates its effectiveness on mature skin. The correlation plot presented dissimilar outcomes in terms of the skin assessment (Figure 10D; Supplementary Table S4). Thus, the true skin age and the score of spots presented a negative association with pores and porphyrins and a significant positive association with texture, spots, and redness. UV spots presented a negative association with most of the evaluated parameters, including texture, pores, and redness. A similar negative correlation with the skin characteristics has also been observed in brown spots and wrinkles. Conversely, a positive association between redness and texture has been obtained. The statistical power for the left side of the face generated an appropriate sample size for porphyrins (power = 0.890) and red areas (power = 0.99), suggesting that for these skin characteristics, the number of allocated individuals proved to be sufficient to assess the outcome of the rosehip oil treatment (Supplementary Table S6).

The topical application of rosehip oil not only improves the overall skin quality but also targets specific concerns related to aging. This effect has also been observed in other studies [63,64]. However, it is important to note that individual skin reactions may vary, and it is important to comprehend the full impact and outcome of the rosehip oil treatment on different skin types and conditions.

4.3. Skin Improvements Evaluated with the VISIA System on the Right Side of the Face

Regarding the right side of the face (Figure 11A), a significant reduction in wrinkles has been regarded after the treatment with rosehip oil. In several individuals, a slight increase in their score has been regarded, which may be due to the individuals sleeping on one side of the face. The PCA scores presented a total variance of 56.1%. As observed in the other evaluated facial parameters, individuals in AG1 tend to have a similar grouping pattern. The subsequent group presents similar outcomes, as also noticed in the previous section and the left side of the face. The following age groups tend to be closely grouped mainly due to the appropriate age and comparable results. Regarding the HCA, the true skin age score has significantly decreased, particularly in the case of age groups 1 and 2. UV and brown spots presented a slightly reduced area in all age groups (Figure 11C). Additionally, the score of spots has also presented a reduced area in all age groups. Texture is among the most debated and popular skin characteristic particularly, among young individuals due to existing skincare trends. Interestingly, a positive influence of the rosehip oil topical application has been regarded as observed by the significant reduction in texture scores. Additionally, Koczka et al. (2018) confirmed the role of *Rosa canina* oil in enhancing skin elasticity and reducing signs of aging through its rich antioxidant profile [33]. The present study supports these findings by demonstrating significant improvements in skin texture and a decrease in the appearance of fine lines, particularly in AG1 (25–35 years old) and AG2 (36–42 years old) [33].

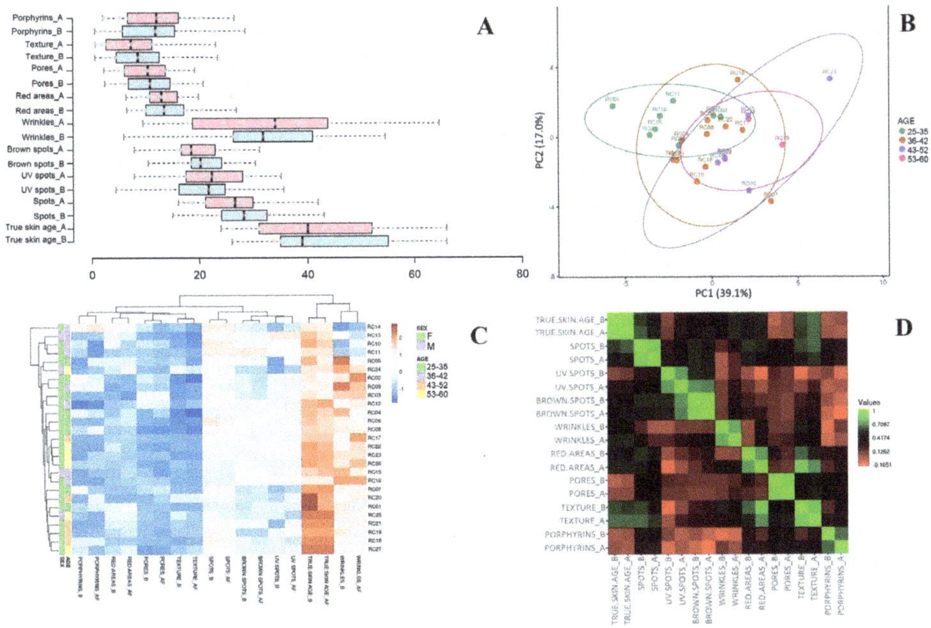

Figure 11. (**A**) A comparison of different skin parameters for the right side of the face, before and after topical applications with rosehip oil at the beginning and after 5 weeks of treatment. (**B**) The principal component analysis of the skin characteristics and their respective cluster position divided by age groups. (**C**) The hierarchical clustering and heatmap visualization of the improved or detrimental outcomes following rosehip oil as a topical face treatment; the figure presents the skin parameters obtained from the right side of the face. (**D**) The Pearson correlation before and after 5 weeks of treatment for the evaluated skin parameters. Correlation coefficients between the evaluated characteristics are displayed in red and green.

Regarding the appearance of pores, the significant reduction in their score is evident, as also observed in the case of texture, particularly in the case of AG2 and AG3. Furthermore, the score of red areas presented several discrepancies among the individuals; an effect was also observed in the other evaluated face positions and may be influenced by the skin type and sensitivity to several compounds present in rosehip oil (i.e., vitamin C). Overall, the results suggest that individuals in AG3 experienced a milder reaction compared to AG4, with only a slight inflammation observed in the former group. Additionally, it is important to note that individual variations in responses were observed within AG4, as seen in the case of patient RC24. Lastly, regarding the score of porphyrins, it can be observed that the right side has comparable results with the left side, which may be due to each individual's skin type (i.e., oily and mixed). The decline in porphyrin levels, often linked to bacterial activity, suggests that Rosa canina oil may possess antimicrobial properties. Previous research by Deliorman Orhan et al. (2007) demonstrated the anti-inflammatory effects of *Rosa canina* extracts, which contribute to the maintenance of a healthy skin microbiome [7]. The results of this study indicate a notable decrease in porphyrins, particularly in AG2 (36–42 years old) and AG4 (53–60 years old), suggesting that the oil may help regulate the bacterial activity on the skin. This aligns with the findings by Oargă et al. (2024), who emphasized the potential of rosehip-based dermatological products in addressing inflammatory skin conditions [4]. Topical treatments using rosehip oil are particularly beneficial for improving the skin tone and texture, especially in older age groups. The reduction in red areas and porphyrins indicate a positive impact on skin health and appearances across all age groups.

The correlation plot presents different outcomes as compared to the front side of the face (Figure 11D; Supplementary Table S5). Thus, the true skin age presented a negative association with pores and porphyrins, whereas the spots presented a positive association with texture and a negative association with wrinkles. UV spots presented a negative association with most of the evaluated parameters, including texture, pores, redness, and wrinkles. A similar negative correlation with the skin characteristics has also been observed in brown spots and wrinkles. Conversely, a positive association with texture has been obtained. The statistical power for the right side of the face generated an appropriate sample size for the porphyrins (power = 0.940) and texture (power = 0.79), which is sufficient to assess the outcome of the rosehip oil treatment, which, among others, might be due to each individual's skin type (Supplementary Table S6).

Overall, the variability in scores across different face positions suggests that individual differences in skin type and sensitivity play a significant role in the effects of rosehip oil. Further research might explore whether this factor impacts the efficacy of rosehip oil in reducing porphyrins and improving skin texture, but also its use as a replacement for vitamin A (natural retinol source). Recent research indicates that rosehip seed oil contains significant levels of compounds that can serve as precursors to vitamin A, commonly referred to as provitamin A. Notably, it was demonstrated that rosehip oil, along with chokeberry oil, shows high concentrations of provitamin A, which includes compounds such as β-carotene that are conventionally converted to active vitamin A in the body [65]. This provitamin A content is associated with skin benefits, as vitamin A derivatives are recognized for their role in promoting skin regeneration, improving cellular turnover, and mitigating signs of aging [66]. Furthermore, as exemplified in Section 3.2, the research on the chemical composition of rosehips confirms the presence of carotenoids in the seeds, which are key precursors for vitamin A activity [34]. Identified compounds like α-carotene and β-carotene in seeds support the notion that these oils provide a natural source of vitamin A precursors. The conversion of these carotenoids into retinol contributes to the oil's antioxidant activity, aiding in reducing oxidative stress in skin cells, a mechanism critical in decreasing fine lines and promoting collagen synthesis, which is essential for skin health [41,66].

Pogostemon cablin Benth. (patchouli) and *Moringa oleifera* (drumstick) oils, both natural antioxidants, prove to be ideal for cosmetic products like body creams due to their capacity to fight free radicals and maintain an optimum skin hydration. The research conducted by Isnaini et al. (2023) evaluated the effectiveness of a body cream formulation using an accelerated stability test technique [67]. The formulation remained stable for at least six months at room temperature, indicating it can be used for a year. After two weeks of applications on dry skin, the formulation increased the moisture content by 30% to 60%. This suggests that the formulation could be used as a commercial cosmetic product in the future and that these natural compounds prove to be valuable for future cosmetic applications [67]. A different study focused on the design and characterization of nanostructured lipid carriers (NLCs) and NLC-based hydrogels using *Passiflora edulis* seed oil. The developed NLC presented a spherical shape, narrow particle sizes, and a high encapsulation efficiency. The nanoparticles showed a superior tyrosinase inhibitory activity and skin retention compared to non-encapsulated oil. Furthermore, the formulations proved to have a non-cytotoxic effect towards HaCat cells and showed suitable viscosity and texture properties for skin applications [68]. The study conducted by Hugo Infante et al. (2023) aimed to assess the permeation depth, antioxidant capacity, and clinical efficacy of *Melaleuca alternifolia* pure essential oil and a nanoemulsion to prevent skin photoaging [69]. Results showed that the nanoemulsion had a lower antioxidant capacity and a higher penetration through the stratum corneum, improving the stratum granulosum morphology. The nanoemulsion also increased the papillary depth, dermis echogenicity, and collagen fibers. *Melaleuca alternifolia* essential oil has the potential to improve photoaged skin, as it can reach deeper skin layers [69].

In a different study, the Visia® Camera System was validated by establishing correlations among measurements, such as the percentile, feature count, and absolute score. The results showed a high level of correlation among the evaluated measurements, with 88.9% of the correlations being statistically significant. The majority of cases showed clear correlations between the variables, with 50% of the absolute values above 0.945; however, UV spots and wrinkles were insignificant as also observed in the present study. The Visia® Camera System has been used to investigate whether skin aspects can be improved by a certain cosmetic product line [70]. In the following year, Henseler continued their research and evaluated the reproducibility and accuracy of facial wrinkles using this system. The standard deviation from frontal captures was about 3%, with an average deviation of 3.36% during the first capture session and 3.4% during the second. The standard deviation of measurements was about 9% when comparing percentiles. The accuracy of the measurements was high, with a correlation coefficient of >0.8 and a statistically significant p-value of <0.001. The TruSkin Age® was slightly higher than the calendrical age by 1.37 years for both facial sides. The study found a satisfactory precision in repeated captures, with absolute scores being preferred over percentiles due to their better precision [1].

These results highlight the potential of the studied *Rosa canina* oil to improve the appearance and health of the skin, supporting its use in skin care products and anti-aging treatments. A detailed investigation of the molecular mechanisms by which *Rosa canina* oil improves skin health could also provide valuable information for the development of more effective and innovative treatments. Future studies should include participants with different skin types to assess the universality and applicability of the observed benefits, thus ensuring that all categories of users can benefit from these effects. In addition, research into the synergistic effects of *R. canina* oil in combination with other bioactive compounds could lead to the development of products with increased efficacy, thus enhancing the therapeutic and cosmetic potential of this valuable natural ingredient. Rosehip oil from Băișoara has considerable therapeutic potential in improving the health and appearance of the skin, supporting its use in anti-aging treatments and high-quality cosmetic products.

In summary, the findings of this study align with the existing literature supporting the dermatological benefits of *Rosa canina* oil. The observed improvements in wrinkle reduction, UV spot mitigation, and bacterial activity regulation reinforce its potential as a valuable ingredient in anti-aging and skincare formulations. Future studies should aim to refine application protocols and explore synergistic formulations to maximize their efficacy across different skin types. Consequently, oxidative stability is another critical consideration regarding the skin sensitivity related to rosehip oil. The oil is susceptible to oxidation when exposed to light and air, which can not only decrease its efficacy but also potentially lead to the formation of irritative byproducts [38]. This degradation can cause a heightened irritation and sensitization upon prolonged exposure, particularly in those not accustomed to using oils topically, which may be further evaluated [71].

5. Limitations of the Pilot Study and Future Perspectives

This pilot study investigating the topical application of rosehip oil presents several limitations that must be acknowledged in the interpretation of its findings. First and foremost, the small sample size compromises the statistical power of this study, making it difficult to generalize the results to a larger population. A limited number of participants may not adequately represent the variability found in a broader demographic, which could affect the overall efficacy and safety conclusions drawn from this study [72]. Future studies should aim to include a larger and more diverse population to strengthen the findings. Additionally, the lack of a control group poses a considerable drawback. Without a placebo or control comparison, it is challenging to determine whether the observed effects on skin conditions can be directly attributed to rosehip oil or if they could result from spontaneous changes over time, a regression to the mean, or participant expectations. Studies have shown that control groups are essential for validating the effectiveness of treatment interventions, as they provide a critical benchmark [73]. Alternative study designs and analytical approaches should be considered to address these issues. For instance, a within-subject or split-face design could be employed, where one side of the face is treated with rosehip oil while the other side serves as a comparator receiving either a placebo or the standard daily skincare routine. This design would mitigate inter-individual variability and allow for a more reliable attribution of treatment effects. Additionally, methods such as historical control analyses or the implementation of a delayed treatment group could provide additional context and serve as pseudo-controls when a randomized control group is not feasible. Such methodological refinements are essential to isolate the treatment effect reliably and enhance the validity of the findings, thus also reinforcing the continued recommendation for appropriate skincare practices, including sun protection, irrespective of the season. Furthermore, variability in participant compliance can introduce additional complexity in the data interpretation. In studies examining subjective outcomes, such as the skin appearance or discomfort, differing levels of adherence to the treatment regimen can lead to inconsistencies in results. Participants may not apply the treatment as instructed, leading to variable exposure levels that could affect their outcomes and the overall efficacy of the rosehip oil. This potential gap in compliance monitoring has been noted in various studies [74]. To mitigate this issue, future studies should monitor adherence closely through methods such as usage diaries or technology-assisted tracking, ensuring participants apply the treatment as directed.

This pilot study's short duration of five weeks limits its ability to capture long-term effects and potential late-onset responses to the treatment; however, it facilitates significant knowledge regarding its use in future studies with longer treatment periods, focused on the oil's stability during storage at higher temperatures that may significantly reduce the polyphenol content and, most importantly, the fatty acids content [75]. Thus, dermatological therapies may require longer periods to demonstrate sustained benefits. The short follow-up may result in an incomplete understanding of the full therapeutic effects and any potential adverse reactions that could manifest over time [76]. Future studies should aim to address these limitations by ensuring longer study durations (i.e., particularly during summer periods), which would help in better understanding the sustained effects of rosehip oil over time and capture any delayed responses. Compliance monitoring could be enhanced by using digital reminders or adherence tracking mechanisms to improve the participant engagement and data integrity. Finally, a double-blind design should be considered to minimize bias and further substantiate the findings regarding the effectiveness of rosehip oil in dermatological applications. By addressing these limitations, subsequent research can better elucidate the therapeutic potential of rosehip oil and provide stronger evidence for its use in topical dermatological applications.

The phenomenon of increased brown spots following the application of rosehip oil—especially during late fall when the UV exposure is minimal—can be attributed to several interrelated factors, including the oil's biological action on the skin, the potential for inflammatory responses, and the physiological tendency of melanocytes (pigment-producing cells in the skin) to react to topical applications. Additionally, while rosehip oil is widely regarded as beneficial for various dermatological conditions, including scarring and inflammatory conditions, individuals may still experience variability in reactions based on their unique skin microbiome and existing skin conditions [77]. Furthermore, individuals with pre-existing pigmentation should exercise caution, as their skin may be more responsive to such stimuli, especially during seasonal transitions when humidity and temperature can fluctuate dramatically. A compromised barrier can also lead to increased transepidermal water loss. If the skin is responding to hydration or repair agents like rosehip oil, this can cause localized sensitivity, resulting in an inflammatory cascade that ultimately leads to pigmentation irregularities. This study's design should account for potential seasonal UV fluctuations by incorporating controls or standardized measures of UV exposure throughout the study period. This is essential to ensure that the observed skin changes are truly due to the topical treatment rather than external UV factors. In doing so, this study avoids the misconception that sun protection might be omitted during the fall; on the contrary, it reinforces that even at periods of lower ambient UV radiation, individuals remain susceptible to photo-induced skin changes [4].

Other investigations should explore the molecular mechanisms underlying the observed effects, particularly regarding collagen synthesis and microbial modulation. For example, the potential of an aerogel-based formula has been investigated in order to naturally conceal shiny facial skin. The aerogel ingredient showed positive in vitro results in measuring the shine induced by a mixture of oleic acid and mineral water. In vivo, two different aerogel formulas (1% and 2% concentrations, in an O/W cosmetic emulsion) were tested on Chinese women under hot and humid conditions known to enhance facial shine. Interestingly, an immediate light scattering effect, masking shine, and a noticeable anti-shine effect in extreme conditions have been observed [78]. As previously highlighted, combining rosehip oil with other bioactive compounds could also be examined to enhance its therapeutic potential. The observed improvements in wrinkle reduction, UV spot mitigation, and bacterial regulation reinforce its potential as a valuable ingredient in anti-aging and skincare formulations [79]. Future-oriented studies should aim to refine application protocols and explore synergistic formulations to maximize their efficacy across different skin types. In this regard, the delivery of hydrophilic tripeptide-3 to the skin using micro-emulsions or nanoemulsions for facial oil reduction has been evaluated. The optimized combination yielded translucent oil-in-water tripeptide-3 nanoemulsions with a high skin penetration and retention. The nanoemulsions not only decreased the sebum production but also enhanced skin moisture levels [80]. Overall, the results suggest that rosehip oil may have a positive impact on skin texture and appearance and further studies could explore the long-term effects of the continued use of this treatment on various kin types (i.e., normal, dry, sensitive, oily, and/or combination).

Rosehip oil is known for its vitamin content, particularly A derivatives (such as trans retinoic acid) and essential fatty acids, which can influence skin pigmentation [20]. Retinoic acid, a well-known active compound in many anti-aging and skin-rejuvenating products, can induce increased cell turnover and desquamation. While this often results in the improvement of skin texture and may clear older pigmented cells, it can also paradoxically stimulate the formation of new pigmentation if not managed properly. In cases where the skin is reacting to a new products or there is excessive exfoliation, transient post-inflammatory hyperpigmentation can occur, where new spots may appear due to the irritation of existing melanocytes [4,81]. To mitigate these outcomes, sunscreen formulations with rosehip oil may be integrated in skincare routines [41].

As highlighted in the present study, rosehip oil contains elevated levels of carotenoids (up to 28.4 μg/mL), which when combined with sunscreen might mitigate photoaging and sunburn [82], but also encourages skin cell renewal and hydration [9].

Lastly, the open-label design of this study introduces a bias that can affect the validity of the findings. Participants knowing they are receiving an active treatment may have heightened expectations, leading to biased self-reports regarding efficacy and satisfaction. Open-label designs can be associated with inflated response rates, which complicate the interpretation of the study and reduce its internal validity. While the study findings contribute to understanding rosehip oil's topical applications, the identified limitations—the small sample size, lack of a control group, compliance variability, short duration, and open-label nature—highlight the need for further research employing rigorous methodologies to validate these results and better define the therapeutic role of rosehip oil in dermatological treatments [83].

6. Conclusions

This pilot study evaluated the therapeutic effects of rosehip oil, mainly *R. canina* from the Băișoara provenance, which presented potential therapeutic purposes. The results showed improvements in skin health, evidenced by a reduction in wrinkles, UV spots, and porphyrins. The analysis with the VISIA system showed a reduction in the depth of wrinkles in several volunteers, confirming the efficacy of rosehip oil in improving the signs of aging. In addition, results demonstrated a slight decrease in UV-induced spots, suggesting that the oil provides effective protection against damage caused by UV exposure.

Evaluations of the rosehip oil treatment showed a noticeable improvement in skin texture and a reduction in redness and spots, suggesting anti-inflammatory and regenerative effects, which may be due to their carotenoids content, including lutein, lycopene, β-carotene, and zeaxanthin, but also their phenolic content and antioxidant activity. Furthermore, a decrease in porphyrins has also been observed, reflecting a decrease in bacterial activity and an overall improvement in skin health. The promising results of this pilot study highlight the need for further research to optimize the application and formulation of rosehip oil-based treatments. These results are the foundation for future and long-term studies with larger sample sizes, which will provide more comprehensive insights into its sustained efficacy. Additional studies should aim to refine application protocols and explore synergistic formulations to maximize their efficacy across different skin types.

Given the positive effects observed on wrinkles and skin texture, it is recommended that *R. canina* oil may be included in cosmetic formulations to reduce the signs of aging. This extended use could capitalize on its antioxidant and regenerative benefits. Collectively, these findings underscore the dual functionality of rosehip seed oil; its richness in carotenoids, particularly provitamin A, aligns with its documented benefits of enhancing skin texture, reducing wrinkles, and improving the overall skin condition. Such evidence supports the continued utilization of rosehip seed oil in dermatological applications and cosmetic formulations targeting anti-aging and skin renewal processes.

Supplementary Materials: The following supporting information can be downloaded at https://www.mdpi.com/article/10.3390/cosmetics12030125/s1: Figure S1: HPLC chromatogram of carotenoids identified from rosehip seed oil. Each peak denotes an identified compound; Figure S2: Assessment of pores evaluated with the VISIA analysis system before (left chart) and after (right chart) 5 weeks of rosehip oil treatment; Table S1: Personal data regarding the volunteers and their assignment group based on age; Table S2: Identification of carotenoids content in *R. canina* oil from the Băișoara area; Table S3: Correlation data showing relationships among the evaluated skin characteristics for the front side of the face; Table S4: Correlation data showing relationships among the evaluated skin characteristics for the right side of the face; Table S5. Correlation data showing relationships among the evaluated skin characteristics for the left side of the face; Table S6: Statistical power and sample size of the pilot study using goal seek. The normality test using the Shapiro–Wilk test and the Mann–Whitney pairwise comparison with Bonferroni corrected p values can be found in the Supplementary Files S1–S3.

Author Contributions: D.P.O.: Conceptualization, Formal Analysis, Data Curation, Investigation, Methodology, Visualization, Writing—Original Draft, Writing—Review and Editing. M.C.-C.: Conceptualization, Data Curation, Formal Analysis, Investigation, Methodology, Validation, Writing—Original Draft, Writing—Review and Editing. S.A.N.: Data Curation, Investigation, Methodology, Visualization, Writing—Review and Editing. M.I.C.: Conceptualization, Validation, Writing—Review and Editing, Supervision. All authors have read and agreed to the published version of the manuscript.

Funding: The authors declare that financial support was not received for the research, authorship, and/or publication of this article.

Institutional Review Board Statement: The present study's protocol was approved by an appropriate Ethics Committee with the reference ID 194/03.10.2024.

Informed Consent Statement: Informed consent was obtained from all subjects involved in the study.

Data Availability Statement: The original contributions presented in this study are included in the article/supplementary material. Further inquiries can be directed to the corresponding authors.

Conflicts of Interest: The authors declare that the research was conducted in the absence of any commercial or financial relationships that could be construed as a potential conflict of interest.

References

1. Henseler, H. Assessment of the reproducibility and accuracy of the Visia® Complexion Analysis Camera System for objective skin analysis of facial wrinkles and skin age. *GMS Interdiscip. Plast. Reconstr. Surg. DGPW* **2023**, *12*, Doc07.
2. Angelov, G.; Boyadzhieva, S.S.; Georgieva, S.S. Rosehip extraction: Process optimization and antioxidant capacity of extracts. *Cent. Eur. J. Chem.* **2014**, *12*, 502–508. [CrossRef]
3. Ayati, Z.; Amiri, M.S.; Ramezani, M.; Delshad, E.; Sahebkar, A.; Emami, S.A. Phytochemistry, traditional uses and pharmacological profile of rose hip: A review. *Curr. Pharm. Des.* **2018**, *24*, 4101–4124. [CrossRef] [PubMed]
4. Oargă, D.P.; Cornea-Cipcigan, M.; Cordea, M. Unveiling the Mechanisms for the Development of Rosehip-Based Dermatological Products: An Updated Review. *Front. Pharmacol.* **2024**, *15*, 1390419. [CrossRef]
5. Jabłońska-Ryś, E.; Zalewska-Korona, M.; Kalbarczyk, J. Antioxidant capacity, ascorbic acid and phenolics content in wild edible fruits. *J. Fruit Ornam. Plant Res* **2009**, *17*, 115–120.
6. Moldovan, I.; Pop, V.C.; Borsai, O.; Lukacs, L.; Ranga, F.; Culea, E.; Damian, G.; Cornea-Cipcigan, M.; Margaoan, R. Dynamics of bioactive compounds under the influence of yellow, blue, and violet light filters on *Hippophae rhamnoides* L. (sea buckthorn) fruits. *Horticulturae* **2023**, *9*, 1312. [CrossRef]
7. Deliorman Orhan, D.; Hartevioğlu, A.; Küpeli, E.; Yesilada, E. In vivo anti-inflammatory and antinociceptive activity of the crude extract and fractions from *Rosa canina* L. fruits. *J. Ethnopharmacol.* **2007**, *112*, 394–400. [CrossRef] [PubMed]
8. Kaur, C.; Kapoor, H.C. Antioxidants in fruits and vegetables–the millennium's health. *Int. J. Food Sci. Technol.* **2001**, *36*, 703–725.
9. Grajzer, M.; Prescha, A.; Korzonek, K.; Wojakowska, A.; Dziadas, M.; Kulma, A.; Grajeta, H. Characteristics of rose hip (*Rosa canina* L.) cold-pressed oil and its oxidative stability studied by the differential scanning calorimetry method. *Food Chem.* **2015**, *188*, 459–466. [CrossRef]
10. Ilyasoğlu, H. Characterization of Rosehip (*Rosa canina* L.) Seed and Seed Oil. *Int. J. Food Prop.* **2014**, *17*, 1591–1598. [CrossRef]
11. Szentmihályi, K.; Vinkler, P.; Lakatos, B.; Illés, V.; Then, M. Rose hip (*Rosa canina* L.) oil obtained from waste hip seeds by different extraction methods. *Bioresour. Technol.* **2002**, *82*, 195–201. [CrossRef] [PubMed]
12. Fromm, M.; Bayha, S.; Kammerer, D.R.; Carle, R. Identification and quantitation of carotenoids and tocopherols in seed oils recovered from different Rosaceae species. *J. Agric. Food Chem.* **2012**, *60*, 10733–10742. [CrossRef]
13. Hodisan, T.; Socaciu, C.; Ropan, I.; Neamtu, G. Carotenoid composition of Rosa canina fruits determined by thin-layer chromatography and high-performance liquid chromatography. *J. Pharm. Biomed. Anal.* **1997**, *16*, 521–528. [CrossRef]

14. Olsson, M.E.; Gustavsson, K.-E.; Andersson, S.; Nilsson, Å.; Duan, R.-D. Inhibition of cancer cell proliferation in vitro by fruit and berry extracts and correlations with antioxidant levels. *J. Agric. Food Chem.* **2004**, *52*, 7264–7271. [CrossRef]
15. Van der Walt, A.C. Transdermal diffusion stability and clinical efficacy of cosmetic formulations containing Rosa rubiginosa rosehip seed oil. Ph.D. Dissertation, North-West University (South Africa), Potchefstroom Campus, Potchefstroom, South Africa, 2016; pp. 1–157.
16. Truong, V.L.; Jeong, W.S. Hair Growth-Promoting Effects of Rosehip Rosa Canina Seed Oil in C57bl/6 Mice. *Prev. Nutr. Food Sci.* **2023**, *28*, 411–417. [CrossRef]
17. Patzelt, A.; Lademann, J.; Richter, H.; Darvin, M.E.; Schanzer, S.; Thiede, G.; Sterry, W.; Vergou, T.; Hauser, M. In vivo investigations on the penetration of various oils and their influence on the skin barrier. *Ski. Res. Technol.* **2012**, *18*, 364–369. [CrossRef]
18. Lin, T.-K.; Zhong, L.; Santiago, J.L. Anti-Inflammatory and Skin Barrier Repair Effects of Topical Application of Some Plant Oils. *Int. J. Mol. Sci.* **2017**, *19*, 70. [CrossRef] [PubMed]
19. Contri, R.V.; Kulkamp-Guerreiro, I.C.; da Silva, S.J.; Frank, L.A.; Pohlmann, A.R.; Guterres, S.S. Nanoencapsulation of rose-hip oil prevents oil oxidation and allows obtainment of gel and film topical formulations. *AAPS PharmSciTech* **2016**, *17*, 863–871. [CrossRef] [PubMed]
20. Zilles, J.C.; Duarte, L.P.; Ruaro, T.C.; Zimmer, A.R.; Külkamp-Guerreiro, I.C.; Contri, R.V. Nanoemulsion Containing Kojic Dipalmitate and Rosehip Oil: A Promising Formulation to Treat Melasma. *Pharmaceutics* **2023**, *15*, 468. [CrossRef]
21. Ojha, P.K.; Poudel, D.K.; Rokaya, A.; Maharjan, S.; Timsina, S.; Poudel, A.; Satyal, R.; Satyal, P.; Setzer, W.N. Chemical Compositions and Essential Fatty Acid Analysis of Selected Vegetable Oils and Fats. *Compounds* **2024**, *4*, 37–70. [CrossRef]
22. Valerón-Almazán, P.; Gómez-Duaso, A.J.; Santana-Molina, N.; García-Bello, M.A.; Carretero, G. Evolution of post-surgical scars treated with pure rosehip seed oil. *J. Cosmet. Dermatol. Sci. Appl.* **2015**, *5*, 161. [CrossRef]
23. Hosny, K.M.; Nahyah, K.S.A.; Alhakamy, N.A. Self-Nanoemulsion Loaded with a Combination of Isotretinoin, an Anti-Acne Drug, and Quercetin: Preparation, Optimization, and in Vivo Assessment. *Pharmaceutics* **2020**, *13*, 46. [CrossRef] [PubMed]
24. Zhou, Y.; Wu, J.; Lin, S.; He, J.; Deng, Y.; He, J.; Cheng, D. The Synergistic Effects of Rosehip Oil and Matrine Against Icerya Aegyptiaca Douglas (Hemiptera: Coccoidea) and the Underlying Mechanisms. *Pest Manag. Sci.* **2022**, *78*, 3424–3432. [CrossRef]
25. Cucu, A.-A.; Pașca, C.; Cucu, A.-B.; Moise, A.R.; Bobiş, O.; Dezsi, Ş.; Blaga Petrean, A.; Dezmirean, D.S. Evaluation of the Main Macro-, Micro- and Trace Elements Found in Fallopia japonica Plants and Their Traceability in Its Honey: A Case Study from the Northwestern and Western Part of Romania. *Plants* **2024**, *13*, 428. [CrossRef]
26. Saini, A.; Kumar, V.; Kaur, R.; Kumar, S.; Gautam, N.; Janghu, S. Conventional and non-conventional approaches for the extraction of rosehip phytocompounds and its bioactive, structural and antimicrobial characterization. *J. Food Meas. Charact.* **2024**, *18*, 2834–2845. [CrossRef]
27. Krist, S. Rose Hip Oil. In *Vegetable Fats and Oils*; Springer: Cham, Switzerland, 2020; Volume 1, pp. 647–650. [CrossRef]
28. Cornea-Cipcigan, M.; Bunea, A.; Bouari, C.; Pamfil, D.; Páll, E.; Urcan, A.C.; Mărgăoan, R. Anthocyanins and Carotenoids Characterization in Flowers and Leaves of Cyclamen Genotypes Linked with Bioactivities Using Multivariate Analysis Techniques. *Antioxidants* **2022**, *11*, 1126. [CrossRef] [PubMed]
29. Attard, E. A rapid microtitre plate Folin-Ciocalteu method for the assessment of polyphenols. *Open Life Sci.* **2013**, *8*, 48–53. [CrossRef]
30. Brand-Williams, W.; Cuvelier, M.E.; Berset, C. Use of a free radical method to evaluate antioxidant activity. *LWT-Food Sci. Technol.* **1995**, *28*, 25–30. [CrossRef]
31. Cohen, J. *Statistical Power Analysis for the Behavioral Sciences*; Routledge: Abingdon-on-Thames, UK, 2013.
32. Roman, I.; Stănilă, A.; Stănilă, S. Bioactive compounds and antioxidant activity of *Rosa canina* L. biotypes from spontaneous flora of Transylvania. *Chem. Cent. J.* **2013**, *7*, 73. [CrossRef]
33. Koczka, N.; Stefanovits-Bányai, É.; Ombódi, A. Total Polyphenol Content and Antioxidant Capacity of Rosehips of Some Rosa Species. *Medicines* **2018**, *5*, 84. [CrossRef] [PubMed]
34. Medveckienė, B.; Kulaitienė, J.; Jarienė, E.; Vaitkevičienė, N.; Hallman, E. Carotenoids, Polyphenols, and Ascorbic Acid in Organic Rosehips (*Rosa* spp.) Cultivated in Lithuania. *Appl. Sci.* **2020**, *10*, 5337. [CrossRef]
35. Demir, N.; Yildiz, O.; Alpaslan, M.; Hayaloglu, A.A. Evaluation of volatiles, phenolic compounds and antioxidant activities of rose hip (*Rosa* L.) fruits in Turkey. *LWT-Food Sci. Technol.* **2014**, *57*, 126–133. [CrossRef]
36. Barros, L.; Carvalho, A.M.; Ferreira, I.C.F.R. Exotic fruits as a source of important phytochemicals: Improving the traditional use of Rosa canina fruits in Portugal. *Food Res. Int.* **2011**, *44*, 2233–2236. [CrossRef]
37. Medveckienė, B.; Kulaitienė, J.; Levickienė, D.; Hallmann, E. The effect of ripening stages on the accumulation of carotenoids, polyphenols and vitamin C in rosehip species/cultivars. *Appl. Sci.* **2021**, *11*, 6761. [CrossRef]
38. Turan, S.; Solak, R.; Kiralan, M.; Ramadan, M.F. Bioactive lipids, antiradical activity and stability of rosehip seed oil under thermal and photo-induced oxidation. *Grasas Y Aceites* **2018**, *69*, e248. [CrossRef]

39. Pećinar, I.; Krstić, D.; Caruso, G.; Popović-Djordjević, J.B. Rapid characterization of hypanthium and seed in wild and cultivated rosehip: Application of Raman microscopy combined with multivariate analysis. *R. Soc. Open Sci.* **2021**, *8*, 202064. [CrossRef]
40. Konopka, I.; Tańska, M.; Dąbrowski, G.; Ogrodowska, D.; Czaplicki, S. Edible Oils from Selected Unconventional Sources—A Comprehensive Review of Fatty Acid Composition and Phytochemicals Content. *Appl. Sci.* **2023**, *13*, 12829. [CrossRef]
41. Chu, C.C.; Nyam, K.L. Application of seed oils and its bioactive compounds in sunscreen formulations. *J. Am. Oil Chem. Soc.* **2021**, *98*, 713–726. [CrossRef]
42. Fujii, T.; Ikeda, K.; Saito, M. Inhibitory effect of rose hip (*Rosa canina* L.) on melanogenesis in mouse melanoma cells and on pigmentation in brown guinea pigs. *Biosci. Biotechnol. Biochem.* **2011**, *75*, 489–495. [CrossRef]
43. Winther, K.; Petcharat, L.; Wongsuphasawat, K. Rose-HIP including seeds and shells reported to reduce symptoms of osteoarthritis, improves quality of the skin by mechanisms which may involve collagen and longevity of cell membranes. *Osteoarthr. Cartil.* **2015**, *23*, A170. [CrossRef]
44. Huang, C.-H.; Chen, S.C.; Wang, Y.C.; Wang, C.-F.; Hung, C.H.; Lee, S.S. Detrimental Correlation Between Air Pollution with Skin Aging in Taiwan Population. *Medicine* **2022**, *101*, e29380. [CrossRef] [PubMed]
45. Kislevitz, M.; Lu, K.B.; Wamsley, C.E.; Hoopman, J.; Kenkel, J.M.; Akgul, Y. Novel Use of Non-Invasive Devices and Microbiopsies to Assess Facial Skin Rejuvenation Following Laser Treatment. *Lasers Surg. Med.* **2020**, *52*, 822–830. [CrossRef]
46. Wang, X.; Shu, X.; Li, Z.; Huo, W.; Zou, L.; Tang, Y.l.; Li, L. Comparison of Two Kinds of Skin Imaging Analysis Software: VISIA® from Canfield and IPP® from Media Cybernetics. *Ski. Res. Technol.* **2018**, *24*, 379–385. [CrossRef]
47. Zuo, Y.; Li, A.; He, H.; Wan, R.; Li, Y.; Li, L. Assessment of Features in Facial Hyperpigmentation: Comparison Study Between VISIA and CSKIN. *Ski. Res. Technol.* **2022**, *28*, 846–850. [CrossRef]
48. Elder, A.; Ring, C.; Heitmiller, K.; Gabriel, Z.; Saedi, N. The Role of Artificial Intelligence in Cosmetic Dermatology—Current, Upcoming, and Future Trends. *J. Cosmet. Dermatol.* **2020**, *20*, 48–52. [CrossRef]
49. Yang, F.; Wang, H.; Guo, M.; Zhou, Z. The Clinical Efficacy of a New Emulsion for Acne and Conspicuous Facial Pore Amelioration. *J. Cosmet. Dermatol.* **2023**, *23*, 958–963. [CrossRef] [PubMed]
50. Zduńska-Pęciak, K.; Dębowska, R.; Kołodziejczak, A.; Rotsztejn, H. Ferulic Acid—A Novel Topical Agent in Reducing Signs of Photoaging. *Dermatol. Ther.* **2022**, *35*, e15543. [CrossRef]
51. Kawałkiewicz, W.; Matthews-Kozanecka, M.; Janus-Kubiak, M.; Kubisz, L.; Hojan-Jezierska, D. Instrumental Diagnosis of Facial Skin—A Necessity or a Pretreatment Recommendation in Esthetic Medicine. *J. Cosmet. Dermatol.* **2020**, *20*, 875–883. [CrossRef]
52. Zawodny, P.; Stój, E.; Kulig, P.; Skonieczna-Żydecka, K.; Sieńko, J. VISIA Skin Analysis System as a Tool to Evaluate the Reduction of Pigmented Skin and Vascular Lesions Using the 532 Nm Laser. *Clin. Cosmet. Investig. Dermatol.* **2022**, *15*, 2187–2195. [CrossRef]
53. Lai, H.T.; Liu, W.; Yi-Chia, W.; Lai, Y.W.; Wen, Z.H.; Wang, H.; Lee, S.-S. The Effect in Topical Use of Lycogen Via Sonophoresis for Anti-Aging on Facial Skin. *Curr. Pharm. Biotechnol.* **2015**, *16*, 1063–1069. [CrossRef]
54. Lintner, K.; Gerstein, F.H.; Solish, N. A Serum Containing Vitamins C & E and a Matrix-repair Tripeptide Reduces Facial Signs of Aging as Evidenced by Primos® Analysis and Frequently Repeated Auto-perception. *J. Cosmet. Dermatol.* **2020**, *19*, 3262–3269. [CrossRef] [PubMed]
55. Verma, K.; Kaushik, P.; Chugh, R.; Kaur, G.; Kathuria, D. Cosmeceutical applications of natural oils and fats. In *Specialized Plant Metabolites as Cosmeceuticals*; Elsevier: Amsterdam, The Netherlands, 2024; pp. 239–256.
56. Ande, S.N.; Bakal, R.L. Potential herbal essential oils: Are they super natural skin protector. *Innov. Pharm. Pharm* **2022**, *10*, 19–24.
57. Khunger, N.; Mehrotra, K. Menopausal Acne—Challenges And Solutions. *Int. J. Women's Health* **2019**, *11*, 555–567. [CrossRef] [PubMed]
58. Zhou, J.; Mehling, A.; Wang, Q.; Wang, X.; Hu, X.; Song, L. Age-related changes in the bacterial composition of healthy female facial skin in Beijing area. *Int. J. Cosmet. Sci.* **2024**, *46*, 982–994. [CrossRef]
59. Nunes, A.; Marto, J.; Gonçalves, L.; Martins, A.M.; Fraga, C.; Ribeiro, H.M. Potential therapeutic of olive oil industry by-products in skin health: A review. *Int. J. Food Sci. Technol.* **2022**, *57*, 173–187. [CrossRef]
60. Bianca Sanabria, M.A.; Berger, L.E.; MrpharmSb, B.B.M.-K.P.; Rao, B.K. Clinical Efficacy of Topical Vitamin C on the Appearance of Wrinkles: A Systematic Literature Review. *J. Drugs Dermatol.* **2023**, *22*, 898–904.
61. Correia, G.; Magina, S. Efficacy of topical vitamin C in melasma and photoaging: A systematic review. *J. Cosmet. Dermatol.* **2023**, *22*, 1938–1945. [CrossRef]
62. Abric, E.; Mathias, J.; Tardieu, A.-S.; Mateos, L.; Eeckhout, C.; Drulhon, F.; Ardiet, N. Anti-Spot, Lightening Effect and Cutaneous Acceptability of a Stable Anhydrous Ecobiological Formulation of 10% L-Ascorbic Acid. *Clin. Cosmet. Investig. Dermatol.* **2024**, *ume 17*, 489–491. [CrossRef]
63. Pereira Oliveira, C.N.; Nani Leite, M.; de Paula, N.A.; Araújo Martins, Y.; Figueiredo, S.A.; Cipriani Frade, M.A.; Lopez, R.F. Nanoemulsions based on sunflower and rosehip oils: The impact of natural and synthetic stabilizers on skin penetration and an ex vivo wound healing model. *Pharmaceutics* **2023**, *15*, 999. [CrossRef]
64. Hailemeskel, B.; Fullas, F. Anti-wrinkle properties of rosehip oil, aloe vera, green tea extract, and frankincense essential oil: Brief review and a survey. *Int. J. Sch. Res. Biol. Pharm.* **2023**, *3*, 1–8. [CrossRef]

65. Michalak, M.; Błońska-Sikora, E.; Dobros, N.; Spałek, O.; Zielińska, A.; Paradowska, K. Bioactive Compounds, Antioxidant Properties, and Cosmetic Applications of Selected Cold-Pressed Plant Oils from Seeds. *Cosmetics* **2024**, *11*, 153. [CrossRef]
66. Rasal, A.V.; Patil, K.; Khot, V.; Laskshapati, L. Formulation and evaluation of an innovative anti-wrinkle cream enriched with bio-retinol. *Indian Drugs* **2024**, *61*, 80–82. [CrossRef]
67. Isnaini, N.; Prajaputra, V.; Maryam, S. Formulation and Evaluation of O/W Body Cream Containing Patchouli Oil (*Pogostemon cablin Benth.*) and Drumstick Oil (*Moringa oleifera*) as Potential Moisturizing Agent. *J. Penelit. Pendidik. IPA* **2023**, *9*, 8001–8007. [CrossRef]
68. Krambeck, K.; Silva, V.; Silva, R.; Fernandes, C.; Cagide, F.; Borges, F.; Santos, D.; Otero-Espinar, F.; Lobo, J.M.S.; Amaral, M.H. Design and characterization of Nanostructured lipid carriers (NLC) and Nanostructured lipid carrier-based hydrogels containing Passiflora edulis seeds oil. *Int. J. Pharm.* **2021**, *600*, 120444. [CrossRef]
69. Hugo Infante, V.; Maria Maia Campos, P.; Darvin, M.; Lohan, S.; Schleusener, J.; Schanzer, S.; Lademann, J.; Meinke, M. Cosmetic Formulations with Melaleuca alternifolia Essential Oil for the Improvement of Photoaged Skin: A Double-Blind, Randomized, Placebo-Controlled Clinical Study. *Photochem. Photobiol.* **2023**, *99*, 176–183. [CrossRef]
70. Henseler, H. Validation of the Visia® Camera System for skin analysis through assessment of the correlations among the three offered measurements–the percentile, feature count and absolute score–as well as the three capture perspectives, from the left, front and right. *GMS Interdiscip. Plast. Reconstr. Surg. DGPW* **2022**, *11*, Doc04.
71. Gavarić, A.; Pastor, K.; Nastić, N.; Vidović, S.; Živanović, N.; Simin, N.; Duarte, A.R.C.; Vladić, J. Recovery of polyphenols from rosehip seed waste using natural deep eutectic solvents and ultrasonic waves simultaneously. *Foods* **2023**, *12*, 3655. [CrossRef]
72. Moreno, I.R.; Ochoa, D.; Román, M.; Cabaleiro, T.; Abad-Santos, F. Utility of Pilot Studies for Predicting Ratios and Intrasubject Variability in High-Variability Drugs. *Basic Clin. Pharmacol. Toxicol.* **2016**, *119*, 215–221. [CrossRef]
73. Etheridge, C.J.; Derbyshire, E. Herbal Infusions and Health. *Nutr. Food Sci.* **2019**, *50*, 969–985. [CrossRef]
74. Rashaan, R.; Arkoumanis, P.T. The Efficacy of Botulinum Toxin A in Treating Palmar Hyperhidrosis–a Literature Review. *Maedica* **2023**, *18*, 712. [CrossRef] [PubMed]
75. Jovanović, A.A.; Balanč, B.; Volić, M.; Pećinar, I.; Živković, J.; Šavikin, K.P. Rosehip extract-loaded liposomes for potential skin application: Physicochemical properties of non- and UV-irradiated liposomes. *Plants* **2023**, *12*, 3063. [CrossRef]
76. Wilsmann-Theis, D.; Kromer, C.; Gerdes, S.; Linker, C.; Magnolo, N.; Sabat, R.; Reich, K.; Mössner, R. A multicentre open-label study of apremilast in palmoplantar pustulosis (APLANTUS). *J. Eur. Acad. Dermatol. Venereol.* **2021**, *35*, 2045–2050. [CrossRef]
77. Belkhelladi, M.; Bougrine, A. Rosehip extract and wound healing: A review. *J. Cosmet. Dermatol.* **2024**, *23*, 62–67. [CrossRef]
78. Cassin, G.; Diridollou, S.; Flament, F.; Adam, A.S.; Pierre, P.; Colomb, L.; Morancais, J.L.; Qiu, H. Concealing a shiny facial skin appearance by an Aerogel-based formula. In vitro and in vivo studies. *Int. J. Cosmet. Sci.* **2018**, *40*, 58–66. [CrossRef] [PubMed]
79. Rattanawiwatpong, P.; Wanitphakdeedecha, R.; Bumrungpert, A.; Maiprasert, M. Anti-aging and brightening effects of a topical treatment containing vitamin C, vitamin E, and raspberry leaf cell culture extract: A split-face, randomized controlled trial. *J. Cosmet. Dermatol.* **2020**, *19*, 671–676. [CrossRef] [PubMed]
80. Magrode, N.; Poomanee, W.; Kiattisin, K.; Ampasavate, C. Microemulsions and Nanoemulsions for Topical Delivery of Tripeptide-3: From Design of Experiment to Anti-Sebum Efficacy on Facial Skin. *Pharmaceutics* **2024**, *16*, 554. [CrossRef]
81. Lei, Z.; Cao, Z.; Yang, Z.; Ao, M.; Jin, W.; Yu, L. Rosehip oil promotes excisional wound healing by accelerating the phenotypic transition of macrophages. *Planta Medica* **2019**, *85*, 563–569. [CrossRef]
82. Mármol, I.; Sánchez-de-Diego, C.; Jiménez-Moreno, N.; Ancín-Azpilicueta, C.; Rodríguez-Yoldi, M.J. Therapeutic applications of rose hips from different *Rosa* species. *Int. J. Mol. Sci.* **2017**, *18*, 1137. [CrossRef]
83. Wirkowska-Wojdyła, M.; Ostrowska-Ligęza, E.; Górska, A.; Brzezińska, R.; Piasecka, I. Assessment of the Nutritional Potential and Resistance to Oxidation of Sea Buckthorn and Rosehip Oils. *Appl. Sci.* **2024**, *14*, 1867. [CrossRef]

cosmetics

MDPI

Article

Systematic Evaluation and Identification of Anti-Inflammatory and Anti-Aging Ginseng Peptides for Skincare Applications

Ze Xia [1,2,3,†], Wei Liu [4,†], Fanmo Zeng [3,5], Sining Kang [3,5], Junxiang Li [3,5], Wenfei Xu [2], Pingxiang Tang [2,3], Xinyi Zheng [2], Dandan Li [1], Xuebin Yang [1], Qing Sheng [1,*] and Xuhui Li [2,3,*]

1 College of Life Sciences and Medicine, Zhejiang Sci-Tech University, Hangzhou 310018, China; zex0909@163.com (Z.X.); lidan05252022@163.com (D.L.); xuebinzznn@163.com (X.Y.)
2 Zhejiang Key Laboratory of Multiomics and Molecular Enzymology, Yangtze Delta Region Institute of Tsinghua University, Zhejiang, Jiaxing 314006, China; xuwenfei.2020@tsinghua.org.cn (W.X.); m220100132@st.shou.edu.cn (P.T.); m220100149@st.shou.edu.cn (X.Z.)
3 AGECODE R&D Center, Yangtze Delta Region Institute of Tsinghua University, Zhejiang, Jiaxing 314000, China; fanmozeng2024@163.com (F.Z.); kangsining@harvbio.com (S.K.); lijunxiang@acrdc.cn (J.L.)
4 Beijing Yiqing Daily Chemical Co., Ltd., Tongzhou 101100, China; liuwei@goldfishtech.cn
5 Harvest Biotech (Zhejiang) Co., Ltd., Jiaxing 314000, China
* Correspondence: csheng@zstu.edu.cn (Q.S.); lixuhui_2013@tsinghua.org.cn (X.L.); Tel.: +86-571-84843302 (Q.S.); +86-18694502360 (X.L.)
† These authors contributed equally to this work.

Abstract: This study explores the potential of ginseng-derived peptides (GPs) as multifunctional bioactive agents for skincare. Unlike previous research into ginseng saponins and polysaccharides, we identified that ginseng extracts containing water-soluble small molecules and polypeptides exhibit potent antioxidant, anti-inflammatory, and anti-aging properties. In vitro assays revealed that ginseng peptide extract (GPE) reduced reactive oxygen species (ROS) and inflammatory cytokines (IL-6, TNF-α, IL-1β) in RAW264.7 macrophages while enhancing collagen synthesis in human skin fibroblasts (HSFs). Validation using 3D epidermal and dermal models further confirmed GPE's ability to mitigate UV-induced damage, restore skin barrier proteins (filaggrin, loricrin), and increase collagen content. In addition, we screened 19 candidate peptides from ginseng extract using machine learning and prioritized their interaction with skin aging and inflammation-related targets. Three peptides (QEGIYPNNDLYRPK, VDCPTDDATDDYRLK, and ADEVVHHPLDKSSEVE) demonstrated significant collagen-promoting, antioxidant, and anti-inflammatory effects in cellular models. These findings highlight the efficacy of computational approaches in identifying natural bioactive ingredients, positioning ginseng peptides as promising candidates for innovative cosmeceutical formulations targeting inflammaging and skin rejuvenation.

Keywords: ginseng peptide; inflammaging; network pharmacology; machine learning; dermatology

1. Introduction

In inflammatory aging in various pathological conditions, senescent cells and macrophages release cytokines such as IL-1β, IL-6, and TNF-α [1–6], which are integral to the senescence-associated secretory phenotype (SASP) and the regulation of inflammatory responses [7–10]. In skin-related diseases, SASPs not only promote the occurrence of chronic inflammation but also induce normal cells to enter the aging process, forming a vicious cycle of inflammation and aging [11,12]. For example, the SASP secreted by senescent skin cells can disrupt the skin microenvironment, exacerbate inflammation, impair epidermal barrier function, and accelerate the aging of fibroblasts [13,14].

Ginseng (*Panax ginseng* C.A. Meyer) has attracted significant attention for its potential anti-inflammatory and anti-aging effects [15–17]. It contains a variety of chemical constituents, including polysaccharides, saponins, sterols, and organic acids, which have been shown to combat fatigue [18], exert anti-inflammatory and anti-aging effects, regulate blood glucose levels [19], enhance cardiac function [20], and possess anti-tumor activities [21,22]. In recent years, with the advancement of biotechnology, there has been a growing interest in exploring the applications of ginseng beyond its traditional medicinal and culinary uses. For instance, based on its role in hair regrowth following chemotherapy, ginseng-derived products have been incorporated into shampoos, which are designed to nourish the hair and mitigate hair loss [23]. Additionally, the antioxidant and anti-aging effects of ginsenosides, along with the moisturizing properties of ginseng polysaccharides and their melanin inhibition, support their use in skincare products targeting aging, hydration, and whitening. However, current research and development efforts have primarily focused on saponins and polysaccharides, with relatively limited and less in-depth studies on ginseng peptides (GPs).

To address the underexplored bioactive components of ginseng, this study systematically investigated the molecular mechanisms underlying the anti-inflammatory and anti-aging effects of ginseng peptide extract (GPE) with in vitro 3D skin models. Additionally, computational biology approaches were employed to identify and predict bioactive peptides within the GPE, with a focus on their potential therapeutic and cosmetic applications in skin anti-aging and anti-inflammatory interventions. This integrated experimental and computational strategy aims to elucidate the functional mechanisms of GPE and expand its utility in biomedical and skincare contexts.

2. Materials and Methods

2.1. Network Pharmacology

2.1.1. Screening Targets of Ginseng Compounds

Using Traditional Chinese Medicine Systems Pharmacology (TCMSP) databases (https://tcmsp-e.com/tcmsp.php, accessed on 13 October 2024) [24] to search for the main components of ginseng, the screening conditions were selected as oral bioavailability OB $\geq$ 30% and DL resistance $\geq$ 0.18, and the main compound components in ginseng were identified and their targets were obtained.

2.1.2. Screening of GPs

The GP database is sourced from the literature on pan ginseng and de novo sequencing. First, we selected peptides with a length $\leq$ 16. Next, relevant target proteins with keywords such as antioxidant, collagen promoting, proliferation promoting, aging, inflammation, and inflammaging in the HSM model were used to predict the interaction probability between GPs and target proteins in this model. The top 50 peptides with the highest probability were selected based on their scores, and each peptide retained the results of three optimal peptide target protein interactions. We used the CAMP model to predict the 150 results obtained below again and selected the results with a score of $\geq$50 based on the results given by the CAMP model (Figure 1).

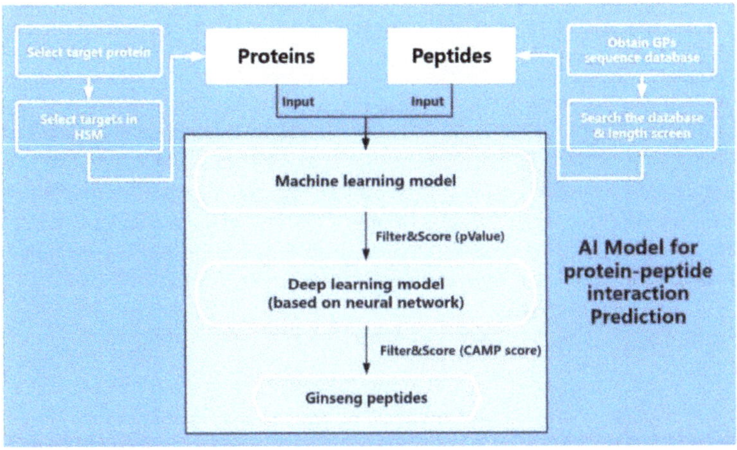

Figure 1. Flowchart of AI prediction to screen for peptides with anti-aging activity. Using HSM and CAMP models to predict protein–peptide interaction.

2.1.3. Disease Targets Screened and Targeted with Network Pharmacological Analysis

We screened inflammaging targets using the GeneCards database (https://www.genecards.org, accessed on 14 October 2024) [25] and the online Mendelian inheritance in man (OMIM) database (https://www.omim.org/, accessed on 14 October 2024) [26], intersecting with targets of ginseng compounds and peptides, and inputting these potential targets into the Search Tool for the Retrieval of Interacting Genes/Proteins (STRING) database (https://cn.string-db.org/, accessed on 20 October 2024) [27] to obtain protein–protein interaction (PPI) network information. The network was visualized using Cytoscape 3.7.2 software. Finally, we analyzed the Kyoto Encyclopedia of Genes and Genomes (KEGG) and gene ontology (GO) functions using the Metascape database (https://metascape.org/gp/index.html, accessed on 21 October 2024) [28,29].

2.2. Preparation of the Ginseng Peptide Extract

Ginseng peptide extract was provided by Harvest Biotech, Zhejiang, Co., Ltd. Black ginseng was ground into a fine powder using a mill. The powder was then dissolved in water at a solid-to-liquid ratio of 1:20. The mixture was soaked overnight at 50 °C with a stirring rate of 120 rpm. After soaking, samples were taken from the solution. Next, sodium hydroxide was added to adjust the pH to 8.2. Subsequently, 2% (w/v) of alkaline protease (1 kg per 50 L of solution) was added in small portions to the black ginseng solution, which was then transferred to a reactor. The mixture was subjected to enzymatic hydrolysis at 50 °C with a stirring rate of 130 rpm for 2.5 h. Post hydrolysis, samples were again taken. Then, 0.4% (w/v) of flocculant (200 g per 50 L of solution) was added to the hydrolyzed solution, followed by static settling overnight at 4 °C. The mixture was then filtered using a plate and frame filter press. After filtration, 2% (w/v) of preservative (a mixture of phenoxyethanol, 1,2-hexanediol, and 1,2-pentanediol) was incorporated into the solution. Finally, the solution was sterilized by filtering it through a 0.22 μm PVDF filter cartridge. The resulting solution was stored at 4 °C away from light until further use.

2.3. Free Radical Scavenging Assessment

2.3.1. DPPH Assay

The 2,2-Diphenyl-1-picrylhydrazyl (DPPH) assay is the most commonly used antioxidant assay for plant extract that measures the ability to act as free radical scavengers or hydrogen donors. Glutathione was used as a standard in the DPPH assay, concentrations were 0, 0.125, 0.25, 0.5, 1, 2, and 4 mg/mL, dissolved and diluted with ddH_2O. The DPPH solution (50 μg/mL) was dissolved with 95% ethanol. For the sample group (As), 150 μL of DPPH solution and 50 μL of glutathione solution were added. For the control group (Ac), 50 μL of glutathione solution and 150 μL of 95% ethanol solution were added. For the blank group (Ab), 150 μL of DPPH solution and 50 μL of ddH_2O were added. When testing the sample, GPE was used instead of the standard, with all other procedures remaining unchanged. After thorough mixing, the samples were reacted in the dark at room temperature for 30 min. The absorbance values were measured at 517 nm with a microplate reader. Finally, the calculation was performed using the following formula (Theorem 1).

$$\text{Free radical scavenging rate (\%) for DPPH} = [1 - (As - Ac)/Ab] \times 100\% \quad (1)$$

Theorem 1. *Calculation formula of DPPH free radical scavenging rate. In this formula, As, Ac, and Ab represent the absorbance values of the sample group, the control group, and the blank group, respectively.*

2.3.2. ABTS Assay

The 2,2′-azino-bis (3-ethylbenzothiazoline-6-sulfonate) diammonium salt (ABTS) radical scavenging assay is also a widely used antioxidant assay for plant extracts or peptides, evaluating the ability to neutralize free radicals via electron donation or radical cation decolorization. Glutathione was used as a standard in the ABTS assay, and concentrations were 0, 0.125, 0.25, 0.5, 1, 2, and 4 mg/mL, dissolved and diluted with ddH_2O. For ABTS working solution preparation, we mixed ABTS solution (7 mM) and potassium persulfate solution (2.45 mM) in equal amounts and allowed the mixture to react in the dark for 12 h. After, we diluted it with PBS (10 mM, pH 7.4) to an absorbance value of 0.7 (734 nm). Next, we mixed 150 μL of ABTS working solution with 10 μL of glutathione. When testing the sample, GPE (different concentrations of GPE were diluted by PBS) was used instead of the standard for the sample group (As), and the PBS group was used as a blank group (Ab). All other procedures remained unchanged. After reacting in the dark at room temperature for 6 min, the absorbance values were measured at 734 nm with a microplate reader. Finally, the calculation was performed using the following formula (Theorem 2).

$$\text{Free radical scavenging rate (\%) for ABTS} = (Ab - As)/Ab \times 100\% \quad (2)$$

Theorem 2. *Calculation formula of ABTS free radical scavenging rate. In this formula, As and Ab represent the absorbance values of the sample group and the blank group.*

2.4. Cell Inflammation Model Based on RAW264.7

2.4.1. Cell Culture of RAW264.7

Normal RAW264.7 cells (Hunan Fenghui Biotechnology Co., Ltd., Changsha, China) were cultured on high-glucose DMEM (Wisent Biotechnology Co., Ltd., Nanjing, China) supplemented with 10% FBS (Gibco, Carlsbad, CA, USA) and 1% penicillin/streptomycin (Shanghai BasalMedia Technologies Co., Ltd., Shanghai, China). The cell cultures were maintained in a humidified incubator at 5% CO_2 and 37 °C. When RAW264.7 cells in T25 cell culture flasks grew to a density above 90%, 1:3 passages were performed.

When the cells reached confluency and were in good condition (with less than 5% differentiated cells), we collected the cells and seeded them at varying cell densities into corresponding culture plates for subsequent experiments.

2.4.2. Cell Viability Assay

After overnight culturing in a 96-well plate (1×10^4 cells/well, 100 μL complete medium/well), the cells were treated with GPE (0, 5, 10, 25, 50, 75, 100 and 150 μg/mL) using a complete medium (containing 5% FBS) for 24 h. Thereafter, 10 μL of cck-8 (Biosharp, Hefei, China) solution was added to each well and incubated at 37 °C for 1 h, and cell-free wells were used as a blank group. Finally, the absorbance of each well was recorded at 450 nm using a multimode plate reader (PerkinElmer, Waltham, MA, USA). The OD values of each group were subtracted from the average value of the blank group, which was the cell viability of each group. The viability of 0 μg/mL GPE group was assessed as 1 (100%).

2.4.3. Nitric Oxide (NO) Production Assay

The RAW264.7 cells were cultured overnight in 24-well plates (5×10^4 cells/well, 500 μL medium/well), then co-stimulated with GPE (0, 50, 100 μg/mL) or GPs (0, 10, 25 μg/mL), and 200 ng/mL of LPS (Sigma-Aldrich, St. Louis, MO, USA) using a complete medium (containing 5% FBS) for 24 h. Untreated cells were used as the blank group, while cells treated with LPS but without GPE or GPs were used as the model group, and cells treated with LPS and positive drug (50 μg/mL Dexamethasone) were used as the positive group. The culture supernatant from each well was collected at the end of the experiments and used to measure NO production. The NO production was determined using a commercial NO assay kit (Beyotime Biotechnology, Shanghai, China) according to the manufacturer's instructions. We mixed 50 μL of cell culture medium and 50 μL of Griess reagents I and II and reacted them in a 96-well plate at room temperature for 10 min. Finally, the absorbance of each well was recorded at 540 nm using a multimode plate reader.

2.4.4. Reactive Oxygen Species (ROS) Measurement

The RAW264.7 cells were cultured overnight in 24-well plates (5×10^4 cells/well, 500 μL medium/well), then co-stimulated with GPE (0, 50, 100 μg/mL) and 200 ng/mL LPS using a complete medium (containing 5% FBS) for 24 h. Then, we removed the medium and washed the cells with 500 μL of PBS at least two times. Next, we used an ROS assay kit (Beijing Solarbio Science & Technology Co., Ltd., Beijing, China) to measure ROS levels. The signal intensity of Ex488/Em525 was then detected using a multimode reader and was observed and photographed under a fluorescence microscope.

2.4.5. RNA Extraction and RT-qPCR

The RAW264.7 cells were cultured overnight in 12-well plates (2×10^5 cells/well, 1 mL medium/well), then co-stimulated with GPE (0, 50, 100 μg/mL) and 200 ng/mL of LPS using a complete medium (containing 5% FBS) for 24 h. After, we removed the DMEM and washed the cells with 1 mL of PBS at least two times. Then, we extracted RNA using a Quick-RNA Microprep Kit (Genstone biotech, Beijing, China) and stored it at −80 °C. Subsequently, reverse transcription was performed using a cDNA first-strand synthesis kit (Tiangen Biotech Co., Ltd., Beijing, China) and diluted with ddH_2O to an appropriate concentration. RT-qPCR was performed using diluted cDNA and qPCR SYBR Green Master Mix (Cat No.11185ES08; YEASEN, Shanghai, China) on Applied Biosystems 3 Fast Real-Time PCR System (Thermo Fisher Scientific, Shanghai, China).

2.5. Histomorphology and Collagen Change Detection Based on In Vitro Skin Model

The in vitro skin model used (Guangdong BioCell Biotechnology Co., Ltd., Dongguan, China) is a skin model which included epidermal and dermal cells. We placed the models into a 6-well plate culture mold, added 3.7 mL of culture medium (#230607, Guangdong

Biocell Biotechnology Co., Ltd.) to each well, incubated in a 37 °C 5% CO_2 incubator (150I, Thermo) and changed the medium every day. After 2 days of cultivation, irradiation and administration, we began according to the grouping and corresponding treatment conditions. The skin models were exposed to a mixture of ultraviolet radiation A (UVA) irradiation at 30 J/cm^2 (365 nm, 16.29 mW/cm^2 for 30 min and 42 s, PL-S 9W/01/2P, Philips, Shanghai, China) and ultraviolet radiation B (UVB) at 50 mJ/cm^2 (311 nm, 2.748 mW/cm^2 for 18 s, PL-S 9W/10/2P, Philips) for 4 consecutive days, with fresh culture medium replaced before each irradiation. The medium was removed after each irradiation, and the skin models were cultured with medium (Guangdong Biocell Biotechnology Co., Ltd.) containing 1% (*v*/*v*) GPE. The positive groups were cultured with 100 μg/mL vitamin C (VC) and 7 μg/mL vitamin E (VE) instead of sample GPE. Control groups were cultured with normal medium. After 4 days of irradiation, the samples were continuously cultured with medium containing 1% (*v*/*v*) GPE or both VC and VE for another 3 days. The medium was replaced with the same fresh medium every day. After the cultivation was completed, we removed the medium in all groups, washed them with PBS solution, and wiped off the residual liquid with a sterile swab. After, we fixed skin models with 4% paraformaldehyde for 24 h.

2.5.1. H&E Staining

After 24 h of fixation, the skin models were washed with PBS and dehydrated by stepwise immersion in solutions of increasing ethanol concentration, embedded in paraffin overnight, then were sealed with dry gum and sliced into 8-μm-thick sections. Sections were stained with hematoxylin and eosin (H&E). Then, the images were taken under the microscope.

2.5.2. Masson Staining

Paraffin sections obtained in Section 2.5.1 were dewaxed and stained using a Masson staining kit (YK2223, Shaanxi Yike biotechnology Service Co., Ltd., Xi'an, China) according to the manufacturer's instructions. Then, the images were taken under a microscope to observe and analyze.

2.5.3. Immunofluorescence

After fixation with 4% paraformaldehyde, tissue embedding and tissue sectioning were performed. After deparaffinization and hydration of the baked slices, the paraffin sections were placed in a 0.01 M sodium citrate antigen repair solution and subjected to high-pressure repair. After cooling, the slices were removed and washed with PBS solution (three times, 5 min each). One drop of 3% H_2O_2 was added to each slice, and they were incubated at room temperature for 30 min, then washed with PBS (three times, 5 min each). We added serum homologous to the secondary antibody and sealed it at 37 °C for 60 min. We added primary antibody working solution dropwise, incubated overnight at 4 °C, and washed with PBS (three times, 5 min each). Next, we added the secondary antibody working solution dropwise and incubated at room temperature for 1 h, before washing with PBS (three times, 5 min each). After the completion of the secondary antibody incubation, we washed with PBS (three times, 5 min each). Then, we shook off the PBS solution attached to the glass slide, added 100 μL of Hoechst working solution dropwise to each slice, and incubated at room temperature for 5 min. We then washed with PBS (three times, 5 min each). We wiped off the PBS solution with absorbent and sealed it with a drop of quenching inhibitor. Finally, we viewed it under a fluorescence microscope (BX43, Olympus, Excitation wavelength: 488 nm, Emission wavelength: 520 nm) and took photos within 24 h. The following primary antibodies were used: anti-collagen I (#72026, Cell Signaling Technology, Danvers, MA, USA, dilution 1:200), anti-collagen III

(22734-1-AP, Proteintech, Wuhan, China, dilution 1:200), anti-collagen IV (ab6311, Abcam, Cambridge, UK, dilution 1:200), anti-collagen XVII (ab186415, Abcam, dilution 1:200) and goat anti-rabbit IgG (ab150077, Abcam, dilution 1:500).

2.6. Tissue Morphological Changes, Detection of Filaggrin (FLG), Loricrin (LOR) and Transglutaminase-1 (TGM1) Content Based on 3D Epidermal Skin Model

The EpiKutis 3D skin model (Guangdong BioCell Biotechnology Co., Ltd.) is another skin model. Unlike in vitro skin models used in Section 2.5, this model only included epidermal cells. It was utilized to evaluate the effects of GPE on cellular barrier function in epidermal layer. 3D skin models were transferred to a 6-well plate with 0.9 mL EpiGrowth medium added in advance according to the grouping. WY14643 is a kind of PPARα agonist, that has anti-inflammatory properties, promoting barrier function in skin cells. In the positive group, 50 μM WY14643 were used instead of sample GPE. Control groups were cultured with normal medium. The positive control group and experimental group were irradiated with UVB (600 mJ/cm^2). After irradiation, the four groups of epidermal models were cultured in a 37 °C 5% CO_2 incubator for 24 h, and then fixed with 4% paraformaldehyde for 24 h. We performed H&E staining and immunofluorescence detection (the same as described in Sections 2.5.2 and 2.5.3), took photos under a microscope for observation, collected images, and conducted analysis. The following primary antibodies were used: anti-FLG (ab218397, Abcam, dilution 1:200), anti-LOR (ab198994, Abcam, dilution 1:500), anti-TGM1 (12912-3-AP, Proteintech, China, dilution 1:200), goat anti-rabbit IgG (ab150077, Abcam, dilution 1:500) and goat anti-mouse IgG (ab150117, Abcam, dilution 1:500).

2.7. Biological Assay and Method Based on HSF Cell Inflammaging Model and UV-Induced Senescence Model

2.7.1. Cell Culture of THP-1 and HSF

Normal HSF cells (NewgainBio, Wuxi, China) were cultured on high-glucose DMEM (Wisent Biotechnology Co., Ltd., Nanjing, China) supplemented with 10% FBS (Gibco, Carlsbad, CA, USA) and 1% penicillin/streptomycin (Shanghai BasalMedia Technologies Co., Ltd., Shanghai, China). When HSF cells in T25 cell culture flasks grew to a density above 90%, 1:3 or 1:2 passages were performed. THP-1 cells (Baidi Biotech Ltd., Jiaxing, China) were cultured on RPMI-1640 medium (Servicebio, Wuhan, China) supplemented with 10% FBS (Gibco, Carlsbad, CA, USA) and 1% penicillin/streptomycin (Shanghai BasalMedia Technologies Co., Ltd.). The cell cultures were maintained in a humidified incubator at 5% CO_2 and 37 °C. When THP-1 cells in T75 cell culture flasks grew to a density above 90%, 1:2 passages were performed.

2.7.2. Induction of Aging to HSF

When THP-1 cells were cultured to at least two T75 cells, we then prepared the RPMI1640 complete medium containing 100 nM PMA, which was used to inoculate the cells into a 6-well plate (1×10^6 cells/well) and cultured them for 24 h. We prepared the RPMI1640 complete culture medium containing 100 nM PMA, 100 ng/mL LPS, and 20 ng/mL IFN-γ, removed the original medium, added the newly prepared complete culture to each well, and continued processing for 24 h to obtain M1 type THP-1 cells. Next, we removed the culture medium, washed it with PBS three times, added new 1640 complete culture medium, incubated it for 48 h, collected the cell culture supernatant, and stored it at −80 °C or proceeded to the next experiment. It was later used to induce the inflammaging model of HSF cells (Figure 2).

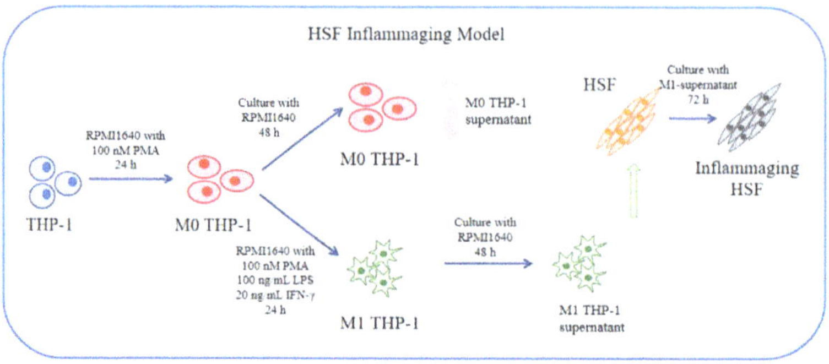

Figure 2. Flowchart of induction of inflammaging model of HSF cells.

For this model, HSF cells were seeded onto a 12-well plate (5 × 10^4 cells/well), and after 24 h, the original culture medium was removed. The THP-1 cell culture supernatant was used to prepare the specified concentration of drug (0, 0.01%, 0.05% GPE, or 0, 10, 25 μg/mL ginseng peptide) or positive drug (50 μg/mL TGF-β1) and stimulated for 72 h. After, we collected the culture supernatant from each well and stored it at −80 °C or performed the next experiment.

For the UV-induced aging model, HSF cells were seeded onto a 12-well plate (4 × 10^4 cells/well). After 24 h, the original culture medium was removed and washed twice with PBS. After 15 J of UVA irradiation, PBS was discarded and replaced with medium (containing 2% FBS) containing the specified concentration of drugs (0, 10, 25 μg/mL ginseng peptide) or positive drug (50 μg/mL TGF-β1) for further incubation for 24 h. After, we collected the culture supernatant from each well and stored it at −80 °C or performed the next experiment.

2.8. Determination of Inflammatory Cytokines (IL-6, TNF-α) and Collagen I by ELISA

The concentration of IL-6, TNF-α and collagen I were determined using IL-6 ELISA kit (EK206, MultiSciences, Hangzhou, China), TNF-α ELISA kit (EK282, MultiSciences, Hangzhou, China) and collagen I ELISA kit (Cat No. 97044ES96, YEASEN, Shanghai, China) according to the manufacturer's instructions and measured the absorbance at 450 nm/630 nm using a microplate reader.

2.9. Statistical Analysis

GraphPad Prism 9.5 software (GraphPad Software, Inc., La Jolla, CA, USA) was used for statistical analyses. Data were expressed as mean values ± SD of almost three independent experiments. A Student's *t*-test was performed to compare the differences between the two groups, and one-way analysis of variance (ANOVA) was performed to compare the differences between multiple groups, followed by Dunnett's test as the post hoc test (* $p < 0.05$, ** $p < 0.01$, *** $p < 0.001$). $p < 0.05$ was regarded as statistically significant, while $p < 0.01$ was regarded as a highly significant difference.

3. Results

3.1. Network Pharmacology Predicts the Anti-Aging and Anti-Inflammatory Efficacy of Ginseng

We first conducted a preliminary network pharmacology analysis of reported ginseng-soluble activities (including small molecules and peptides) and collected targets for various skin diseases. Specifically, we used the TCMSP [24] database for screening the small molecules and their regulatory targets in ginseng, while the identified GPs and their potential corresponding targets were obtained using mass spectrum data for ginseng peptides [30] and regulatory target prediction through HSM and CAMP models. We retrieved skin anti-aging and repair-related targets through the GeneCards [25] and OMIM databases [26], removed duplicates, and intersected them with the potential targets of ginseng, resulting in a total of 121 potential targets.

After integrating the potential targets of ginseng compounds and GPs, a network relationship was constructed using Cytoscape 3.7.2, resulting in a network comprising 159 nodes and 256 interaction edges. Then, we used the STRING database [27] to construct a key target network, which consisted of 134 nodes and 1895 interacting edges (Figure 3A). Using Cytoscape 3.7.2 software to analyze the PPI network, the top five targets with the highest degree values were obtained: TNF, AKT1, IL-6, INS, and IL-1β. This suggests that ginseng may alleviate inflammaging through these main targets (Figure 3B).

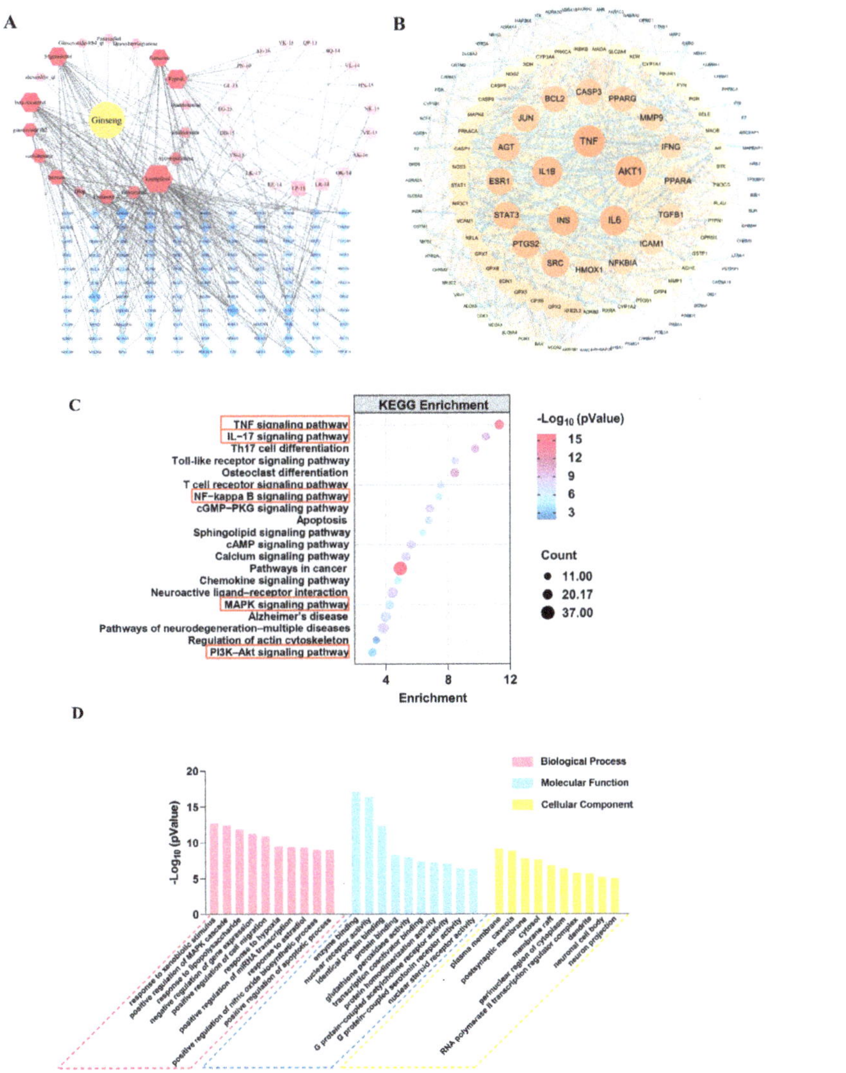

Figure 3. Visualization of interacting targets in the active components of ginseng and senescence: (**A**) complex target pathway network of ginseng components against senescence. The deep-red hexagonal

nodes represent the small molecule components in ginseng, the pink square nodes represent the calculated ginseng peptides, and the blue node is the potential target; (**B**) PPI network of the selected core objectives; (**C**) KEGG pathway analysis of senescence targets. The horizontal axis represents fold enrichment, the depth of the bubble color represents the enrichment degree of the pathway, and the bubble size represents the number of genes enriched in this pathway. The red boxes represent the main signaling pathways; (**D**) GO enrichment of ginseng against senescence. The ordinate indicates the degree of enrichment of the pathway, expressed in the form of $-\text{Log}_{10}$ (p Value).

We performed GO functional annotation and KEGG pathway analysis on 159 targets of the PPI network using the Metascape database [28]. Then, the top 20 targets were visualized as bubble plots (Figure 3C), and it was found that ginseng may alleviate inflammaging through the TNF, IL-17, NF kappa B, MAPK, and PI3K-Akt signaling pathways. In the biological processes (BPs), ginseng mainly has a great influence on MAPK cascade reaction, response to lipopolysaccharide (LPS), and negative regulation of gene expression and nitric oxide biosynthesis process. Regarding the molecular functions (MFs), the functions of ginseng pharmaceutical components are mainly related to enzyme binding, protein binding, and transcription coactivator binding. The targets in the cellular components (CCs) are closely related to the plasma membrane and cytosol (Figure 3D).

Based on the analysis of the results from network pharmacology and computational biology, we hypothesize that ginseng may exert anti-inflammatory and anti-aging effects by down-regulating the expression of pro-inflammatory cytokines such as IL-6, IL-1β, and TNF-α.

3.2. *In Vitro Evaluation of the Antioxidative and Anti-Inflammatory Effects of GPE*

To verify the anti-inflammatory effects of GPE, we first assessed its antioxidant activity and anti-inflammatory activities. We initially obtained GPE through enzymatic hydrolysis, which mainly included water-soluble small molecules and water-soluble polypeptides. The results showed that the half-maximal scavenging rates were 5.13% and 4.91%, respectively, indicating that the GPE possesses good free radical scavenging ability (Figure 4A,B).

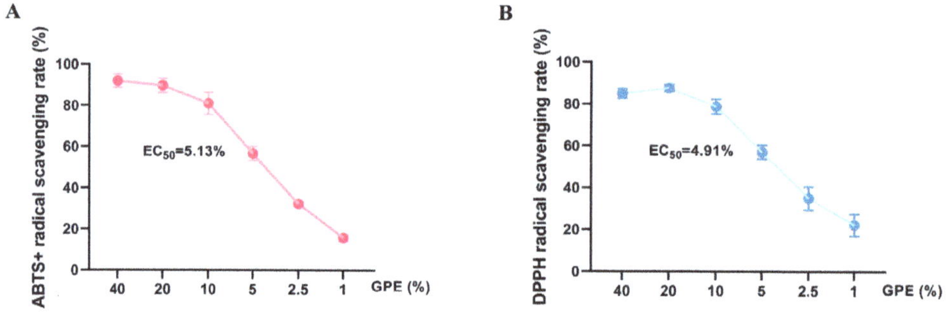

Figure 4. *Cont.*

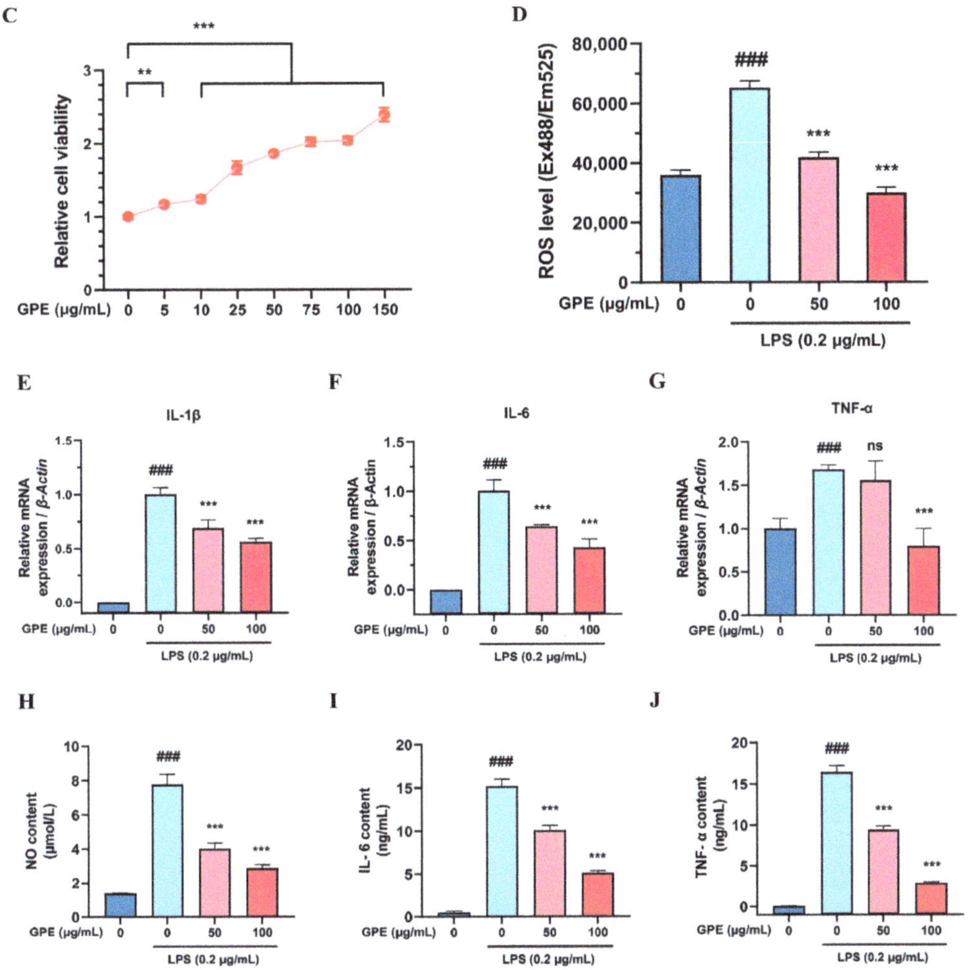

Figure 4. In vitro evaluation of the antioxidative and anti-inflammatory effects of GPE in RAW264.7 cell model. (**A**) ABTS+ radical scavenging rate; (**B**) DPPH radical scavenging rate; EC_{50}: half-maximal effective concentration; (**C**) cell viability calculated by the cck-8 assay after incubation of 24 h with different concentrations of GPE. The viability of untreated group was assessed as 1 (100%); (**D**) ROS levels after 24 h of treatment with different concentration of GPE; (**E–G**) changes in IL-1β, IL-6, and TNF-α gene expression evaluated by RT-qPCR; (**H**) changes in NO content; (**I**,**J**) release of IL-6 and TNF-α obtained by ELISA. Values are expressed as mean ± SD (n = 3). ANOVA was used for all statistical analyses and Dunnett's test was used for the post hoc test. Untreated cells were used as the blank group, while cells treated with LPS but without GPE were used as the model group. ### $p < 0.001$ vs. blank group; ** $p < 0.01$ vs. control group; *** $p < 0.001$ vs. control group; ns, not significant between the indicated groups.

Next, we evaluated the effects of GPE on the viability of RAW264.7 cells. The results showed that as the concentration of GPE increased (0–150 μg/mL), cell viability increased in a dose-dependent manner. This indicates that GPE promotes the viability of RAW264.7 cells within a certain concentration range and may have a positive impact on cell growth (Figure 4C). Subsequently, we further determined the antioxidative and anti-inflammatory effects of GPE in the RAW264.7 model. The ROS experimental results indicated that GPE could effectively reduce the intracellular ROS levels in RAW264.7 cells, which may also be one of the important mechanisms of its anti-inflammatory action (Figure 4D). Gene expression analysis found that GPE significantly inhibited the gene expression levels of IL-6, IL-1β, and TNF-α (Figure 4E–G). Additionally, the cytokine analysis results showed that the levels of IL-6 and TNF-α in cells treated with GPE were significantly reduced, while the levels of NO were also significantly decreased (Figure 4H–J). The experimental results indicated that GPE has significant antioxidative and anti-inflammatory effects.

3.3. Systematic Evaluation of the Anti-Inflammaging Effects of GPE

To verify whether GPE can alleviate skin cell senescence by reducing inflammatory responses, we utilized the HSF inflammaging model to assess the anti-aging efficacy of GPE. The HSF inflammaging model is an HSF cell senescence model induced by inflammatory cytokines produced by M1-type THP-1 cells. Compared with the use of a single inflammatory cytokine to induce an aging model, it can simulate cellular aging caused by changes in complex inflammatory environments in the skin more realistically [31,32]. In this inflammaging model, we found that the content of collagen I in the supernatant of the model group (cells treated with M1 THP-1 supernatant without GPE) significantly decreased, while the content of MMP-1 significantly increased. However, after the addition of GPE, the levels of collagen I and MMP-1 in the supernatant tended to return to normal (Figure 5A,B).

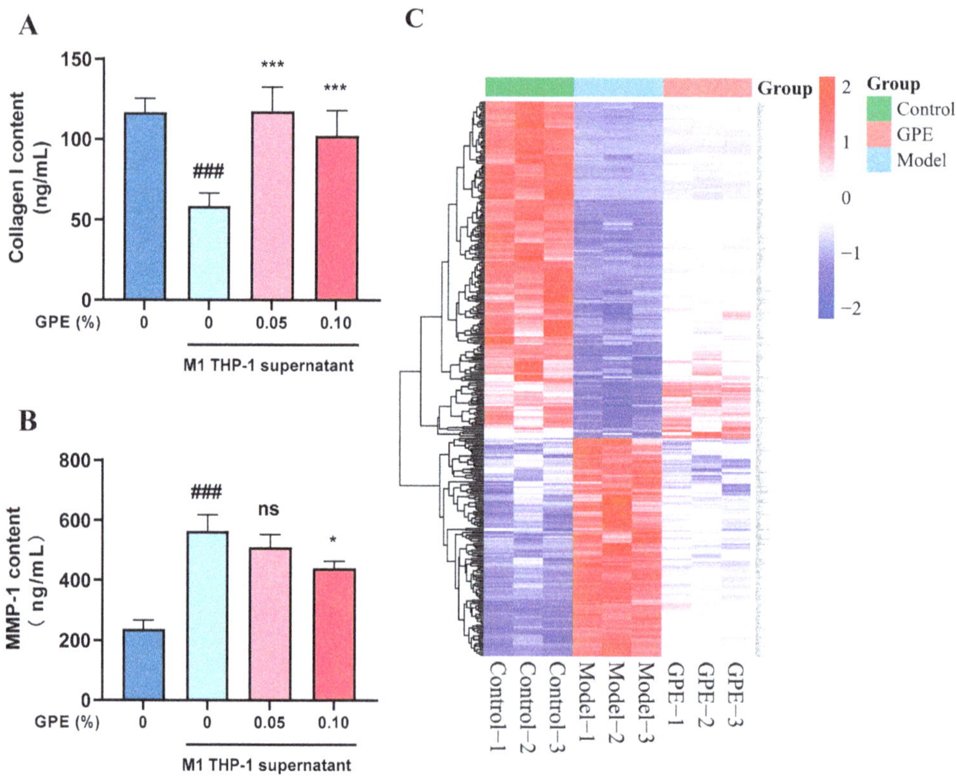

Figure 5. *Cont.*

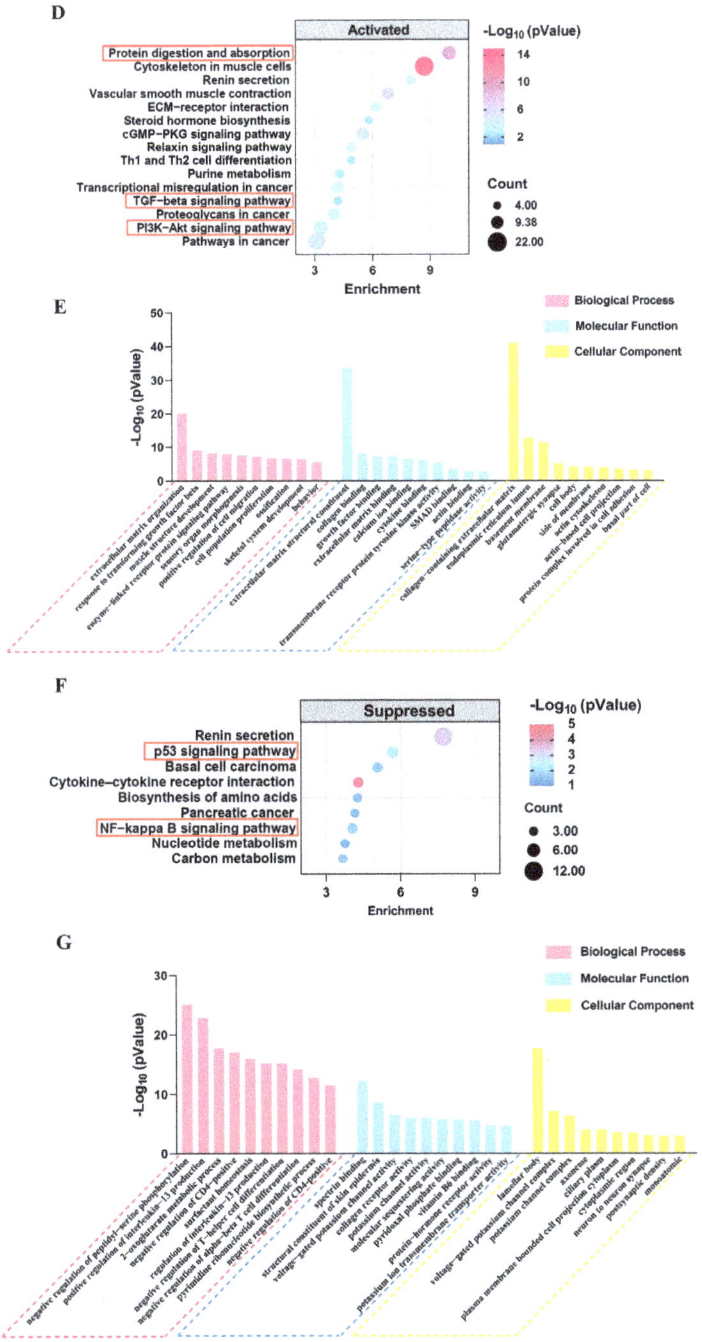

Figure 5. Evaluation of the anti-aging effect of GPE on inflammaging HSF cells. (**A**,**B**) inflammaging HSF cells were treated with different doses of GPE for 24 h and ELISA was used to detect collagen I and MMP-1 content change. Untreated cells were used as the blank group, while cells treated with THP-1 supernatant but without GPE were used as the model group. Values are expressed as mean ± SD (n = 3). ANOVA was used for all statistical analyses and Dunnett's test was used for the

post hoc test. ### $p < 0.001$ vs. Control group; * $p < 0.05$, *** $p < 0.001$ vs. Model group; ns, not significant between the indicated groups (vs. Model group); (**C**) heat map of differentially expressed genes in transcriptome sequencing. The red and blue blocks represent high and low expression levels of genes, respectively; (**D**,**E**) KEGG pathway analysis and GO enrichment of activated genes; (**F**,**G**) KEGG pathway analysis and GO enrichment of suppressed genes. In KEGG enrichment analysis, the horizontal axis represents fold enrichment, the depth of bubble color represents the enrichment degree of the pathway, and the bubble size represents the number of genes enriched in this pathway. The red boxes represent the main signaling pathways. In GO enrichment analysis, the ordinate indicates the degree of enrichment of the pathway, expressed in the form of $-\mathrm{Log}_{10}$ (p Value).

To comprehensively investigate the differential gene expression levels of GPE in HSF cells, we conducted transcriptome sequencing. After filtering the results, a total of 203 activated genes and 132 inhibited genes were identified (Figure 5C). Using the Metascape database for GO functional annotation and KEGG pathway analysis, it was found that the activated genes were primarily enriched in protein digestion and absorption and the TGF-β and PI3K-Akt signaling pathways (Figure 5D). In the GO enrichment analysis, the enrichment items for GPE included cell population proliferation, muscle structure development, response to TGF-β, growth factor binding, cytokine binding, collagen binding, and collagen-containing extracellular matrix (Figure 5E). The suppressed genes were mainly enriched in the p53 and NF-kappa B signaling pathways (Figure 5F). In the GO enrichment analysis, the enrichment items for GPE included a structural constituent of the skin epidermis and collagen receptor activity (Figure 5G). Based on the comprehensive analysis of the transcriptome sequencing results, we determined that the anti-aging effects of GPE on HSF cells are primarily manifested in the increased synthesis of collagen. The increase in collagen content further indicates that GPE has a protective effect against aging in HSF cells induced by inflammatory factors.

3.4. Systematic Evaluation of Anti-Inflammatory Aging in Dermatological Application Using Multi-Layer Skin Models

To confirm whether GPE can protect epidermal skin cells from UV damage, we first used UV irradiation on a 3D epidermal skin model to induce damage and detected changes in tissue morphology, filaggrin (FLG), loricrin (LOR), and transglutaminase 1 (TGM1) content. H&E staining showed that the number of sunburn cells increased after UVB irradiation but significantly decreased after GPE treatment (Figure 6A,B). Immunofluorescence was then used to detect the expression levels of FLG, LOR, and TGM1. The results showed that the expression levels of FLG, LOR, and TGM1 were significantly lower in cells after UVB irradiation compared to the control group. In contrast, the levels of these proteins were significantly higher in the GPE-treated group, with enhancement rates of 95.56%, 146.15%, and 207.69%, respectively (Figure 6C–E). These data indicate that GPE can upregulate the expression of FLG, LOR, and TGM1 proteins, protect epidermal skin cells from UVB-induced damage, and possess reparative effects.

After confirming the efficacy of GPE in the epidermal cells, we further sought to verify whether GPE also has anti-aging effects on dermal skin cells. Therefore, we utilized a combination of UVA (30 J/cm^2) and UVB (50 mJ/cm^2) irradiation to induce an in vitro skin model. H&E staining was used to observe histological changes, revealing that the thickness of the viable skin cell layer significantly decreased after UV irradiation but increased after GPE treatment. Masson staining also showed that GPE treatment could improve the UV-induced decrease in collagen fiber content (Figure 7A–C). Subsequently, we used immunofluorescence to detect changes in collagen content in the in vitro skin model. After UV induction, the content of collagen I, collagen III, collagen IV, and collagen XVII all significantly decreased, but GPE treatment significantly increased the collagen content, with enhancement rates of 77.36%, 213.33%, 176.92%, and 146.15%, respectively (Figure 7A,D–G). This indicates that skin function improved after treatment with 1% (v/v) GPE. Based on the above results, we determined that GPE applied to the in vitro skin model significantly increases collagen content and improves skin senescence induced by UV.

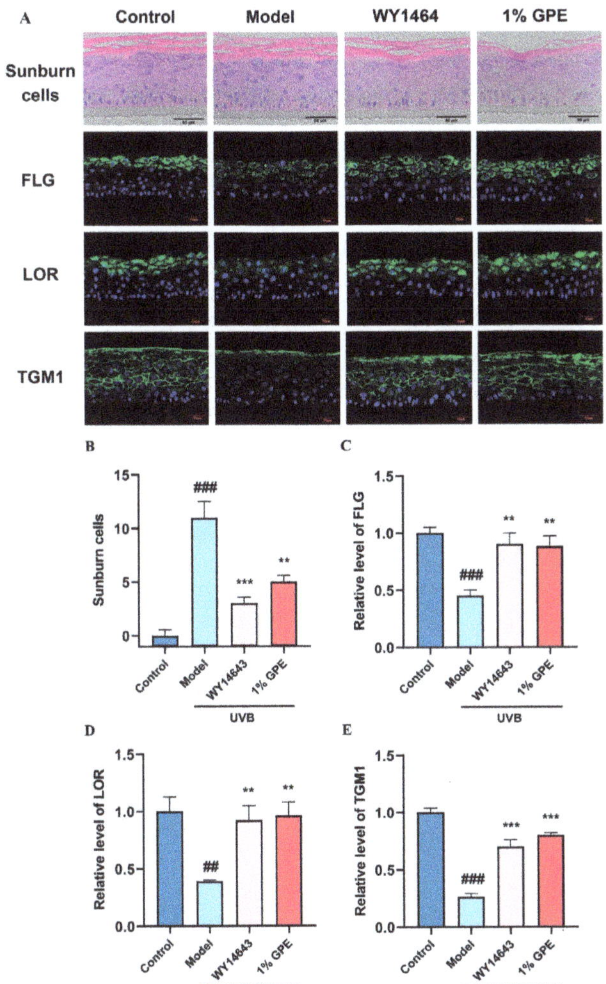

Figure 6. Effects of GPE treatment on UV-induced EpiKutis 3D epidermal skin model damage as assessed via immunofluorescence and H&E staining. All groups (except the control group) were treated with UVB. WY14643 (pirinixic acid) was used as the positive control drug in the experiment. The results show that 1% (v/v) GPE could protect epidermal skin cells from UVB-induced damage and possess reparative effects. (**A**) sunburn cells were calculated and expression of filaggrin (FLG), loricrin (LOR), and transglutaminase-1 (TGM1) proteins were analyzed after treatment with 1% (v/v) GPE in 3D human skin cells using a confocal microscope. Scale bar = 100 μm; (**B**–**E**) sunburn cells and relative expression level of FLG, LOR, and TGM1 proteins were analyzed. Values are expressed as mean ± SD (n = 3). Comparisons between groups were performed using a Student's t-test. ANOVA was used for all statistical analyses and Dunnett's test was used for the post hoc test. $^{##}$ $p < 0.01$, $^{###}$ $p < 0.001$ vs. Control group; ** $p < 0.01$, *** $p < 0.001$ vs. Model group.

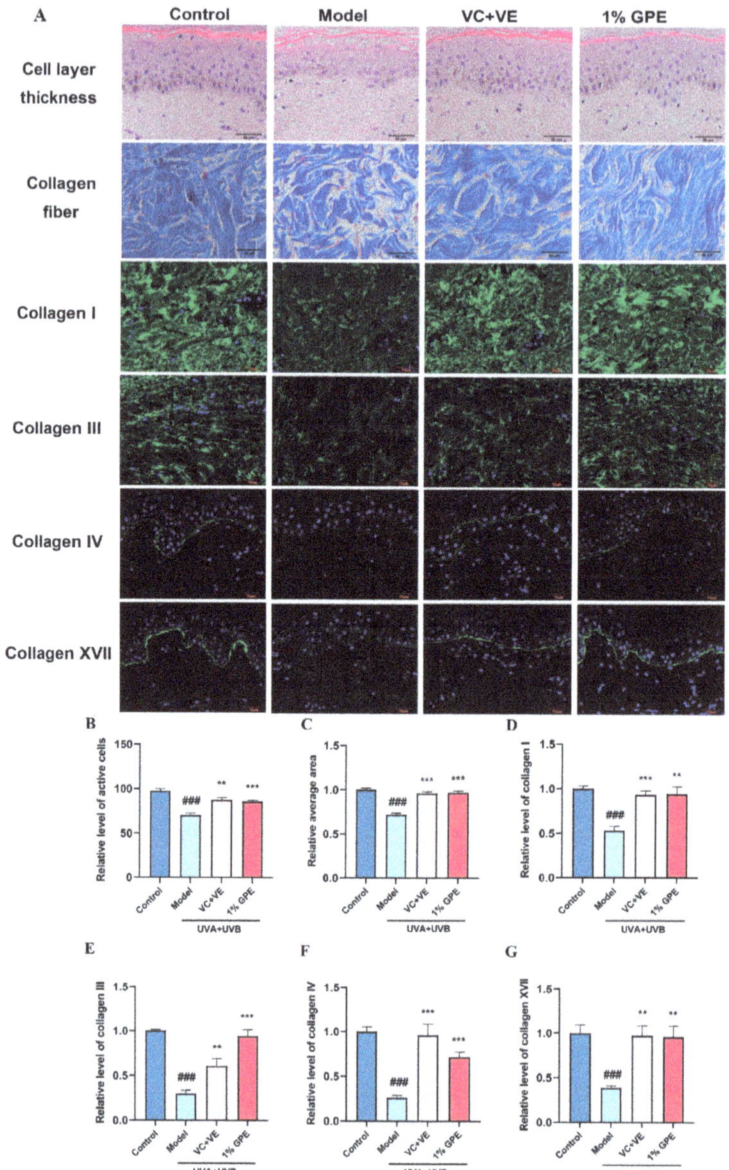

Figure 7. Protective effect of GPE on the UV-induced in vitro skin model. All groups (except the control group) were treated with UVA and UVB. VC (vitamin C) and VE (vitamin E) were used as the positive control drug in the experiment. The results indicated that impaired skin function due to UV exposure was markedly reversed after 1% (*v*/*v*) GPE treatment. (**A**) protective effect of GPE was evaluated through H&E staining, Masson staining, and immunofluorescence on in vitro skin models. Scale bar = 100 μm; (**B**) cell layer thickness, i.e., the relative level of active cells; (**C**) relative content of collagen fiber; (**D**–**G**) relative content of collagen I, collagen III, collagen IV, and collagen XVII was analyzed. Values are expressed as mean ± SD (n = 3). Comparisons between groups were performed using a Student's *t*-test. ANOVA was used for all statistical analyses and Dunnett's test was used for the post hoc test. ### $p < 0.001$ vs. Control group; ** $p < 0.01$, *** $p < 0.001$ vs. Model group.

3.5. Prediction of GPs Based on Protein Interaction Model

Next, we utilized the in-house customized machine learning model [33], which predicts the probability of protein–peptide interactions, to screen for bioactive peptides in GPE. We first applied the modified HSM model to the GP database [30]. Based on the scores output by the model, we selected the top 50 GPs with the highest interaction probabilities with the targets related to inflammaging in the modified HSM model [33,34]. This step provided us with a list of high-potential peptide candidates. Subsequently, we used the modified CAMP model to further predict the results from the HSM model [33,35]. We conducted an in-depth analysis of these 50 peptides using the CAMP model and, based on the predicted scores, ultimately selected peptides with interaction scores greater than 0.5. Through the predictions of these two models, we identified 19 peptides and 14 potential target proteins, providing a list of GPs with potential anti-aging and anti-inflammatory effects (Appendix A Table A1). After summarizing and screening the results, we synthesized the top five GPs with different sequences (QK-14, EG-15, VL-14, VK-15, AE-16) for subsequent experimental validation (Figure 8).

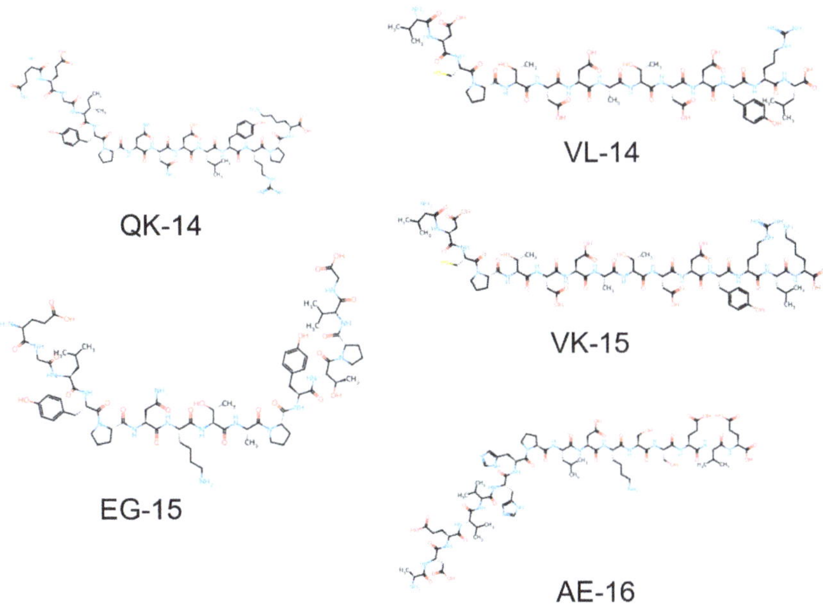

Figure 8. Structure of the obtained GPs screened. QK-14 (QEGIYPNNDLYRPK), VL-14 (VDCPTDDATDDYRL), EG-15 (EGLYPNKTAPYTPVG), VK-15 (VDCPTDDATDDYRLK), and AE-16 (ADEVVHHPLDKSSEVE).

3.6. Evaluation of Anti-Inflammatory and Anti-Aging Effects of GPs

To verify whether the peptides we screened have anti-inflammatory and anti-aging effects, we selected the top five GPs (QK-14, EG-15, VL-14, VK-15, and AE-16) and conducted preliminary evaluation using a UVA-induced UV aging model. It was found that the three peptides QK-14, VK-15, and AE-16 can increase the content of collagen I in senescent cells (Figure 9A). After in-depth exploration, we found that it has anti-aging effects on cell aging caused by UV and inflammatory factors and can enhance the expression of collagen I in cells (Figure 9B,C). Applied to the RAW264.7 cell inflammation model, it was found through ELISA detection that three GPs can significantly reduce the levels of IL-6 and TNF-α produced by M1 type RAW264.7 cells, and NO detection also showed the same effect (Figure 9D–F). Based on the above results, we determined that the GPs we calculated have excellent anti-inflammatory effects and can alleviate the decrease in collagen content caused by UV or inflammatory factors in skin cells with anti-aging effects.

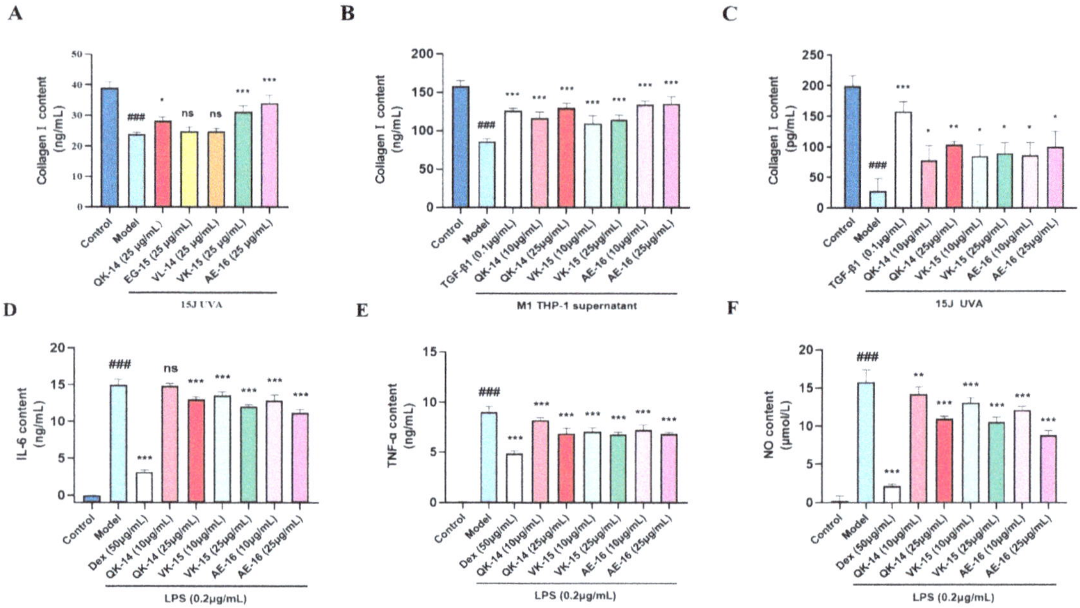

Figure 9. Verification of the anti-inflammatory and anti-aging effects of GPs. (**A**) preliminary evaluation of GPs using UVA-induced HSF senescence model; (**B**) changes in collagen I content in the HSF cell inflammaging model after treatment with QK-14, VK-15, or AE-16 (10 or 25 μg/mL); (**C**) changes in collagen I content in the HSF cell UVA-induced senescence model after treatment with GPs, TGF-β1 was used as the positive control drug; (**D–F**) changes in IL-6, TNF-α, and NO levels in the RAW264.7 cell inflammation model treated with GPs, Dex (Dexamethasone) was used as the positive control drug in the experiment. Values are expressed as mean ± SD (n = 3). ANOVA was used for all statistical analyses and Dunnett's test was used for the post hoc test. ### $p < 0.001$ vs. Control group; * $p < 0.05$, ** $p < 0.01$, *** $p < 0.001$ vs. Model group. ns, not significant between the indicated groups (vs. Model group).

4. Discussion

This study explores the effects of GPE on anti-inflammatory aging activity and skin protection, revealing potential mechanisms at the cellular and molecular levels. Cellular assays demonstrated that GPE enhanced cell viability within a defined concentration range, ameliorating inflammatory responses and supporting tissue repair. While in an inflammatory aging model, GPE significantly elevated collagen deposition and reduced MMP-1 expression. In vitro skin models exposed to UV irradiation showed that GPE attenuated epidermal damage and enhanced collagen synthesis. Machine learning-driven screening identified three bioactive peptides (QK-14, VK-15, AE-16) from GPE that demonstrated dual anti-inflammatory and collagen-promoting activities in both cellular senescence and inflammation models, collectively indicating GPE's therapeutic potential in dermatoprotection and anti-aging interventions.

Ginseng (*Panax ginseng* C.A. Meyer) has long been recognized for its medicinal properties, including anti-inflammatory, anti-aging, and immunomodulatory effects. These properties are primarily attributed to its saponins and polysaccharides, which have been extensively studied for their therapeutic potential in various health conditions [36–39].

However, the role of ginseng peptides (GPs) in skincare has remained relatively underexplored compared to saponins and polysaccharides [30]. Our study utilizes network pharmacology and machine learning to identify and validate the bioactive peptides within GPE, demonstrating their potential to reduce inflammation and promote collagen synthesis in both in vitro and ex vivo models. This approach aligns with recent trends in computational biology, which emphasize the use of predictive models to identify bioactive compounds from natural sources [40–45].

Compared to previous studies, our work provides a more comprehensive evaluation of GPE's effects on skin health. For instance, while earlier research has focused on the antioxidant properties of ginseng saponins [46–48], our study extends this by demonstrating the anti-inflammatory and collagen-promoting effects of GPE. Additionally, our use of multi-layer skin models to validate GPE's reparative effects on UV-induced damage offers a more realistic assessment of its potential in skincare applications [49,50]. Our findings on GPE's bioactivity are consistent with studies on other herbal peptides. For example, Ganoderma lucidum (*G. lucidum, Lingzhi*), commonly known as Reishi mushroom, is another herbal source that has been extensively studied for its bioactive peptides. Reishi peptides have been reported to exhibit significant anti-inflammatory, antioxidant, and immunomodulatory effects [51,52]. For example, a study identified several bioactive peptides from Reishi mushroom that demonstrated potent anti-inflammatory effects by inhibiting the production of pro-inflammatory cytokines such as IL-6 and TNF-α [51]. Another study reported that ginseng peptides could promote collagen synthesis in human skin models, thereby enhancing skin elasticity and reducing signs of aging [53]. Similarly, peptides derived from soybeans have been shown to exhibit anti-inflammatory and anti-aging effects, particularly in reducing the expression of inflammatory cytokines such as IL-6 and TNF-α [54]. Likewise, peptides from green tea or other plants have demonstrated significant antioxidant properties, which are crucial for mitigating oxidative stress-induced skin aging [55,56]. These parallels suggest that herbal peptides may share common mechanisms in promoting skin health through anti-inflammatory and antioxidative pathways.

Traditional methods of peptide discovery from natural sources often involve labor-intensive and time-consuming processes, such as extraction, purification, and in vitro and in vivo testing. These methods typically require extensive experimentation and can be limited by the complexity of natural products and the variability in their biological activities. In contrast, our study employs a combination of network pharmacology and machine learning to predict and identify bioactive peptides from ginseng extract. This computational approach offers an advantage over traditional methods: by using predictive models, we can rapidly screen and prioritize potential bioactive peptides, significantly reducing the time and resources required for experimental validation [57], and network pharmacology allows us to integrate multiple data sources, providing a more holistic view of their therapeutic potential [58].

Notably, compared with other computational screening methods, our computational approach has several unique features and advantages: (1) We combined multiple algorithms, including the HSM model [34] and the CAMP model [35], to enhance the accuracy and reliability of peptide prediction. The HSM model provided a preliminary screening of potential peptides, while the CAMP model further refined the selection by predicting protein–peptide interactions [45]. (2) Our approach not only identified bioactive peptides but also predicted their potential targets and pathways involved in anti-inflammatory and anti-aging effects. This comprehensive analysis can identify potential targets and pathways involved in the bioactivity of various peptides, facilitating a deeper understanding of their mechanisms of action [59]. (3) Our computational framework can be easily adapted to screen peptides from other herbal sources, making it a versatile tool for identifying bioactive compounds from natural products.

However, our study also has some limitations: (1) the bioactivity of the identified peptides was only suggested in vitro, and our 3D experimental models do not fully replicate the complexity of human skin [60]. Therefore, clinical trials are also needed to confirm GPE's efficacy and safety in human subjects, and more comprehensive models are needed to validate the bioactivity of ginseng peptides; (2) the discovery of bioactive peptides in GPE offers new possibilities for cosmeceuticals against inflammaging. Future research should explore the specific molecular mechanisms of GPE's effects, such as related signaling pathways; (3) additionally, studying the synergistic effects of GPE with other bioactive compounds could enhance its therapeutic potential, which can determine optimal combination and dosages for skincare formulations [61].

As computational biology evolves, the virtual screening methods used here can be improved. Better machine learning algorithms and predictive models could increase the accuracy and efficiency of identifying bioactive peptides from natural sources [62]. This would allow our research approach to be applied more broadly to the discovery of bioactive compounds in various natural ingredients. Moreover, future research could focus on the structural optimization and modification of the identified ginseng peptides using various computational approaches to enhance their efficacy, stability, and bioavailability for practical applications in skincare and therapeutic formulations [63–65].

5. Conclusions

In conclusion, our study provides a comprehensive evaluation of GPE's potential as an anti-inflammatory and anti-aging agent for skincare. By integrating network pharmacology, machine learning, and experimental validation, we have identified key ginseng peptides that demonstrate significant bioactivity. These findings not only expand the current understanding of ginseng's therapeutic potential but also offer new directions for the development of innovative skincare formulations. Future research should aim to address the limitations identified here and further explore the clinical applications of ginseng peptides in skin health.

Author Contributions: Conceptualization, Z.X., W.L., F.Z., S.K. and J.L.; methodology, F.Z., W.X., Z.X. and P.T.; software, W.X. and D.L.; validation, Z.X., P.T. and X.Z.; formal analysis, D.L. and X.Y.; investigation, Z.X.; resources, J.L., Q.S. and X.L.; data curation, Z.X. and F.Z.; writing—original draft preparation, Z.X.; writing—review and editing, J.L.; visualization, Z.X.; supervision, S.K. and W.L.; project administration, S.K. and W.L.; funding acquisition, Q.S., X.L. and W.L. All authors have read and agreed to the published version of the manuscript.

Funding: This research was funded by The Zhejiang Provincial Natural Science Foundation of China (LQ23H160002); The Collaborative Project with China Resources Research Institute of Science & Technology (ILH24001); The fund of Yangtze Delta Region Institute of Tsinghua University, Zhejiang (LZZLX23C002); and The Key Research and Development Program of Shihezi City (No. 2024SF08).

Institutional Review Board Statement: Not applicable.

Informed Consent Statement: Not applicable.

Data Availability Statement: The data presented in this study are available in this article. Other data that support the findings of this study are available upon request from the corresponding author.

Conflicts of Interest: W.L. was employed by Beijing Yiqing Daily Chemical Co., Ltd. F.Z., S.K., and J.L. were employed by Harvest Biotech (Zhejiang) Co., Ltd. The remaining authors declare that the research was conducted in the absence of any commercial or financial relationships that could be construed as potential conflicts of interest.

Abbreviations

The following abbreviations are used in this article:

ABTS+	2,2′-azino-bis (3-ethylbenzothiazoline-6-sulfonate) diammonium salt
BP	biological processes
CC	cellular components
COL1A1	Collagen type I alpha 1
Dex	Dexamethasone
DPPH	2,2-diphenyl-1-picrylhydrazyl
ELISA	Enzyme-linked immunosorbent assay
FLG	Filaggrin
GO	Gene ontology
GP	Ginseng peptides
GPE	Ginseng peptide extract
HSM	Hierarchical statistical mechanical modeling
KEGG	Kyoto Encyclopedia of Genes and Genomes
LOR	Loricrin
MF	molecular functions
NO	Nitric oxide
OMIM	Online Mendelian inheritance in man
PPI	Protein–protein interaction
ROS	Reactive oxygen species
RT-qPCR	Quantitative real-time polymerase chain reaction
SASP	senescence-associated secretory phenotype
STRING	Search Tool for the Retrieval of Interacting Genes/Proteins
TGM1	Transglutaminase-1
TCMSP	Traditional Chinese Medicine Systems Pharmacology

Appendix A

Table A1. Sequence of GPs, keywords of targets, and predicted scores of modified HSM and CAMP models.

Peptide_Sequence	Keywords	Protein	HSM Score (p_Value)	CAMP Score
LNQEGIYPNNDLYRP	proliferation	BIN1	0.000222856	0.99666637
LNQEGIYPNNDLYR	Inflammation, proliferation	FGR	0.000571771	0.9769991
QEGIYPNNDLYRPK	Proliferation, proliferation	FGR	0.0013086	0.97212684
LNQEGIYPNNDLYRP	inflammation	PSTPIP1	0.000271744	0.94917411
ANLLHKLEETLGMNDK	Inflammation, proliferation, oxidat	TP53BP2	0.0000043	0.908577681
LNQEGIYPNNDLYRP	inflammation	BTK	0.000251922	0.89676905
VDCPTDDATDDYR	proliferation	FYN	0.000111128	0.8797974
NQEGIYPNNDLYRPK	Inflammation, proliferation	FGR	0.00017643	0.86943609
HLNAVPEIDFTKNEN	inflammation	PSTPIP1	0.000113509	0.86510551
LNQEGIYPNNDLYRP	Inflammation, oxidat	PIK3R1	0.000212876	0.84001458
VDCPTDDATDDYRL	proliferation	FYN	0.000110712	0.83449197

Table A1. *Cont.*

Peptide_Sequence	Keywords	Protein	HSM Score (p_Value)	CAMP Score
LNQEGIYPNNDLYR	inflammation	BTK	0.00057089	0.8144055
LNQEGIYPNNDLYR	oxidat	NCF4	0.000371933	0.79551464
SEYVLTDINVCVNQ	inflammation	BTK	0.000883059	0.79394358
VDCPTDDATDDYR	Inflammation, proliferation	FGR	0.00016603	0.7798081
LNQEGIYPNNDLYR	Inflammation, oxidat	PIK3R1	0.000545315	0.77899241
QEGIYPNNDLYRP	collagen	WWC1	0.000382538	0.76215649
VDCPTDDATDDYRLK	proliferation	FYN	0.0000267	0.73349589
ADEVVHHPLDKSSEVE	Inflammation, proliferation, oxidat	JIP1	0.0000021	0.730936706
PEIDFTKNEN	inflammation	BTK	0.000421379	0.72908759
VDCPTDDATDDYRL	Inflammation, proliferation	FGR	0.000133978	0.72363502
NQEGIYPNNDLYRPK	oxidat	NCF4	0.000182778	0.71487856
GVQKTEVEATSTVPAQKL	Inflammation, proliferation, oxidat	JIP1	0.0000027	0.708695173
EGLYPNKTAPYTPVG	Inflammation, proliferation	FGR	0.000300853	0.69255173
DKSSEVETTDRGLFD	proliferation	VAV1	0.0000597	0.685672045
VQVLEGNGGVGTIKN	proliferation	ARHGAP26	0.0000169	0.6738382
LNQEGIYPNNDLYRP	oxidat	NCF4	0.000162998	0.64596671
QEGIYPNNDLYRPK	Inflammation, proliferation	ITK	0.001193209	0.62135625
LTVTPEEPVVVEK	Inflammation, proliferation	FGR	0.000264268	0.59293336
VQVLEGNGGVGTIKN	Inflammation, proliferation	FGR	0.0000117	0.5275951
EGLYPNKTAPYTPVG	oxidat	SGR	0.000258785	0.52582133
EHTNTEDKQFWEHE	inflammation	BTK	0.0000917	0.50152653

References

1. Hofseth, L.J.; Ying, L. Identifying and Defusing Weapons of Mass Inflammation in Carcinogenesis. *Biochim. Biophys. Acta (BBA)-Rev. Cancer* **2006**, *1765*, 74–84. [CrossRef] [PubMed]
2. Barton, G.M. A Calculated Response: Control of Inflammation by the Innate Immune System. *J. Clin. Investig.* **2008**, *118*, 413–420. [CrossRef]
3. Davies, L.C.; Jenkins, S.J.; Allen, J.E.; Taylor, P.R. Tissue-Resident Macrophages. *Nat. Immunol.* **2013**, *14*, 986–995. [CrossRef] [PubMed]
4. Zhao, Y.; Zhao, N.; Kollie, L.; Yang, D.; Zhang, X.; Zhang, H.; Liang, Z. Sasanquasaponin from Camellia Oleifera Abel Exerts an Anti-Inflammatory Effect in RAW 264.7 Cells via Inhibition of the NF-κB/MAPK Signaling Pathways. *Int. J. Mol. Sci.* **2024**, *25*, 2149. [CrossRef]
5. Wu, T.; Zhang, J.; Geng, M.; Tang, S.-J.; Zhang, W.; Shu, J. Nucleoside Reverse Transcriptase Inhibitors (NRTIs) Induce Proinflammatory Cytokines in the CNS via Wnt5a Signaling. *Sci. Rep.* **2017**, *7*, 4117. [CrossRef] [PubMed]
6. Wang, Y.; Miao, X.; Jiang, Y.; Wu, Z.; Zhu, X.; Liu, H.; Wu, X.; Cai, J.; Ding, X.; Gong, W. The Synergistic Antitumor Effect of IL-6 Neutralization with NVP-BEZ235 in Hepatocellular Carcinoma. *Cell Death Dis.* **2022**, *13*, 146. [CrossRef]
7. Ross, E.A.; Devitt, A.; Johnson, J.R. Macrophages: The Good, the Bad, and the Gluttony. *Front. Immunol.* **2021**, *12*, 708186. [CrossRef]
8. van Furth, R.; Cohn, Z.A.; Hirsch, J.G.; Humphrey, J.H.; Spector, W.G.; Langevoort, H.L. The Mononuclear Phagocyte System: A New Classification of Macrophages, Monocytes, and Their Precursor Cells. *Bull. World Health Organ.* **1972**, *46*, 845–852.
9. Fu, J.; Lu, Z.; Wu, G.; Yang, Z.; Wu, X.; Wang, D.; Nie, Z.; Sheng, Q. *Gastrodia elata* Specific miRNA Attenuates Neuroinflammation via Modulating NF-κB Signaling Pathway. *Int. J. Neurosci.* **2024**, *134*, 1652–1662. [CrossRef]
10. Chen, Q.; Kang, J.; Fu, C. The Independence of and Associations among Apoptosis, Autophagy, and Necrosis. *Signal Transduct. Target. Ther.* **2018**, *3*, 18. [CrossRef]

11. Acosta, J.C.; Banito, A.; Wuestefeld, T.; Georgilis, A.; Janich, P.; Morton, J.P.; Athineos, D.; Kang, T.-W.; Lasitschka, F.; Andrulis, M.; et al. A Complex Secretory Program Orchestrated by the Inflammasome Controls Paracrine Senescence. *Nat. Cell Biol.* **2013**, *15*, 978–990. [CrossRef] [PubMed]
12. Barbé-Tuana, F.; Funchal, G.; Schmitz, C.R.R.; Maurmann, R.M.; Bauer, M.E. The Interplay between Immunosenescence and Age-Related Diseases. *Semin. Immunopathol.* **2020**, *42*, 545–557. [CrossRef]
13. Li, X.; Li, C.; Zhang, W.; Wang, Y.; Qian, P.; Huang, H. Inflammation and Aging: Signaling Pathways and Intervention Therapies. *Signal Transduct. Target. Ther.* **2023**, *8*, 239. [CrossRef]
14. Li, Y.; Tian, X.; Luo, J.; Bao, T.; Wang, S.; Wu, X. Molecular Mechanisms of Aging and Anti-Aging Strategies. *Cell Commun. Signal.* **2024**, *22*, 285. [CrossRef] [PubMed]
15. Attele, A.S.; Wu, J.A.; Yuan, C.-S. Ginseng Pharmacology. *Biochem. Pharmacol.* **1999**, *58*, 1685–1693. [CrossRef] [PubMed]
16. Yun, T.K. Brief Introduction of Panax Ginseng CA Meyer. *J. Korean Med. Sci.* **2001**, *16*, S3. [CrossRef]
17. Xia, P.; Mao, Y.; Liang, Z. Two Important Ginseng Plants in Southeast Asia: A Systematic Review of Their Traditional Uses, Botany, Phytochemistry, and Pharmacology. *Acta Physiol. Plant.* **2022**, *44*, 105. [CrossRef]
18. Lu, G.; Liu, Z.; Wang, X.; Wang, C. Recent Advances in Panax Ginseng C.A. Meyer as a Herb for Anti-Fatigue: An Effects and Mechanisms Review. *Foods* **2021**, *10*, 1030. [CrossRef]
19. Hernández-García, D.; Granado-Serrano, A.B.; Martín-Gari, M.; Naudí, A.; Serrano, J.C. Efficacy of Panax Ginseng Supplementation on Blood Lipid Profile. A Meta-Analysis and Systematic Review of Clinical Randomized Trials. *J. Ethnopharmacol.* **2019**, *243*, 112090. [CrossRef]
20. Fan, W.; Huang, Y.; Zheng, H.; Li, S.; Li, Z.; Yuan, L.; Cheng, X.; He, C.; Sun, J. Ginsenosides for the Treatment of Metabolic Syndrome and Cardiovascular Diseases: Pharmacology and Mechanisms. *Biomed. Pharmacother.* **2020**, *132*, 110915. [CrossRef]
21. Li, J.; Li, F.; Jin, D. Ginsenosides Are Promising Medicine for Tumor and Inflammation: A Review. *Am. J. Chin. Med.* **2023**, *51*, 883–908. [CrossRef] [PubMed]
22. Kim, Y.-J.; Zhang, D.; Yang, D.-C. Biosynthesis and Biotechnological Production of Ginsenosides. *Biotechnol. Adv.* **2015**, *33*, 717–735. [CrossRef]
23. Truong, V.-L.; Keum, Y.-S.; Jeong, W.-S. Red Ginseng Oil Promotes Hair Growth and Protects Skin against UVC Radiation. *J. Ginseng Res.* **2021**, *45*, 498–509. [CrossRef]
24. Ru, J.; Li, P.; Wang, J.; Zhou, W.; Li, B.; Huang, C.; Li, P.; Guo, Z.; Tao, W.; Yang, Y.; et al. TCMSP: A Database of Systems Pharmacology for Drug Discovery from Herbal Medicines. *J. Cheminform.* **2014**, *6*, 13. [CrossRef] [PubMed]
25. Stelzer, G.; Rosen, N.; Plaschkes, I.; Zimmerman, S.; Twik, M.; Fishilevich, S.; Stein, T.I.; Nudel, R.; Lieder, I.; Mazor, Y.; et al. The GeneCards Suite: From Gene Data Mining to Disease Genome Sequence Analyses. *Curr. Protoc. Bioinform.* **2016**, *54*, 1–30. [CrossRef]
26. Hamosh, A.; Amberger, J.S.; Bocchini, C.; Scott, A.F.; Rasmussen, S.A. Online Mendelian Inheritance in Man (OMIM®): Victor McKusick's Magnum Opus. *Am. J. Med. Genet. Part A* **2021**, *185*, 3259–3265. [CrossRef] [PubMed]
27. Szklarczyk, D.; Kirsch, R.; Koutrouli, M.; Nastou, K.; Mehryary, F.; Hachilif, R.; Gable, A.L.; Fang, T.; Doncheva, N.T.; Pyysalo, S.; et al. The STRING Database in 2023: Protein–Protein Association Networks and Functional Enrichment Analyses for Any Sequenced Genome of Interest. *Nucleic Acids Res.* **2023**, *51*, D638–D646. [CrossRef]
28. Zhou, Y.; Zhou, B.; Pache, L.; Chang, M.; Khodabakhshi, A.H.; Tanaseichuk, O.; Benner, C.; Chanda, S.K. Metascape Provides a Biologist-Oriented Resource for the Analysis of Systems-Level Datasets. *Nat. Commun.* **2019**, *10*, 1523. [CrossRef]
29. Lu, M.; Xu, X.; Xi, B.; Dai, Q.; Li, C.; Su, L.; Zhou, X.; Tang, M.; Yao, Y.; Yang, J. Molecular Network-Based Identification of Competing Endogenous RNAs in Thyroid Carcinoma. *Genes* **2018**, *9*, 44. [CrossRef]
30. Ye, X.; Zhao, N.; Yu, X.; Han, X.; Gao, H.; Zhang, X. Extensive Characterization of Peptides from *Panax ginseng* C.A. Meyer Using Mass Spectrometric Approach. *Proteomics* **2016**, *16*, 2788–2791. [CrossRef]
31. Horiba, S.; Kami, R.; Tsutsui, T.; Hosoi, J. IL-34 Downregulation—Associated M1/M2 Macrophage Imbalance Is Related to Inflammaging in Sun-Exposed Human Skin. *JID Innov.* **2022**, *2*, 100112. [CrossRef] [PubMed]
32. Xu, Y.; Liu, Y.; Li, J.; Li, Y.; Xu, L.; Dong, K.; Lin, X.; Zhang, T. Unveiling a Novel In-Vitro Model of Skin Inflammaging. *Front. Med.* **2025**, *12*, 1556680. [CrossRef] [PubMed]
33. Li, X.-H.; Su, W.-R.; Wang, F.-F.; Li, K.; Zhu, J.-Y.; Zhu, S.-Y.; Kang, S.-N.; He, C.-F.; Li, J.-X.; Lin, X. Computational Biology in Topical Bioactive Peptide Discovery for Cosmeceutical Application: A Concise Review. *Biomed. Eng. Commun.* **2023**, *2*, 14. [CrossRef]
34. Cunningham, J.M.; Koytiger, G.; Sorger, P.K.; AlQuraishi, M. Biophysical Prediction of Protein–Peptide Interactions and Signaling Networks Using Machine Learning. *Nat. Methods* **2020**, *17*, 175–183. [CrossRef]
35. Lei, Y.; Li, S.; Liu, Z.; Wan, F.; Tian, T.; Li, S.; Zhao, D.; Zeng, J. A Deep-Learning Framework for Multi-Level Peptide–Protein Interaction Prediction. *Nat. Commun.* **2021**, *12*, 5465. [CrossRef]
36. Zheng, Y.; Shao, R.; Xia, P.; Liang, Z.; Yan, K. Activity and Function Studies of the Promoter Cis-Acting Elements of the Key Enzymes in Saponins Biosynthesis of DS from Panax Notoginseng. *Protoplasma* **2022**, *259*, 163–171. [CrossRef]

37. Su, J.; Su, Q.; Hu, S.; Ruan, X.; Ouyang, S. Research Progress on the Anti-Aging Potential of the Active Components of Ginseng. *Nutrients* **2023**, *15*, 3286. [CrossRef]
38. Fan, W.; Fan, L.; Wang, Z.; Mei, Y.; Liu, L.; Li, L.; Yang, L.; Wang, Z. Rare Ginsenosides: A Unique Perspective of Ginseng Research. *J. Adv. Res.* **2024**, *66*, 303–328. [CrossRef]
39. Zhou, G.; Wang, C.-Z.; Mohammadi, S.; Sawadogo, W.R.; Ma, Q.; Yuan, C.-S. Pharmacological Effects of Ginseng: Multiple Constituents and Multiple Actions on Humans. *Am. J. Chin. Med.* **2023**, *51*, 1085–1104. [CrossRef]
40. Xie, W.-Y.; Ji, Z.-H.; Ren, W.-Z.; Zhao, P.-S.; Wei, F.-H.; Hu, J.; Yuan, B.; Gao, W. Wheat Peptide Alleviates DSS-Induced Colitis by Activating the Keap1–Nrf2 Signaling Pathway and Maintaining the Integrity of the Gut Barrier. *Food Funct.* **2024**, *15*, 5466–5484. [CrossRef]
41. Wang, W.; Lin, H.; Shen, W.; Qin, X.; Gao, J.; Cao, W.; Zheng, H.; Chen, Z.; Zhang, Z. Optimization of a Novel Tyrosinase Inhibitory Peptide from Atrina Pectinata Mantle and Its Molecular Inhibitory Mechanism. *Foods* **2023**, *12*, 3884. [CrossRef] [PubMed]
42. Yin, H.; Zhang, S.; Yue, H.; Wang, M.; Zeng, J.; Wu, W.; Wang, J.; Zheng, H.; Xue, C.; Zhao, Y.-T. Isolation, Identification and in Silico Analysis of Two Novel Cytoprotective Peptides from Tilapia Skin against Oxidative Stress-Induced Ovarian Granulosa Cell Damage. *J. Funct. Foods* **2023**, *107*, 105629. [CrossRef]
43. Zhu, Q.; Xue, J.; Wang, P.; Wang, X.; Zhang, J.; Fang, X.; He, Z.; Wu, F. Identification of a Novel ACE Inhibitory Hexapeptide from Camellia Seed Cake and Evaluation of Its Stability. *Foods* **2023**, *12*, 501. [CrossRef]
44. Yuxiu, Z.; Haisheng, L.; Lei, D.; Jialong, G.; Wenhong, C.; Xiaoming, Q.; Zhongqin, C.; Huina, Z.; Saiyi, Z. A Novel Tyrosinase Inhibitory Peptide Obtained from Sipunculus Nudus Gelatin Hydrolysate: Preparation, Identification, and Action Mechanism. *LWT* **2024**, *202*, 116293. [CrossRef]
45. Feifei, W.; Wenrou, S.; Sining, K.; Siyu, Z.; Xiaolei, F.; Junxiang, L.; Congfen, H.; Xuhui, L. A Novel Functional Peptide, Named EQ-9 (ESETRILLQ), Identified by Virtual Screening from Regenerative Cell Secretome and Its Potential Anti-Aging and Restoration Effects in Topical Applications. *Peptides* **2023**, *169*, 171078. [CrossRef]
46. Baik, I.-H.; Kim, K.-H.; Lee, K.-A. Antioxidant, Anti-Inflammatory and Antithrombotic Effects of Ginsenoside Compound K Enriched Extract Derived from Ginseng Sprouts. *Molecules* **2021**, *26*, 4102. [CrossRef]
47. Ratan, Z.A.; Haidere, M.F.; Hong, Y.H.; Park, S.H.; Lee, J.-O.; Lee, J.; Cho, J.Y. Pharmacological Potential of Ginseng and Its Major Component Ginsenosides. *J. Ginseng Res.* **2021**, *45*, 199–210. [CrossRef]
48. Wu, H.; Pei, H.; Liu, J.; Zeng, J.; Liu, S.; Chen, W.; He, Z.; Du, R. Protective Effect of Total Saponins of Ginseng Stems and Leaves (GSLS) on Chlorpyrifos-Induced Brain Toxicity in Mice through the PTEN/PI3K/AKT Axis. *Aging* **2022**, *14*, 8982–8999. [CrossRef]
49. Yang, J.-E.; Ngo, H.T.T.; Hwang, E.; Seo, S.A.; Park, S.W.; Yi, T.-H. Dietary Enzyme-Treated Hibiscus Syriacus L. Protects Skin against Chronic UVB-Induced Photoaging via Enhancement of Skin Hydration and Collagen Synthesis. *Arch. Biochem. Biophys.* **2019**, *662*, 190–200. [CrossRef]
50. Bai, D.; Wang, Z.; Xiao, Y.; Liu, T.; Pu, Y.; Sun, H.; Wang, M.; Guo, C.; Zhang, J. Transdermal Delivery of Elastin Peptide Assisted by Betaine-Based Deep Eutectic Solvent to Ameliorate Skin Photoaging. *Biomater. Adv.* **2024**, *163*, 213965. [CrossRef]
51. Aursuwanna, T.; Noitang, S.; Sangtanoo, P.; Srimongkol, P.; Saisavoey, T.; Puthong, S.; Reamtong, O.; Karnchanatat, A. Investigating the Cellular Antioxidant and Anti-Inflammatory Effects of the Novel Peptides in Lingzhi Mushrooms. *Heliyon* **2022**, *8*, e11067. [CrossRef] [PubMed]
52. Meng, M.; Wang, L.; Yao, Y.; Lin, D.; Wang, C.; Yao, J.; Sun, H.; Liu, M. Ganoderma Lucidum Polysaccharide Peptide (GLPP) Attenuates Rheumatic Arthritis in Rats through Inactivating NF-κB and MAPK Signaling Pathways. *Phytomedicine* **2023**, *119*, 155010. [CrossRef]
53. Zhu, N.; Xu, M.-H.; Li, Y. Bioactive Oligopeptides from Ginseng (Panax Ginseng Meyer) Suppress Oxidative Stress-Induced Senescence in Fibroblasts via NAD+/SIRT1/PGC-1α Signaling Pathway. *Nutrients* **2022**, *14*, 5289. [CrossRef]
54. Wen, L.; Bi, H.; Zhou, X.; Jiang, Y.; Zhu, H.; Fu, X.; Yang, B. Structure Characterization of Soybean Peptides and Their Protective Activity against Intestinal Inflammation. *Food Chem.* **2022**, *387*, 132868. [CrossRef]
55. Kanlayavattanakul, M.; Khongkow, M.; Klinngam, W.; Chaikul, P.; Lourith, N.; Chueamchaitrakun, P. Recent Insights into Catechins-Rich Assam Tea Extract for Photoaging and Senescent Ageing. *Sci. Rep.* **2024**, *14*, 2253. [CrossRef] [PubMed]
56. Mo, Q.; You, S.; Fu, H.; Wang, D.; Zhang, J.; Wang, C.; Li, M. Purification and Identification of Antioxidant Peptides from Rice Fermentation of Lactobacillus Plantarum and Their Protective Effects on UVA−Induced Oxidative Stress in Skin. *Antioxidants* **2022**, *11*, 2333. [CrossRef] [PubMed]
57. Zhang, H.; Saravanan, K.M.; Wei, Y.; Jiao, Y.; Yang, Y.; Pan, Y.; Wu, X.; Zhang, J.Z.H. Deep Learning-Based Bioactive Therapeutic Peptide Generation and Screening. *J. Chem. Inf. Model.* **2023**, *63*, 835–845. [CrossRef]
58. Feifei, W.; Wenrou, S.; Jinyue, S.; Qiaochu, D.; Jingjing, L.; Jin, L.; Junxiang, L.; Xuhui, L.; Xiao, L.; Congfen, H. Anti-ageing Mechanism of Topical Bioactive Ingredient Composition on Skin Based on Network Pharmacology. *Int. J. Cosmet. Sci.* **2025**, *47*, 134–154. [CrossRef]

59. Zhang, M.; Zhou, J.; Wang, X.; Wang, X.; Ge, F. DeepBP: Ensemble Deep Learning Strategy for Bioactive Peptide Prediction. *BMC Bioinform.* **2024**, *25*, 352. [CrossRef]
60. Jing, C.; Guo, J.; Li, Z.; Xu, X.; Wang, J.; Zhai, L.; Liu, J.; Sun, G.; Wang, F.; Xu, Y.; et al. Screening and Research on Skin Barrier Damage Protective Efficacy of Different Mannosylerythritol Lipids. *Molecules* **2022**, *27*, 4648. [CrossRef]
61. Kim, J.-H.; Lee, R.; Hwang, S.-H.; Choi, S.-H.; Kim, J.-H.; Cho, I.-H.; Lee, J.I.; Nah, S.-Y. Ginseng and Ginseng Byproducts for Skincare and Skin Health. *J. Ginseng Res.* **2024**, *48*, 525–534. [CrossRef] [PubMed]
62. Gupta, R.; Srivastava, D.; Sahu, M.; Tiwari, S.; Ambasta, R.K.; Kumar, P. Artificial Intelligence to Deep Learning: Machine Intelligence Approach for Drug Discovery. *Mol. Divers.* **2021**, *25*, 1315–1360. [CrossRef] [PubMed]
63. Wang, L.; Wang, N.; Zhang, W.; Cheng, X.; Yan, Z.; Shao, G.; Wang, X.; Wang, R.; Fu, C. Therapeutic Peptides: Current Applications and Future Directions. *Signal Transduct. Target. Ther.* **2022**, *7*, 48. [CrossRef]
64. Zhang, Q.-Y.; Yan, Z.-B.; Meng, Y.-M.; Hong, X.-Y.; Shao, G.; Ma, J.-J.; Cheng, X.-R.; Liu, J.; Kang, J.; Fu, C.-Y. Antimicrobial Peptides: Mechanism of Action, Activity and Clinical Potential. *Mil. Med. Res.* **2021**, *8*, 48. [CrossRef]
65. Song, G.; Sun, Y.; Liu, T.; Zhang, X.; Zeng, Z.; Wang, R.; Li, P.; Li, C.; Jiang, G. Transdermal Delivery of Cu-Doped Polydopamine Using Microneedles for Photothermal and Chemodynamic Synergistic Therapy against Skin Melanoma. *Chem. Eng. J.* **2021**, *426*, 130790. [CrossRef]

cosmetics

MDPI

Article

Investigation of the Anti-Aging Effects of Composite Nanocarriers Based on Autophagy Regulation and Oxidative Stress Inhibition

Min Liu [1], Lei Ye [1], Lingling Jiang [1,*], Xi Wang [2,3], Cui Sun [1], Jiuyan Zheng [1] and Wei Liu [2,3]

[1] Suzhou Misifu Cosmetic Co., Ltd., Suzhou 215299, China; liumin@misifu.com (M.L.); suncui@misifu.com (C.S.)
[2] National Engineering Research Center for Nanomedicine, Huazhong University of Science and Technology, Wuhan 430075, China; wliu@hust.edu.cn (W.L.)
[3] College of Life Science and Technology, Huazhong University of Science and Technology, Wuhan 430074, China
* Correspondence: jianglingling@misifu.com; Tel.: +86-18262283403

Abstract: Focusing on the anti-aging mechanism and efficient utilization of anti-aging active ingredients in the skin is an excellent strategy to mitigate aging. In this study, ribose/collagen/decarboxylated carnitine hydrochloride/palmitoyl tripeptide-1 composite nanocarriers (RCDP NCs) were synthesized using transdermal drug delivery nanotechnology. The drug delivery of composite nanocarriers and the anti-aging mechanism of RCDP NCs were studied through transdermal behavior, cell uptake, cell proliferation, antioxidant enzyme activity, lipid oxidation product expression, β—galactosidase content, autophagy vesicle number, autophagy-related protein expression, and other indicators. The results showed that the composite nanocarriers on the skin could reach a dermal depth of 460.0 μm (4 h). The uptake of RCDP NCs by keratinocytes and fibroblasts increased by 47.37% and 89.11% (4 h), respectively. RCDP NCs promoted cell proliferation, enhanced the activities of the main antioxidant enzymes, and reduced the production of the lipid oxidation product malondialdehyde (MDA). Sequestosome-1 protein (p62) decreased, whereas both the ratio of microtubule-associated protein light chain 3 II/microtubule-associated protein light chain 3 I (LC3II/LC3I) and the number of autophagosomes increased, indicating that RCDP NCs promoted autophagy. The drug delivery nanotechnology in this study achieved better transdermal application of active ingredients, which could mitigate skin aging faster and more effectively.

Keywords: composite nanocarriers; percutaneous penetration; cell uptake; anti-aging; autophagy

1. Introduction

Skin aging is a physiological phenomenon characterized by degenerative alterations in cutaneous morphology and functionality resulting from various factors. As human longevity increases and an aging society emerges, public awareness regarding aesthetics has progressively heightened. Consequently, the investigation of the fundamental mechanisms underlying skin aging and methods to mitigate this process has become a prominent focus within cosmetic research [1]. The complexity of skin aging involves multiple concurrent mechanisms; therefore, a singular efficacy approach cannot adequately address all manifestations of aging [2]. Thus, a comprehensive examination of the mechanisms involved in skin aging, combined with the development of skincare materials that exhibit multiple efficacy pathways, can facilitate the creation of more potent anti-aging cosmetic ingredients.

Free radical accumulation induces oxidative stress, which can potentially damage lipids, proteins, nucleic acids, and cellular organelles. The intracellular buildup of non-functional or deteriorated organelles and proteins adversely impacts cellular integrity, thereby promoting cellular senescence [3]. When the accumulation of damage from free radicals exceeds the body's reparative capacity, alterations in cellular differentiation occur or may be completely abolished. This results in cellular rupture, apoptosis, and subsequent skin aging [4]. Autophagy plays a crucial role in maintaining intracellular homeostasis under various stress conditions; it functions as a survival mechanism against both endogenous and exogenous stressors [5]. However, cellular senescence is associated with reduced autophagic activity. This decline allows for the continuous accumulation of damaged organelles and proteins, further accelerating the process of senescence. Therefore, moderate activation and regulation of autophagy are beneficial for delaying senescence [6].

Ribose, a pentose sugar, plays a crucial role in stimulating the synthesis of adenosine triphosphate, rejuvenating the cellular energy supply, enhancing cellular respiration, and improving cutaneous elasticity as well as the appearance of wrinkles [7–9]. Recombinant human collagen III (hereafter referred to as collagen, sequence: GERGAPGFRGPAGPNGIPGEKGPAGERGAP) represents a functional domain fragment derived from human collagen III. This recombinant form exhibits high activity and affinity, with notable transdermal performance. The curved triple helical structure of recombinant human collagen III facilitates superior adhesion to melanocytes, fibroblasts, macrophages, and other cell types, resulting in enhanced cellular attachment. Furthermore, this structural configuration promotes self-assembly into collagen fibers. The abundant integrin binding sites within collagen interact with cellular surface receptors, thereby augmenting integrin binding and subsequently regulating processes such as cellular recognition, adhesion, and migration [10,11]. Decarboxy carnosine HCL demonstrates multiple functionalities, including antioxidant effects, anti-glycation properties, pollution protection capabilities, and anti-aging benefits. This compound protects cutaneous cells, diminishes wrinkles, and exerts protective effects on enzymatic activity and collagen integrity [12]. Palmitoyl tripeptide-1 promotes extracellular matrix synthesis, including collagen and glycosaminoglycan production, thereby reducing wrinkle depth, volume, and density [13].

Nanocarriers possess the capability to encapsulate both hydrophilic and lipophilic active substances. Their nanoscale dimensions and favorable deformability confer an enhanced capacity to penetrate physiological tissue barriers [14]. When incorporated into cosmetic or transdermal drug delivery formulations, nanocarriers significantly improve the cutaneous penetration of active ingredients. The simultaneous encapsulation of multiple mechanistically distinct active components within a single nanocarrier achieves sustained release, prolonged efficacy, and synergistic effects through multi-target mechanisms, substantially enhancing overall effectiveness [15,16]. Consequently, this research focused on the development of ribose/collagen/decarboxy carnosine HCL/palmitoyl tripeptide-1 composite nanocarriers (RCDP NCs), which simultaneously incorporate ribose, collagen, decarboxy carnosine, and palmitoyl tripeptide-1. Free RCDP was employed as a control to evaluate the transdermal penetration, cellular uptake, anti-oxidative capacity, anti-aging effects, and autophagic regulation of RCDP NCs.

The present investigation is predicated on the efficient utilization of established active ingredients that demonstrate proven anti-aging efficacy. A comprehensive examination of the anti-aging mechanisms associated with these ingredient combinations has been conducted. The selection criteria for active ingredient combinations encompassed: small molecular collagen active fragments capable of effective cellular recognition, adhesion promotion, and migration enhancement; ribose for cellular autophagy stimulation; decarboxy carnosine HCL, which exhibits potent antioxidant and anti-glycation properties; and palmitoyl tripeptide-1, which enhances extracellular matrix component synthesis, including collagen and glycosaminoglycan. These bioactive constituents are encapsulated within nanocarriers that possess a bilayer vesicular structure for optimized delivery.

The efficiency optimization approach incorporates bioavailability enhancement through penetration-promoting technology, achieves multi-mechanism coordination and synergistic anti-aging effects via a scientific combination of ingredients, augments the stability of anti-aging active ingredients through nanocarrier encapsulation, prevents efficacy reduction through multidimensional optimization, and ultimately maximizes the cutaneous anti-aging effects of these active ingredients.

2. Materials and Methods

2.1. Main Materials, Reagents, and Instruments

Bama pig skin was procured from the Zhifu District Yourong Biological Studio, Yantai, Shandong province, China. Human immortalized keratinocytes (HaCaT) and human skin fibroblasts (HSF) were obtained from the Kunming Cell Bank at the Chinese Academy of Sciences, Kunming, Yunnan province, China.

Ribose was acquired from Shanghai Zhirou Chemical Co., Ltd., Shanghai, China. Decarboxy carnosine HCL and palmitoyl tripeptide-1 were purchased from Nanjing Leon Biotechnology Co., Ltd., Nanjing, Jiangsu province, China. Recombinant human collagen III (trade name: micoreCol.III) was sourced from Shanxi Jinbo Bio-Pharmaceutical Co., Ltd., Taiyuan, Shanxi province, China. Dulbecco's modified eagle medium (DMEM), trypsin, double antibody, fetal bovine serum, and phosphate-buffered saline (PBS) solution were purchased from Thermo Fisher Scientific, Waltham, MA, USA. Cell Counting Kit-8 (CCK-8), reactive oxygen species detection kit, β-galactosidase staining kit, CAT, GSH-Px, SOD, and MDA detection kits were acquired from Beyotime Biotechnology Co., Ltd., Shanghai, China. The bicinchoninic acid protein concentration assay kit, glyceraldehyde-3-phosphate dehydrogenase (GAPDH) primary antibody (rabbit source), microtubule-associated protein 1 light chain 3 (LC3) primary antibody (rabbit source), Sequestosome-1 (also known as P62) primary antibody (rabbit source), and horseradish peroxidase-conjugated affinipure goat anti-rabbit IgG (H&L) were purchased from Servicebio Biotechnology Co., Ltd., Wuhan, Hubei province, China. All other reagents were supplied by Sinopharm Group, Shanghai, China.

The SHJ-6A digital magnetic stirring water bath was acquired from Changzhou Yineng Experimental Instrument Factory, Changzhou, China. The BB150 carbon dioxide cell incubator was purchased from Thermo Fisher Scientific, Waltham, MA, USA. The SW-CJ-2FD ultra-clean worktable was obtained from Suzhou Purification Equipment Co., Ltd., Suzhou, Jiangsu province, China. The Zetasizer/Nano-ZS90 Particle Size Analyzer was procured from Malvern Panalytical, Marvin, UK. The multi-label microplate reader was supplied by Perkin Elmer, Waltham, MA, USA. The FV3000 laser confocal microscope and flow cytometer were purchased from Olympus, Tokyo, Japan, and Beckman, Brea, CA, USA, respectively. The MF52-N inverted microscope was obtained from Guangzhou Mingmei Optoelectronic Technology Co., Ltd., Guangzhou, Guangdong province, China. The CT15RE desktop high-speed refrigerated centrifuge and HT7700 transmission electron microscope were supplied by Hitachi, Tokyo, Japan. The WB electrophoresis apparatus and chemiluminescence imaging system were acquired from Servicebio Biotechnology Co., Ltd., Wuhan, Hubei province, China.

2.2. Experimental Methods

2.2.1. Preparation Method of RCDP NCs

For the preparation of 1000.00 g of RCDP NCs, 20.00 g of ribose, 0.30 g of palmitoyl tripeptide-1, 100.00 g of glycerol, 150.00 g of pentylene glycol, 100.00 g of ethoxydiglycol, 100.00 g of polysorbate 80, 100.00 g of PEG-40 hydrogenated castor oil, and 30.00 g of lecithin were accurately weighed. These components were subsequently stirred at 55 °C until a homogeneous mixed solution A was formed. In a separate container, 10.00 g of decarboxy carnosine HCl, 1.00 g of collagen, 5.00 g of sodium bisulfite, and 383.70 g of water were accurately weighed and stirred at 55 °C to create a homogeneous mixed solution B. Mixed solution A was then gradually incorporated into mixed solution B and thoroughly stirred at 55 °C until a uniform mixed solution C was achieved. This final mixture was subjected to homogenization twice in a high-pressure homogenizer at 850 bar, resulting in the formation of RCDP NCs containing numerous vesicles with bilayer structures. The resultant RCDP NCs were diluted to 0.1 g/mL with pure water without filtration. The particle size, polydispersity index (PDI), and Zeta potential of these nanocarriers were subsequently measured using a Malvern particle size analyzer. Measurements were conducted at a test temperature of 25 °C with an equilibration time of 120 s.

2.2.2. Percutaneous Penetration

Rhodamine B (RhoB) was utilized as a fluorescent marker in place of the functional component to prepare rhodamine B composite nanocarriers (RhoB NCs), while an equivalent concentration (0.20 g/L) of rhodamine B solution (Free RhoB) was formulated. For the preparation of 1000.00 g of RhoB NCs, 100.00 g of glycerol, 150.00 g of pentylene glycol, 100.00 g of ethoxydiglycol, 100.00 g of polysorbate 80, 100.00 g of PEG-40 hydrogenated castor oil, and 30.00 g of lecithin were accurately weighed. These components were subsequently stirred at 55 °C until a homogeneous mixed solution A was formed. In a separate container, 0.20 g of rhodamine B, 5.00 g of sodium bisulfite, and 414.80 g of water were accurately weighed and stirred at 55 °C to create a homogeneous mixed solution B. Mixed solution A was then gradually incorporated into mixed solution B and thoroughly stirred at 55 °C until a uniform mixed solution C was achieved. The final mixture was homogenized twice in an 850-bar high-pressure homogenizer to obtain RhoB NCs.

Panama pig-isolated skin was secured between the receiving and supply chambers of the diffusion cell. Subsequently, 0.5 g of RhoB NCs essence (5% RhoB NCs) and Free RhoB essence (containing an equivalent RhoB concentration as in RhoB NCs) were separately applied to the supply chamber, with PBS serving as the receiving solution. After 2 and 4 h of penetration and diffusion, residual samples were gently removed from the skin surface, and the target area of the skin was excised and thoroughly rinsed, and excess moisture was eliminated. The specimens were cryosectioned, and the resultant slices were examined via laser confocal microscopy, with representative areas selected for imaging. Fluorescence signals were quantitatively analyzed using ImageJ software (Version 1.54) to determine penetration depth and fluorescence intensity.

2.2.3. Quantification of Cellular Uptake

Cellular uptake behavior was evaluated using laser confocal microscopy. HaCaT cells were cultured in 35 mm confocal dishes at a density of 2.0×10^5 cells per dish, while HSF cells were seeded at 1.6×10^5 cells per dish. After 24 h of cultivation, DMEM medium containing Free RhoB or RhoB NCs was introduced for subsequent 2 h and 4 h incubation periods. Following incubation, the medium was aspirated, and cells were washed three times with PBS solution, fixed with paraformaldehyde, and stained with 4′,6-diamidino-2-phenylindole solution for 15 min. Observations were conducted under a 60-fold objective using laser confocal microscopy. Flow cytometry was employed for the quantitative assessment of cellular uptake. HaCaT cells were seeded in 6-well plates at 2.0×10^5 cells per well, whereas HSF cells were plated at 1.6×10^5 cells per well. The treatment protocol remained consistent with previous methods. After 2 h and 4 h cultivation periods, cell pellets were harvested, resuspended in 0.5 mL of cold PBS solution, and mean fluorescence intensity within cells was determined via flow cytometry.

2.2.4. Cell Proliferation Activity

HaCaT and HSF cells were seeded in 96-well plates at densities of 8×10^3 and 4×10^3 cells per well, respectively, followed by 24 h of cultivation. Each well received 100 μL of DMEM complete medium containing either RCDP NCs (at concentrations of 200, 400, and 800 mg/L) or Free RCDP (with functional component concentrations equivalent to those in RCDP NCs). Control wells were supplemented with 100 μL of DMEM complete medium only. Three replicates were maintained for each experimental group. After 48 h of cultivation, cell proliferation rates were assessed using the CCK-8 method.

2.2.5. β-Galactosidase Staining Observation

HSF cells at passage 15 were seeded in 6-well plates at 1.6×10^5 cells per well. Following 24 h of cultivation, the control group received only DMEM complete medium. Treatment groups were administered DMEM complete medium containing either RCDP NCs (400 mg/L) or Free RCDP (with functional component concentrations equivalent to those in RCDP NCs). After 48 h of cultivation, cells from each group were washed three times with PBS according to kit instructions. Subsequently, 1.0 mL of β-galactosidase staining fixative was added to each well, followed by room temperature fixation for 15 min and three PBS washes. Thereafter, 1.0 mL of β-galactosidase staining working solution was applied, and samples were incubated overnight at 37 °C. Staining outcomes were visualized microscopically.

2.2.6. Inhibition of Oxidative Stress

HSF cells were seeded in 24-well plates at 8×10^4 cells per well in 500 μL of medium. After 24 h of cultivation, the supernatant was removed. The model group received DMEM medium containing 0.6 mmol/L H_2O_2. Treatment groups were administered DMEM complete medium containing both 0.6 mmol/L H_2O_2 and RCDP NCs (at concentrations of 200, 400, and 800 mg/L) or various concentrations of Free RCDP (with functional component concentrations equivalent to those in RCDP NCs). The control group received H_2O_2-free DMEM medium. Each experimental group comprised three replicate wells and was maintained for 24 h. Cellular CAT activity, SOD activity, GSH-Px activity, and MDA content were determined according to the respective kit protocols.

2.2.7. Autophagy-Related Protein Expression in Cells

HSF cells were seeded into 10 mm cell culture dishes at a density of 8×10^5 cells per dish, with 10 mL of medium in each dish. Following 24 h of incubation, the supernatant was aspirated. Subsequently, DMEM complete medium containing either 400 mg/L or 800 mg/L of RCDP NCs or Free RCDP (at concentrations equivalent to the functional components in RCDP NCs) was introduced, and the cells were cultured for an additional 48 h. Cellular protein was extracted from each experimental group using protein lysate, and protein concentration was quantified using a protein determination kit. For electrophoretic analysis, 60 μg of protein per lane was loaded. Following sodium dodecyl sulfate-polyacrylamide gel electrophoresis, proteins were transferred to polyvinylidene fluoride membranes using a semi-dry transfer methodology. The membranes were incubated with primary antibodies against P62, LC3I, LC3II, and GAPDH at appropriate dilutions overnight at 4 °C. After three 10 min washes with tris-buffered saline containing tween 20, horseradish peroxidase-conjugated goat anti-rabbit secondary antibody was applied. A 1:1 chemiluminescence solution was prepared and applied to the membrane positioned on the exposure plate. Images were captured and analyzed using an automated fluorescence and visible light gel imaging analysis system.

2.2.8. Autophagosomes

HSF cells were seeded into 10 mm cell culture dishes at a density of 8×10^5 cells per dish, with 10.0 mL of medium per dish. After 24 h of cultivation, the supernatant was removed. DMEM complete medium containing 400 mg/L of either RCDP NCs or Free RCDP was then introduced, and the cells were maintained in culture for 48 h. The cells were subsequently harvested using trypsin digestion and centrifuged until aggregation occurred. The cellular pellet obtained through centrifugation was dissociated and fixed with 2.5% glutaraldehyde in phosphoric acid buffer, then stored at 4 °C overnight. Following multiple rinses with PBS, the specimens were post-fixed with 1% osmic acid for 1–2 h. The cellular samples underwent gradient dehydration through a series of increasing ethanol and acetone concentrations. The samples were then embedded, sectioned, and examined for autophagic structures via transmission electron microscopy.

2.2.9. Data Processing

Statistical analysis was performed using the Statistical Package for the Social Sciences version 20.0 software. All quantitative data were presented as mean ± standard deviation (SD). Comparisons between experimental groups were conducted using one-way analysis of variance. Statistical significance was established at $p < 0.05$.

3. Results

3.1. Particle Size, PDI, and Zeta Potential

The RCDP NCs synthesized in this study were observed as a light yellow transparent liquid, characterized by an average particle size of 33.1 nm, a PDI of 0.292, and a Zeta potential of (−34.2 ± 0.7) mV. As illustrated in Figure 1, the size distribution by number displays a narrow range, indicating uniform vesicle dimensions, with nearly all particles measuring below 50 nm in diameter. Figure 2 demonstrates a similarly narrow range for Zeta potential distribution, with vesicle surfaces predominantly exhibiting negative charges, thereby confirming the stability of the RCDP NCs.

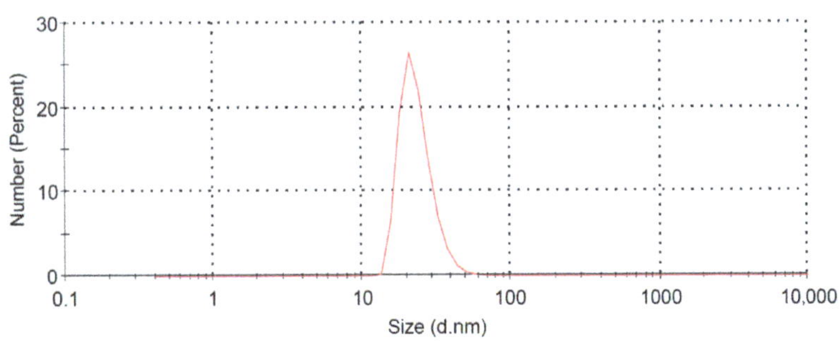

Figure 1. Size distribution by number of RCDP NCs.

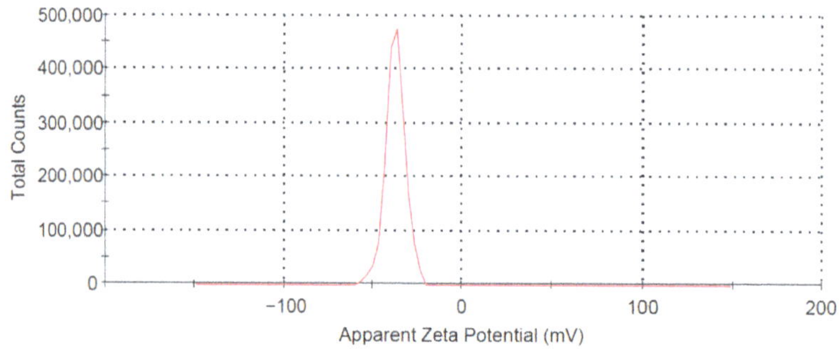

Figure 2. Zeta potential distribution of RCDP NCs.

3.2. In Vitro Skin Penetration

The stratum corneum constitutes a formidable epidermal barrier against environmental elements, resulting in limited transdermal penetration of therapeutic agents and impeding their delivery to the primary anti-aging target site—the dermis. To assess the dermal penetration capability of RCDP NCs, RhoB was encapsulated within the nanocarriers, and both penetration depth and fluorescence intensity were assessed at various time intervals utilizing the vertical Franz diffusion cell methodology. The experimental results are presented in Figure 3.

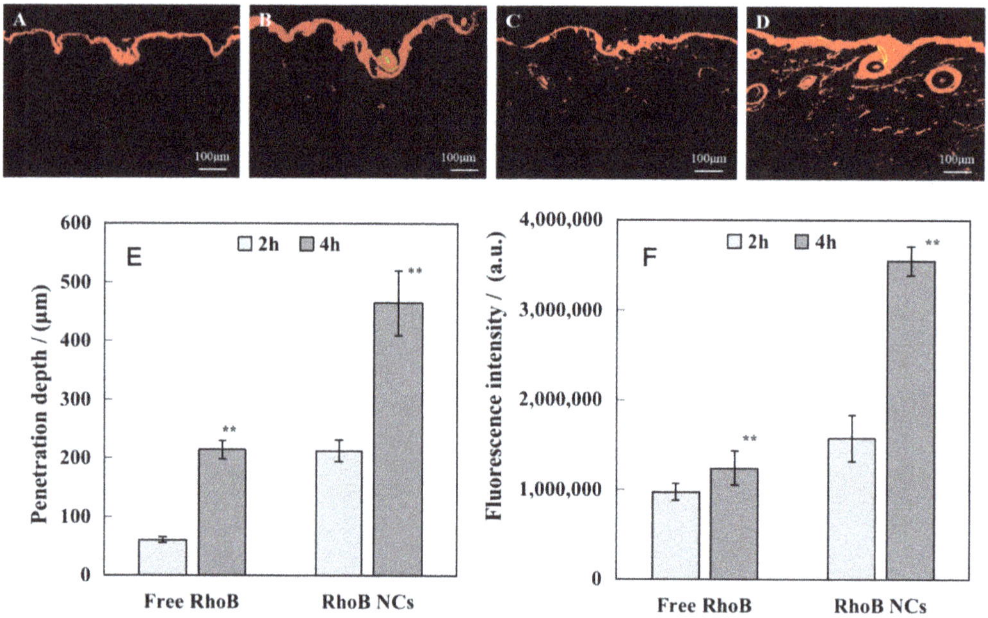

Figure 3. Percutaneous penetration of Free RhoB and RhoB NCs, (**A**). Free RhoB infiltration for 2 h (10×); (**B**). Free RhoB permeates for 4 h (10×); (**C**). RhoB NCs permeate for 2 h (10×); (**D**). RhoB NCs permeate for 4 h (10×); (**E**). Penetration depth of Free RhoB and RhoB NCs at 2 h and 4 h of penetration; (**F**). Fluorescence intensity of Free RhoB and RhoB NCs at 2 h and 4 h of penetration. (Note: compared with Free RhoB at the same time, ** $p < 0.01$).

As observed in Figure 3, Free RhoB within 2 h was predominantly localized in the stratum corneum and was unable to penetrate this barrier. In contrast, RhoB NCs successfully penetrated the stratum corneum barrier within the same timeframe. With extended temporal exposure, the fluorescence penetration depth of RhoB NCs in the skin significantly increased within 4 h, ultimately reaching a depth of 460.0 μm. The fluorescence intensity and penetration depth of RhoB NCs in the skin were significantly greater than those observed for Free RhoB. These findings illustrate that nanocarriers can facilitate the rapid and efficient delivery of encapsulated components to deep dermal tissues.

3.3. Cell Proliferation

The rapid remodeling of the skin epidermis and dermis significantly impacts the metabolic state of skin renewal. Consequently, we systematically investigated the proliferative effects of Free RCDP and RCDP NCs on HaCaT and HSF cells. The resulting data are presented in Figure 4.

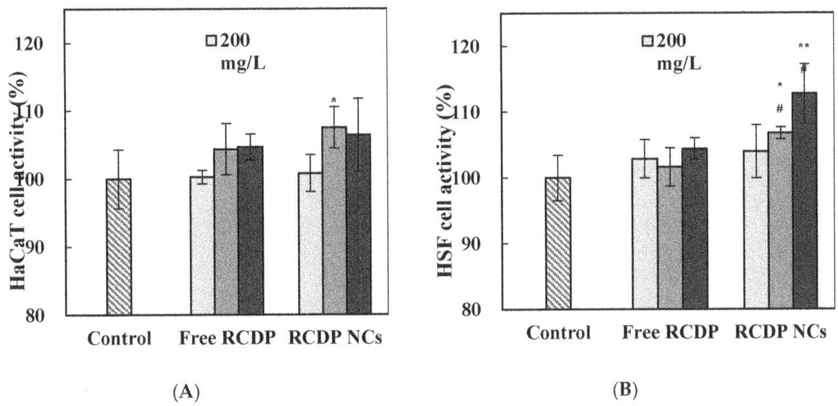

Figure 4. Effects of Free RCDP and RCDP NCs on the proliferation of HaCaT (**A**); Effects of Free RCDP and RCDP NCs on HSF proliferation (**B**). (Note: compared with the control, ** $p < 0.01$, * $p < 0.05$; compared with the same concentration of Free RCDP, # $p < 0.05$.).

As illustrated in Figure 4A, RCDP NCs at a concentration of 400 mg/L significantly enhanced the proliferation of HaCaT cells compared to the control group ($p < 0.05$). Figure 4B indicates that RCDP NCs at concentrations of both 400 mg/L and 800 mg/L markedly stimulated HSF cell proliferation relative to the control group ($p < 0.01$ or $p < 0.05$). Furthermore, when compared with Free RCDP, HSF cell proliferation was significantly increased at RCDP NC concentrations of 400 mg/L and 800 mg/L ($p < 0.05$).

3.4. The Uptake of RCDP NCs by Cells

To assess the efficacy of RCDP NCs in delivering functional components into HaCaT and HSF cells, it was deemed essential to investigate the cellular entry behavior of these nanocarrier-encapsulated components [17]. The cellular uptake patterns of Free RhoB and RhoB NCs by HaCaT and HSF cells were analyzed using laser confocal microscopy and flow cytometry. RhoB NCs were utilized for both qualitative and quantitative assessments. The experimental results are presented in Figure 5.

With extended incubation periods, the fluorescence intensity within both HaCaT and HSF cells was significantly enhanced. As illustrated in Figure 5A,B, minimal fluorescence intensity was observed in the Free RhoB solution group after 2 h of incubation. In contrast, the RhoB NCs group exhibited significantly stronger fluorescence intensity compared to the Free RhoB group at the same time point. After 4 h of incubation, the fluorescence intensity in the RhoB NCs group was further increased. According to Figure 5C, following incubation periods of 2 h and 4 h, the mean fluorescence intensity of HaCaT cells treated with RhoB

NCs measured 4213 and 6894, respectively. These values represented increases of 53.73% and 47.37%, respectively, compared with the Free RhoB group. Figure 5D demonstrates that after incubation periods of 2 h and 4 h, the mean fluorescence intensity of HSF cells treated with RhoB NCs was quantified at 2451 and 4635, respectively. These measurements indicated increases of 27.60% and 89.11%, respectively, compared with the Free RhoB group. These data confirm that RCDP NCs can effectively deliver encapsulated bioactive substances into targeted skin cells.

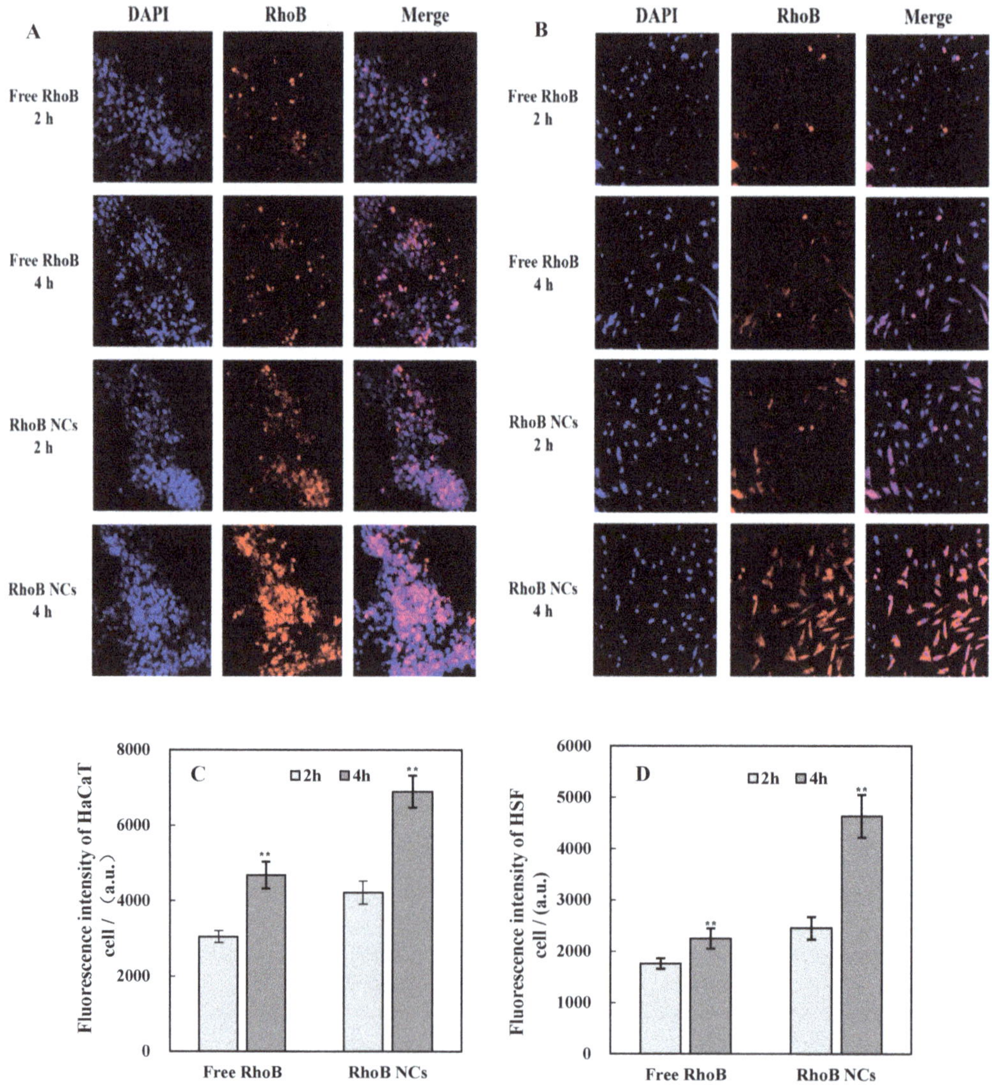

Figure 5. Cell uptake of Free RhoB and RhoB NCs. (**A**) The uptake of Free RhoB and RhoB NCs by HaCaT was observed using laser confocal microscopy (60×). (**B**) The uptake of Free RhoB and RhoB NCs by HSF was observed using laser confocal microscopy (60×). (**C**) Quantitative analysis of HaCaT cell uptake using flow cytometry; (**D**) Quantitative analysis of HSF cell uptake using flow cytometry. (Note: compared with Free RhoB at the same time, ** $p < 0.01$).

3.5. Effect of RCDP NCs on Cellular Oxidative Stress

Oxidative stress is widely recognized as a significant contributor to skin aging. This phenomenon can lead to the degradation and alteration of various intracellular molecules, including lipids, nucleic acids, and proteins, ultimately resulting in apoptosis and the progression of skin aging [18]. The effects of Free RCDP and RCDP NCs on oxidative parameters within the oxidized cell model are illustrated in Figure 6.

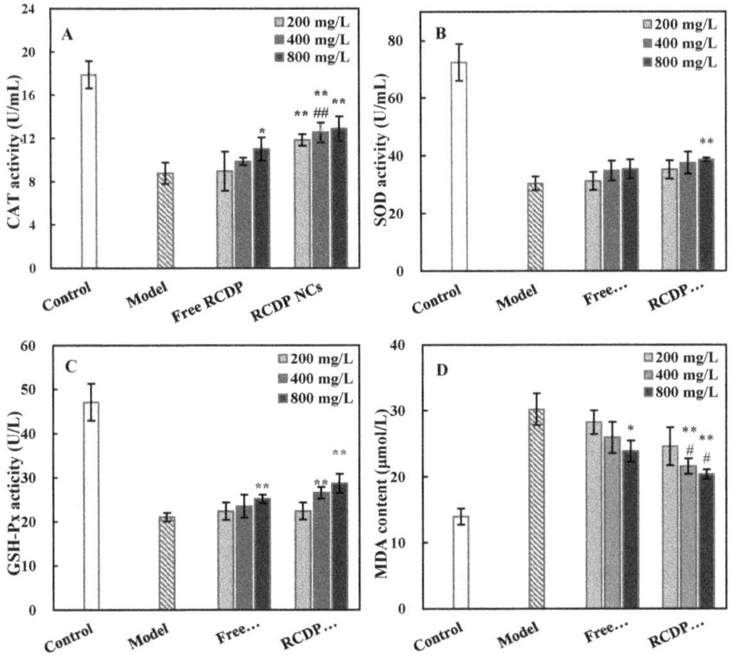

Figure 6. Effects of Free RCDP and RCDP NCs on oxidation indices of HSF cells, (**A**) CAT activity; (**B**) SOD activity; (**C**) GSH-Px activity; (**D**) MDA content. (Note: compared with the control, ** $p < 0.01$, * $p < 0.05$; compared with the same concentration of Free RCDP, ## $p < 0.01$, # $p < 0.05$).

As illustrated in Figure 6, HSF cells in the model group exhibited a significant increase in the activities of CAT, SOD, and GSH-Px, accompanied by a decrease in MDA content following exposure to 0.6 mmol/L H_2O_2. The CAT activity analysis revealed that 800 mg/L Free RCDP and all three concentrations of RCDP NCs (200 mg/L, 400 mg/L, 800 mg/L) significantly enhanced CAT activity compared to the model group ($p < 0.05$ or $p < 0.01$). Furthermore, at a concentration of 400 mg/L, RCDP NCs demonstrated significantly higher CAT activity in oxidatively damaged cells compared to Free RCDP ($p < 0.01$). In terms of SOD activity, RCDP NCs at 800 mg/L significantly increased SOD activity relative to the model group ($p < 0.01$). The assessment of GSH-Px activity demonstrated that Free RCDP at 800 mg/L and RCDP NCs at concentrations of 400 mg/L and 800 mg/L significantly increased GSH-Px activity compared to the model group ($p < 0.01$). The measurement of MDA content indicated that Free RCDP at 800 mg/L and RCDP NCs at concentrations of 400 mg/L and 800 mg/L significantly reduced MDA levels compared to the model group ($p < 0.05$ or $p < 0.01$). Additionally, RCDP NCs at 400 mg/L and 800 mg/L demonstrated significantly lower MDA content compared to Free RCDP ($p < 0.05$). These findings suggest that RCDP NCs possess superior antioxidant properties compared to Free RCDP at equivalent concentrations.

3.6. β-Galactosidase

β-galactosidase serves as a classical biomarker for cellular senescence. This enzyme interacts with the 5-bromo-4-chloro-3-indolyl-β-D-galactopyranoside substrate, generating dark blue products that are commonly utilized to identify senescent cells [19]. The results of β-galactosidase staining in HSF cells treated with Free RCDP and RCDP NCs are presented in Figure 7.

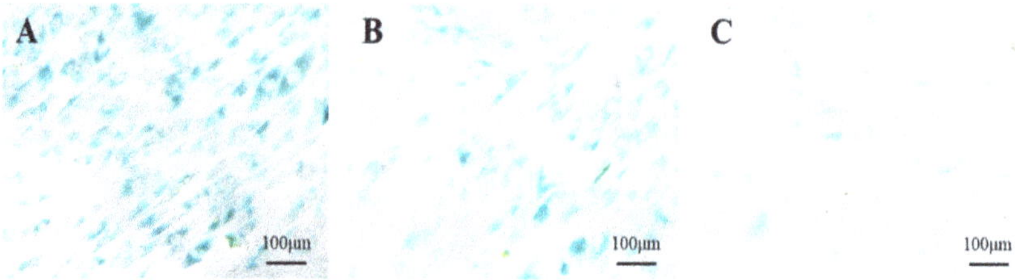

Figure 7. β-galactosidase staining of HSF cells, (**A**) The 15th generation HSF cells in the control group (10×); (**B**) Free RCDP group of the 15th generation HSF cells (10×); (**C**) RCDP NC group of the 15th generation HSF cells (10×).

As illustrated in Figure 7, the cells of the 15th generation displayed positive β-galactosidase staining with a prominent dark blue coloration, indicating substantial cellular senescence. Following treatment with Free RCDP and RCDP NCs, the blue coloration of the cells was noticeably attenuated. The RCDP NC treatment resulted in markedly lighter coloration compared to Free RCDP treatment, suggesting a more pronounced reduction in the senescent state of HSF cells after RCDP NCs administration.

3.7. Expression of Autophagy-Related Proteins in Cells

LC3, a homologous protein associated with autophagy-related genes, typically exists in two forms: LC3I and LC3II. During autophagy activation, LC3I is converted into LC3II [20]. P62 serves as a specific substrate that can be degraded through its binding to ubiquitinated proteins, subsequently forming a complex with the LC3II protein and undergoing degradation via the autophagic lysosomal pathway. Elevated levels of P62 expression are correlated with reduced autophagy activity [21]. The expression profiles of the autophagy-related proteins LC3I and LC3II, along with the autophagy degradation-related protein P62, as determined by western blot analysis, are presented in Figure 8.

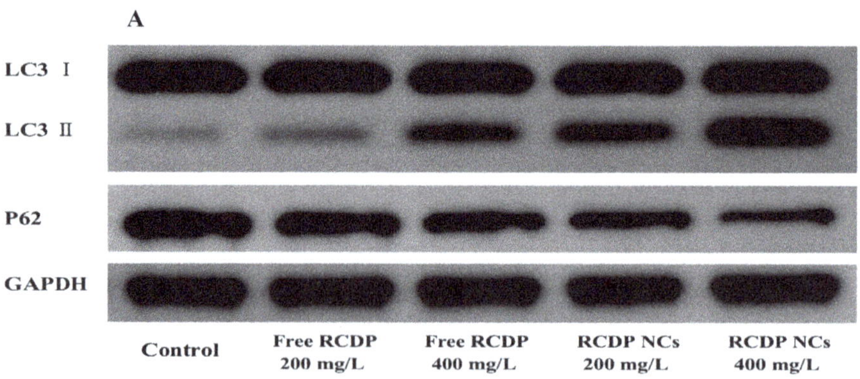

Figure 8. *Cont.*

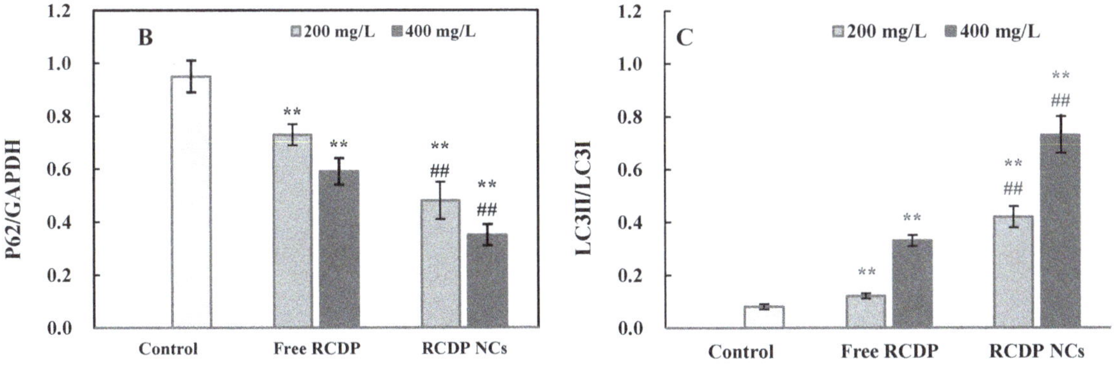

Figure 8. Expression of autophagy-related proteins in HSF cells treated with Free RCDP and RCDP NCs. (**A**) Expression of LC3I, LC3II, and P62 proteins; (**B**) P62/GAPDH protein expression ratio; (**C**) LC3II/LC3I protein expression ratio. (Note: compared with the control, ** $p < 0.01$; compared with the same concentration of Free RCDP, ## $p < 0.01$.).

As illustrated in Figure 8, the autophagy-related protein P62 content was significantly reduced in both the Free RCDP and RCDP NC groups compared to the control group ($p < 0.01$). Concurrently, there was a substantial increase in the ratio of autophagy-related proteins LC3II/LC3I ($p < 0.01$). When compared with equivalent concentrations of Free RCDP, the RCDP NCs group exhibited a more pronounced decrease in P62 protein expression ($p < 0.01$) and a greater increase in the LC3II/LC3I ratio ($p < 0.01$). These findings indicate that RCDP NCs effectively enhance the autophagic capacity of HSF cells with superior efficacy relative to their free components.

3.8. Observation of Autophagosomes Using Transmission Electron Microscopy

Cellular autophagy involves the encapsulation of damaged or misfolded proteins and organelles within membranous structures, forming double-membrane vesicles that subsequently fuse with lysosomes for content degradation [22]. Autophagosomes within cellular structures were visualized using transmission electron microscopy, and the results are presented in Figure 9.

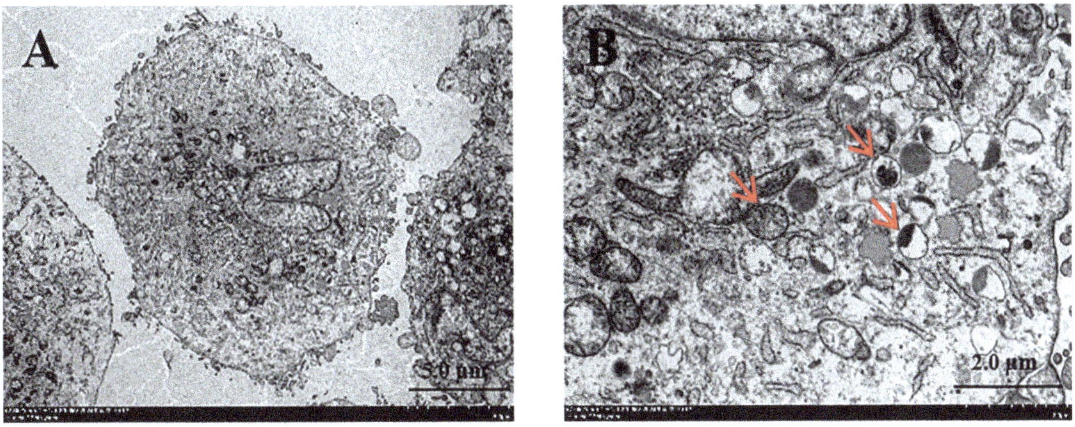

Figure 9. *Cont.*

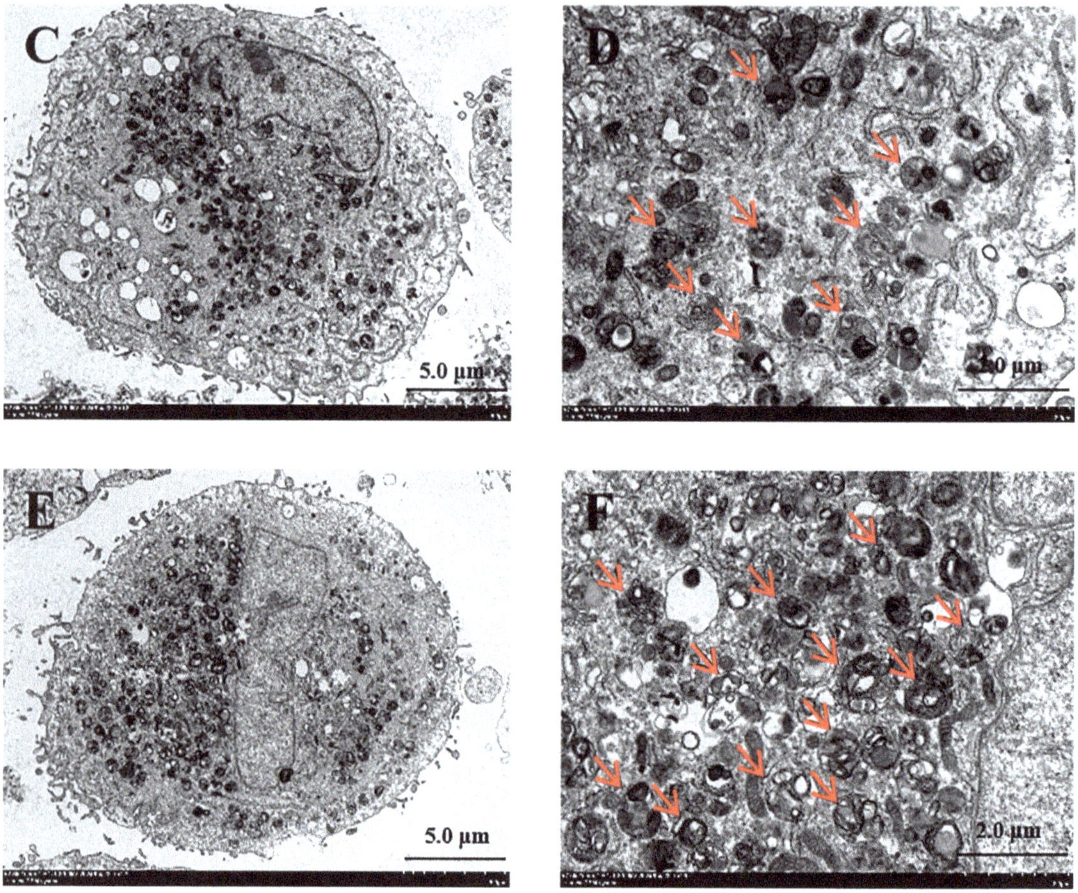

Figure 9. Autophagic lysosomes of HSF cells observed via transmission electron microscopy: (**A**) represents the control group (2500×), (**B**) represents the control group (7000×), (**C**) represents the Free RCDP group (2500×), (**D**) represents the Free RCDP group (7000×), (**E**) represents the RCDP NCs group (2500×), and (**F**) represents the RCDP NCs group (7000×). The black arrows indicate autophagic lysosomes.

Transmission electron microscopy observations, as illustrated in Figure 9, revealed a minimal presence of autophagosomes in the cytoplasm of cells from the control group (A,B). In contrast, the cytoplasm of cells treated with Free RCDP exhibited an increased number of autophagosomes compared to the control group (C,D). Notably, the highest density of autophagosomes was observed in the cytoplasm of cells treated with RCDP NCs (E,F). These microscopic findings further substantiate that RCDP NCs enhance the autophagic capacity of HSF cells more effectively than their free components.

4. Discussion

Human aging is regulated by genetically predetermined programs and is simultaneously influenced by cumulative internal and external "wear and tear". Both mechanisms manifest distinctly at the cellular and molecular levels. The skin, which functions as the outermost organ of the human body, undergoes aging processes that follow comparable patterns. Moreover, skin aging characteristics—such as wrinkles, sagging, pigmentation, and textural roughness—are more visible and measurable [23,24]. As quality of life standards elevate and aesthetic consciousness intensifies, the societal desire to delay cutaneous aging and preserve youthfulness has strengthened considerably. Through advanced research methodologies and technological innovations, scientists have conducted comprehensive investigations into skin aging mechanisms and developed targeted anti-aging interventions [25–27]. Although numerous active ingredients with demonstrated anti-aging properties are incorporated into cosmetic formulations and exhibit efficacy through single or multiple pathways, the outcomes frequently fail to meet consumer expectations. The fundamental challenge in addressing skin aging resides in elucidating the anti-aging mechanisms and optimizing the utilization efficiency of active ingredients [28–31]. In this research context, efficient utilization encompasses maximizing the effectiveness of anti-aging compounds on the skin through enhanced bioavailability via penetration technologies, achieving synergistic anti-aging effects through scientifically validated ingredient combinations, and augmenting the stability of active compounds to prevent efficacy degradation.

The present study was designed based on the optimization of the utilization efficiency of established anti-aging active ingredients. The formulation incorporates small molecular collagen active fragments selected for their superior cellular recognition and adhesion properties, promoting cellular migration [10,11]; ribose for enhancing cellular autophagy [7–9]; decarboxylcarnosine HCl for its exceptional antioxidant and anti-glycation capabilities [12]; and palmitoyl tripeptide-1 for facilitating the synthesis of extracellular matrix components, including collagen and glycosaminoglycan [13]. These active compounds were encapsulated within nanocarriers possessing bilayer vesicular structures for targeted delivery. Transdermal penetration assessments and cellular uptake investigations of RCDP NCs demonstrated that these carriers significantly enhanced penetration efficiency, depth, and cellular utilization of the anti-aging compounds.

The experimental results revealed that RCDP NCs increased the activities of SOD, CAT, and GSH-Px enzymes, thereby reducing free radical production in cells under oxidative stress. The observed decrease in MDA levels indicated that RCDP NCs inhibited oxidative stress reactions and minimized cellular damage. Furthermore, RCDP NCs decreased the autophagy degradation-related protein p62 content, elevated the ratio of autophagy-related protein LC3II/LC3I, and increased autophagosome formation. During autophagy, p62 serves as a key ligating protein. p62 is selectively wrapped into the autophagosome and subsequently degraded by protein hydrolases in the autophagic lysosome, and its expression level negatively correlates with autophagic activity [32]. LC3I animalizes or binds to autophagosome membranes and is converted into LC3II, so that an increase in the LC3II/LC3I ratio indicates an increase in autophagy [33,34]. Through autophagy activation, RCDP NCs facilitated the clearance of damaged proteins and senescent organelles, consequently delaying cellular aging processes. Additionally, RCDP NCs stimulated proliferation in keratinocytes and fibroblasts, thereby enhancing metabolic functions in aging skin. After treating 15 generations of HSF aging cells with RCDP NCs, a significant reduction in β-galactosidase, a cell aging biomarker, was observed, indicating substantial improvement in cellular senescence status.

In this study, RCDP NCs primarily delayed cellular senescence through potent antioxidant activity and autophagy promotion. The antioxidant properties enhanced inherent cellular defense mechanisms, mitigating damage to proteins, lipids, and organelles caused by reactive oxygen species and other free radicals. Autophagy functioned as a critical survival mechanism for aging cells responding to internal and external stressors [35]. By eliminating damaged proteins and organelles, reducing oxidative stress-induced damage from various factors, and removing damaged proteins and senescent mitochondria among other organelles, RCDP NCs comprehensively delayed cellular senescence, promoted cellular proliferation, and retarded skin aging progression. Compared with previous studies on anti-aging active ingredients [36,37], this research not only explored the anti-aging mechanisms but also emphasized the efficient utilization of these compounds. This integrated approach presents greater potential for achieving desired anti-aging effects in subsequent skin aging treatments, thereby yielding enhanced economic benefits and market value.

5. Conclusions

In this study, RCDP NCs were successfully synthesized using transdermal drug delivery nanotechnology. The evaluation of transdermal behavior revealed that RCDP NCs exhibited superior penetration efficiency, reaching a dermal depth of 460.0 μm within 4 h, significantly outperforming free components. Enhanced cellular uptake was observed in both HaCaT and HSF, with fluorescence intensity increasing by 47.37% and 89.11%, respectively, after 4 h of incubation, confirming the nanocarriers' ability to effectively deliver bioactive compounds into target skin cells.

RCDP NCs demonstrated potent anti-aging effects by promoting cell proliferation in HaCaT and HSF cells, elevating antioxidant enzyme activities (CAT, SOD, GSH-Px), and reducing lipid oxidation product (MDA) levels. Notably, β-galactosidase staining indicated a marked reduction in senescent HSF cells following RCDP NC treatment. Furthermore, autophagy regulation studies revealed that RCDP NCs decreased the expression of the autophagy degradation-related protein P62, increased the LC3II/LC3I ratio, and enhanced autophagosome formation, indicating robust activation of autophagy pathways. These mechanisms collectively contributed to delaying cellular senescence and mitigating oxidative stress-induced damage.

Compared to Free RCDP, RCDP NCs exhibited enhanced stability, bioavailability, and synergistic anti-aging efficacy through multi-target mechanisms. The integration of nanotechnology with scientifically formulated active ingredients optimized transdermal delivery, cellular uptake, and functional performance. These findings underscore the potential of RCDP NCs as promising candidates for anti-aging cosmetic formulations, offering significant advancements in addressing skin aging through improved ingredient utilization and mechanistic coordination. Further research and clinical validation are warranted to translate these innovations into practical skincare applications.

Author Contributions: Conceptualization, L.Y. and C.S.; data curation, M.L., L.Y., X.W. and W.L.; formal analysis, M.L.; investigation, M.L. and L.J.; methodology, M.L.; project administration, M.L. and J.Z.; resources, L.Y., J.Z. and W.L.; supervision, L.Y. and C.S.; validation, L.J.; writing—original draft, L.J. and X.W.; writing—review and editing, M.L. and L.J. All authors have read and agreed to the published version of the manuscript.

Funding: This research received no external funding.

Institutional Review Board Statement: Not applicable.

Informed Consent Statement: Not applicable.

Data Availability Statement: The datasets supporting this article's conclusions are available from the corresponding author upon reasonable request.

Acknowledgments: The authors thank Jiuyan Zheng for testing funding and Wei Liu for polishing the English in the manuscript.

Conflicts of Interest: Min Liu, Lei Ye, Lingling Jiang, Cui Sun and Jiuyan Zheng were all employed by Suzhou Misifu Cosmetic Co., Ltd. The remaining authors declare that the research was conducted in the absence of any commercial or financial relationships that could be construed as a potential conflict of interest.

References

1. Warsito, M.F.; Kusumawati, I. The impact of herbal products in the prevention, regeneration and delay of skin aging. *Adv. Exp. Med. Biol.* **2019**, *1178*, 155–174. [CrossRef] [PubMed]
2. Zhao, X.J.; Hong, Y.H.; Liu, W. Nanocarrier technology-from transdermal drug delivery to functional cosmetics. *Deterg. Cosmet.* **2016**, *44*, 12–16. [CrossRef]
3. Martinez-Lopez, N.; Athonvarangku, D.; Singh, R. Autophagy and aging. *Adv. Exp. Med. Biol.* **2015**, *847*, 73–87. [CrossRef] [PubMed]
4. Palmer, D.M.; Kitchin, J.S. Oxidative damage, skin aging, antioxidants and a novel antioxidant rating system. *J. Drugs Dermatol.* **2010**, *9*, 11–15. [PubMed]
5. Gu, Y.; Han, J.; Jiang, C.; Zhang, Y. Biomarkers, oxidative stress and autophagy in skin aging. *Ageing Res. Rev.* **2020**, *59*, 101036. [CrossRef]
6. Mariño, G.; Niso-Santano, M.; Baehrecke, E.H.; Kroemer, G. Self-consumption: The interplay of autophagy and apoptosis. *Nat. Rev. Mol. Cell Biol.* **2014**, *15*, 81–94. [CrossRef] [PubMed]
7. Roupe, G. Skin of the aging human being. *Lakartidningen* **2001**, *98*, 1091–1095. [PubMed]
8. Muggleton-Harris, A.L.; Defuria, R. Age-dependent metabolic changes in cultured human fibroblasts. *In Vitro Cell Dev. Biol.* **1985**, *21*, 271–276. [CrossRef]
9. Shecterle, L.M.; St Cyr, J.A. Dermal benefits of topical D-ribose. *Clin. Cosmet. Investig. Dermatol.* **2009**, *2*, 151–152. [CrossRef]
10. Wang, J.; Hu, H.; Wang, J.; Qiu, H.; Gao, Y.; Xu, Y.; Liu, Z.; Tang, Y.; Song, L.; Ramshaw, J.; et al. Characterization of recombinant humanized collagen type III and its influence on cell behavior and phenotype. *Collagen Leather* **2023**, *5*, 84–96. [CrossRef]
11. Hua, C.; Zhu, Y.; Xu, W.; Ye, S.; Zhang, R.; Lu, L.; Jiang, S. Characterization by high-resolution crystal structure analysis of a triple-helix region of human collagen type III with potent cell adhesion activity. *Biochem. Biophys. Res. Commun.* **2019**, *508*, 1018–1023. [CrossRef] [PubMed]
12. Baek, J.H.; Kong, U.S.; Lee, G.M. Cosmetic Composition for Preventing Skin Wrinkles, Containing Carcinine(Decarboxy Carnosine 2HCl) Having Collagen Synthetic and Antioxidant Effects. GB Patent Application No. KR20030010259, 11 September 2024.
13. Yang, F.; Zhang, X.; Wang, H.; Guo, M.; Zhang, J.; Feng, X.; Yu, J.; Yang, J.; Zhu, J.; Wang, Y. Comprehensive evaluation of the efficacy and safety of a new multi-component anti-aging topical eye cream. *Skin Res. Technol.* **2024**, *30*, e13790. [CrossRef]
14. Zhang, S.; Zhou, H.; Chen, X.; Zhu, S.; Chen, D.; Luo, D.; Chen, S.; Liu, W. Microneedle Delivery Platform Integrated with Codelivery Nanoliposomes for Effective and Safe Androgenetic Alopecia Treatment. *ACS Appl. Mater. Interfaces* **2024**, *16*, 15701–15717. [CrossRef]
15. Zhou, H.; Luo, D.; Chen, D.; Tan, X.; Bai, X.; Liu, Z.; Yang, X.; Liu, W. Current Advances of Nanocarrier Technology-Based Active Cosmetic Ingredients for Beauty Applications. *Clin. Cosmet. Investig. Dermatol.* **2021**, *14*, 867–887. [CrossRef] [PubMed]
16. Wang, Z.P.; Shen, H.H.; Luo, D.; Chen, D.; Sheng, J.; Liu, W. Preparation and efficacy evaluation of nanoliposomes for co-delivery of anti-alopecia agents. *Chin. Surf. Deter. Cos.* **2020**, *50*, 396–401. [CrossRef]
17. Kim, B.S.; Na, Y.G.; Choi, J.H.; Kim, I.; Lee, E.; Kim, S.-Y.; Lee, J.-Y.; Cho, C.-W. The Improvement of Skin Whitening of Phenylethyl Resorcinol by Nanostructured Lipid Carriers. *Nanomaterials* **2017**, *7*, 241. [CrossRef]
18. Wang, M.; Charareh, P.; Lei, X.; Zhong, J.L. Autophagy: Multiple Mechanisms to Protect Skin from Ultraviolet Radiation-Driven Photoaging. *Oxidative Med. Cell. Longev.* **2019**, *2019*, 8135985. [CrossRef]
19. Han, S.; Li, H.; Luo, F.; Chen, X.; Cen, Y.; Liu, P.; Chen, Z.; Lan, T.; Lin, J. Inhibitory Effect of Seawater Pearl Hydrolysate on UVA-Induced Photoaging of Human Skin Fibroblasts. *Evid. Based. Complement. Alternat. Med.* **2022**, *2022*, 1558288. [CrossRef]
20. Misovic, M.; Milenkovic, D.; Martinovic, T.; Ciric, D.; Bumbasirevic, V.; Kravic-Stevovic, T. Short-term exposure to UV-A, UV-B, and UV-C irradiation induces alteration in cytoskeleton and autophagy in human keratinocytes. *Ultrastruct. Pathol.* **2013**, *37*, 241–248. [CrossRef]
21. Jung, C.H.; Ro, S.H.; Cao, J.; Otto, N.M.; Kim, D.-H. mTOR regulation of autophagy. *FEBS Lett.* **2010**, *584*, 1287–1295. [CrossRef]
22. Klionsky, D.J.; Emr, S.D. Autophagy as a regulated pathway of cellular degradation. *Science* **2000**, *290*, 1717–1721. [CrossRef]
23. Gilchrest, B.A. A review of skin aging and its medical therapy. *Br. J. Dermatol.* **1996**, *135*, 867–875. [CrossRef] [PubMed]
24. Gilchrest, B.A.; Garmys, M.; Yaar, M.A. Aging and photoaging effectgene expression in cultured human Keratinocyte. *Arch. Dermatol.* **1994**, *130*, 82–86. [CrossRef]
25. Boismal, F.; Serror, K.; Dobos, G.; Zuelgaray, E.; Bensussan, A.; Michel, L. Skin aging: Pathophysiology and innovative therapies. *Med. Sci.* **2020**, *36*, 1163–1172. [CrossRef]
26. Csekes, E.; Račková, L. Skin Aging, Cellular senescence and natural polyphenols. *Int. J. Mol. Sci.* **2021**, *22*, 12641. [CrossRef] [PubMed]
27. Fisher, G.J.; Kang, S.; Varani, J.; Bata-Csorgo, Z.; Wan, Y.; Datta, S.; Voorhees, J.J. Mechanisms of photoaging and chronological skin aging. *Arch. Dermatol.* **2002**, *138*, 1462–1470. [CrossRef] [PubMed]
28. Pozos-Nonato, S.; Domínguez-Delgado, C.L.; Campos-Santander, K.A.; Benavides, A.A.; Pacheco-Ortin, S.M.; Higuera-Piedrahita, R.I.; Resendiz-González, G.; Molina-Trinidad, E.M. Novel nanotechnological strategies for skin anti-aging. *Curr. Pharm. Biotechnol.* **2023**, *24*, 1397–1419. [CrossRef]

29. He, X.; Gao, X.; Guo, Y.; Xie, W. Research progress on bioactive factors against skin aging. *Int. J. Mol. Sci.* **2024**, *25*, 3797. [CrossRef]
30. He, X.; Wan, F.; Su, W.; Xie, W. Research progress on skin aging and active Ingredients. *Molecules* **2023**, *28*, 5556. [CrossRef]
31. Pintea, A.; Manea, A.; Pintea, C.; Vlad, R.-A.; Bîrsan, M.; Antonoaea, P.; Rédai, E.M.; Ciurba, A. Peptides: Emerging candidates for the prevention and treatment of skin senescence: A review. *Biomolecules* **2025**, *15*, 88. [CrossRef]
32. Alotaibi, M.R.; As, S.H.; Alaqil, F.A.; Almutairi, M.; Alhazzani, K.; Sulaiman, A.A.; Isab, A.A.; Alotaibi, N.H. A newly synthesizedplatinum-based compound (PBC-II) increases chemosensitivity of HeLa ovarian cancer cells via inhibition of autophagy. *Saudi Pharm. J.* **2019**, *27*, 1203–1209. [CrossRef] [PubMed]
33. Chen, S.M.; Gong, Y.C.; Sui, L.; Chu, M.; Dong, Y. Role of tangeretin on autophagy in human gastric cancer AGS cells and its mechanism. *Chin. Pharmacol. Bull.* **2019**, *35*, 1671–1676. [CrossRef]
34. Qi, M.; Tan, B.E. Molecular mechanism of autophagy regulating oxidative stress in animals. *Chin. J. Anim. Nut.* **2020**, *32*, 10.
35. Sun, D.; Wu, R.; Zheng, J.; Li, P.; Yu, L. Polyubiquitin chain-induced p62 phase separation drives autophagic cargo segregation. *Cell Res.* **2018**, *28*, 405–415. [CrossRef]
36. Ferreira, M.S.; Magalhães, M.C.; Oliveira, R.; Sousa-Lobo, J.M.; Almeida, I.F. Trends in the Use of Botanicals in Anti-Aging Cosmetics. *Molecules* **2021**, *26*, 3584. [CrossRef]
37. Cronin, H.; Draelos, Z.D. Top 10 botanical ingredients in 2010 anti-aging creams. *J. Cosmet. Dermatol.* **2010**, *9*, 218–225. [CrossRef]

cosmetics

MDPI

Article

The Oral Intake of Specific Bovine-Derived Bioactive Collagen Peptides Has a Stimulatory Effect on Dermal Matrix Synthesis and Improves Various Clinical Skin Parameters

Ehrhardt Proksch [1], Denise Zdzieblik [2] and Steffen Oesser [2,*]

[1] Department of Dermatology, University Hospital of Schleswig-Holstein, Campus Kiel, 24105 Kiel, Germany; eproksch@dermatology.uni-kiel.de
[2] CRI, Collagen Research Institute, Schauenburgerstr 116, 24118 Kiel, Germany; denise.zdzieblik@cri-mail.org
* Correspondence: steffen.oesser@cri-mail.org; Tel.: +49-431979987310

Abstract: Collagen products are widely marketed for skin improvement. This study evaluated the efficacy of VERISOL B in relation to key skin aging parameters. In a double-blind, placebo-controlled trial, 66 women (aged 35–55) were randomized to receive either 2.5 g of bovine-derived bioactive collagen peptides (SCPs) (n = 33) or a placebo (n = 33) daily for 8 weeks. Their eye wrinkle volume, skin elasticity, and hydration were objectively measured at baseline (X_0), 4 weeks (X_4), and 8 weeks (X_8). Additionally, the SCPs' impact on type I collagen, elastin, and proteoglycan biosynthesis was assessed in human dermal fibroblasts. The SCP supplementation significantly ($p < 0.05$) reduced their eye wrinkle volume and improved their skin elasticity and hydration within 4 weeks. After 8 weeks of treatment, the positive effects were even more pronounced for all of the clinical parameters measured ($p < 0.05$). The fibroblast experiments confirmed the SCPs' stimulatory impact on dermal metabolism ($p < 0.05$). In conclusion, oral SCP supplementation effectively reduced wrinkles and enhanced skin elasticity and hydration, likely by promoting extracellular matrix biosynthesis.

Keywords: bovine-derived collagen peptides; collagen peptides; anti-aging; wrinkles; skin hydration; cutometry; corneometry

1. Introduction

The appearance and integrity of the skin deteriorate with age due to the synergistic effects of chronological aging and photoaging, hormonal deficiency, and environmental influences [1]. A decrease in various metabolic activities, such as quantitative and qualitative changes in dermal collagen and elastin, leads to changes in skin texture and the typical signs of aging. The loss of connective tissue during skin aging leads to reduced elasticity, a loss of skin tone, and the progressive development and deepening of facial wrinkles and folds [2].

Facial wrinkles, accompanied by an age-related decrease in the skin's elasticity and a leathery appearance of the affected skin areas, are the best-known signs of skin aging. Certain facial areas, such as the corners of the eyes, commonly known as crow's feet, are particularly susceptible to wrinkling [3,4].

The function and healthy appearance of the skin depend on an adequate supply of essential nutrients. The relationship between nutrition and skin has become an issue of current interest worldwide. Intervention studies suggest dietary supplements can modulate or delay skin aging and improve the skin's integrity [5].

In addition to topical applications, a significant trend in skin care is the consumption of various supplements, such as vitamins, antioxidants, fatty acids, and hydrolyzed proteins, to enhance the appearance and texture of the skin [6–16]. As a result, there is increasing focus on the efficacy of these products, and several clinical studies suggest that dietary supplements can influence skin health [17–21].

Collagen peptides have been used for some time in dietary supplements to improve the skin's properties and appearance. Collagen peptides are partially absorbed intact from the digestive tract [22,23] and accumulate in skin tissue [24]. In vitro studies have shown that specific collagen peptides have a stimulatory effect on molecules of the dermal extracellular matrix (ECM), such as collagen, elastin, and proteoglycans [25–28]. The ECM is produced by dermal fibroblasts, cells responsible for synthesizing and maintaining the matrix molecules, thus counteracting age-related degradation processes.

Collagen hydrolysates have antioxidant properties and promote cell regeneration and synthesis of the extracellular matrix in the dermal connective tissue [6]. Additionally, orally administered collagen peptides enhance the skin's elasticity, hydration, and dermal density [8,29] while also reducing wrinkles and improving the skin's firmness [30], as described in systematic reviews of the literature on the effects of collagen supplementation on skin health [6,30,31]. The benefits of orally administered collagen peptides include significant improvements in skin elasticity and hydration and wrinkle reduction, as well as wound healing, ulcer treatment, and skin regeneration. The efficacy of collagen peptides has been demonstrated in several clinical trials, with the daily dosages ranging from 2.5 g to 10 g and the treatment periods ranging from 4 to 24 weeks.

Clinical trials on healthy women have shown that daily supplementation of 2.5 g of specific bioactive collagen peptides (SCPs) over 8 weeks significantly improves skin elasticity [32]. Another study [25] showed a positive impact on wrinkle reduction after 8 weeks of SCP treatment. In addition, a significant increase in type I collagen and elastin contents was observed in the skin of the women who had consumed 2.5 g SCPs/day for 8 weeks.

Most of these clinical trials have been conducted using porcine- or fish-derived specific bioactive collagen peptides [25,32,33], but there is also an increasing number of studies on the effects of bovine-derived collagen products on various skin parameters [34,35]. Based on the existing studies on specific products made from porcine-based collagen peptides (VERISOL P) [25,32], the aim of the current study was to investigate the efficacy of a specific bovine-derived bioactive collagen peptide product (VERISOL B) in relation to various clinically relevant skin parameters. Moreover, to gain a better understanding of the potential mechanisms of action, the biosynthesis of dermal matrix molecules in human dermal fibroblast cultures was investigated following the supplementation of this specific bovine-derived product.

2. Materials and Methods

2.1. The Test Product

The test product used in this study (SCPs) was a specific bioactive collagen peptide composition derived from the special enzymatic hydrolysis of bovine type I collagen. The peptide composition was produced and provided by GELITA AG (Eberbach, Germany) and is commercially available under the brand name VERISOL® B. It has a mean molecular weight of approximately 2.0 kDa. It is free of allergens. The recommended dosage ranges from 2.5 g to 5 g per day, with no concerns of overdose. This product is classified as a nutritional supplement and was classified by the United States Food and Drug Administration (FDA) as generally recognized as safe (GRAS) [36].

2.2. In Vitro Tests

We have used the fibroblast cells NHDF (PromoCell, Heidelberg, Germany). The cells were cultured in an optimized medium (HAM's F12) supplemented with 10% FBS, 20 U/mL penicillin–streptomycin, 50 µg/mL partricin, 0.05 mg/mL ascorbic acid, and 1 mM glutamine. The primary dermal fibroblasts were then seeded into 12-well culture plates following the proteolytic detachment of the adherent cells using Accutase solution and incubated at 37 °C with 5% CO_2. The culture medium was refreshed every 48 h, and the cells were maintained until they reached 80% confluence. Subsequently, the regular medium was replaced with medium containing the test product at a concentration of 0.5 mg/mL. Control experiments were conducted by culturing the cells in the initial medium without any additional treatment. To assess the mRNA expression profiles, the fibroblasts were stimulated for 24 h. The biosynthesis of the extracellular matrix molecules was evaluated after a 2-week treatment. The results are expressed relative to those for the untreated control.

2.3. Gene Expression Studies

Gene expression studies were conducted by treating fibroblast monolayer cultures with the test product for 24 h and comparing them to untreated controls. RNA was extracted using the phenol–chloroform method (PeqGoldTriFast, VWR, Erlangen, Germany). The matrix molecule RNA expression was measured semi-quantitatively using RT-PCR (DyNAmo Flash SYBR Green, Thermo Scientific, Waltham, MA, USA) after determining the RNA concentration photometrically.

Each reaction used 10–40 ng of transcribed cDNA (optimized in preliminary experiments), 12.5 µL of distilled water, 5.0 µL of DyNAmo Flash Master Mix, and 1.25 µL of the primers described in Table 1. The reaction profile included 7′ at 95 °C, followed by 35–45 cycles of denaturation (10″ at 95 °C), annealing (15″ at temperatures in Table 1), and elongation (20″ at 72 °C), with a final step at 60 °C for 60″. The RNA expression was determined relative to GAPDH.

Table 1. Primer sequences used for RT-PCR analysis.

Gene	Forward Sequences (5′-3′)	Reverse Sequences (3′-5′)	Annealing (°C)	Accession
GAPDH	GCTCTCTGCTCCTCCTGTTC	ACTCCGACCTTCACCTTCC	63.0	NG_007073.2
Collagen type I	AATGGTGCTCCTGGTATTGC	ACCAGGTTCACCGCTGTTAC	59.0	NM_000088
Decorin	TGATTTGGGTCTGGACAAAG	TGCCCAGTTCTATGACAATC	60.0	AF_491944.2
Biglycan	CCTCCAGGTGGTCTATCTGC	CATCAGGATGTGTGGCTGTG	58.0	AH_002674.2
Elastin	AAGGTGGCTGCCAAAGC	ACTCCTCCAAGTGGGAACTG	60.0	NM_00501

The primer annealing temperatures were optimized between 50 and 70 °C using gradient PCR. The products were separated on 2% agarose gel (3 g of agarose in 150 mL of buffer of 0.05 M EDTA, 40 mM Tris, and 57.1 mL of acetic acid), stained with 10 µL of GelRed (41003, BIOTREND Chemikalien GmbH, Cologne, Germany), and analyzed using an Alpha Innotech Fluorchem Imager (Bio-Rad Laboratories GmbH, Munich, Germany). Products were specified using a 50 bp DNA fluoro-ladder (8263.1, Carl Roth GmbH, Karlsruhe, Germany).

2.4. Measurement of the ECM Macromolecules

Commercially available assay kits were used to quantitatively analyze the extracellular matrix molecules investigated. These methods were used for all of the cell culture studies, including the batch-to-batch analysis.

Newly formed collagen was determined after three weeks of stimulation. Synthesized collagen was isolated using the Sircol assay (tebu-bio, Offenbach, Germany) according to the manufacturer's instructions. For this purpose, the culture medium was discarded, and the adherent cell layers were digested in 0.1 mg of pepsin solution in 0.5 M acetic acid overnight at 4 °C. The cell suspensions were neutralized by adding 100 µL of acidic neutral reagent, and the synthesized collagen was then separated by adding 200 µL of the isolation and concentration kit solution and vigorously shaking it overnight at 4 °C. After centrifugation (12,000 rpm, 10′) and disposal of the supernatant, the isolated collagen was resuspended in 1 mL of Sircol dye reagent. Following 30 min of shaking and further centrifugation, the collagen pellet was overlaid with 750 µL of cold acid wash reagent. After centrifugation, the supernatant was discarded, and the enriched collagen was dissolved in 250 µL of alkaline solution. From these sample solutions, 200 µL was taken for photometric quantification of the synthesized collagen. The absorbance was measured at 492 nm. The amount was determined from the measured standard collagen solutions.

For the detection of proteoglycans, the Blyscan glycosaminoglycan assay (tebu-bio, Offenbach, Germany) was used. The biosynthesis of the proteoglycans was determined after two weeks of stimulation. Following the manufacturer's instructions, the cell layers were coated with 1 mL of papain extraction solution after discarding the cell culture medium and incubated for 3 h at 65 °C, with vigorous shaking. The cell suspensions were centrifuged (10,000× *g*, 10′), and the supernatants were collected. After adding 1 mL of Blyscan dye solution and shaking it (30′), the supernatants were again centrifuged (12,000 rpm, 10′) and discarded. The isolated proteoglycan pellets were resuspended in 500 µL of dissociation solution. A total of 200 µL per sample was subjected to photometric determination at 656 nm and compared to the untreated control experiments.

Elastin synthesis was determined using the Fastin-Elastin assay kit (tebu-bio, Offenbach, Germany). The fibroblasts were stimulated with the test product for 2 weeks or left untreated in the control experiment. After discarding the culture medium, the cell monolayers were dissolved in 250 µL of trypsin. The trypsin solution was removed after centrifugation (3000 rpm, 10′). The cell pellets were dissolved in 100 µL of oxalic acid solution (1 M) and boiled for 1 h in a water bath. After cooling, 300 µL of the elastin precipitation solution was added to precipitate the dissolved elastin within 15 min. After centrifugation (10′), the supernatant was discarded, and the elastin pellet was dissolved in 1 mL of dye solution through shaking for 90 min. Following centrifugation and disposal of the supernatant, the elastin pellet was resuspended in 260 µL of dye dissociation solution. A total of 200 µL of the resolvent was used for the photometric elastin measurements at a wavelength of 492 nm.

2.5. The Study Design of the Clinical Trial

The trial was a monocentric, double-blind, randomized, placebo-controlled trial (RCT) studying the effects of specific bioactive collagen peptides (SCPs) on various skin parameters and was conducted in accordance with the guidelines for Good Clinical Practice (GCP) and in accordance with the Declaration of Helsinki at the SGS SIT GmbH in Hamburg, Germany. This study's primary objective was defined as a change in eye wrinkle volume after 8 weeks of treatment. In addition, the effects on the skin's elasticity and hydration (as secondary objectives) were examined after 4 and 8 weeks.

The examination was approved by the Institutional Review Board of the Ethics Committee of the Hamburg Medical Association (PV5894) and registered in the German Clinical Trials Register (DRKS00036187). All subjects received detailed information listing all relevant individual parameters of the study. After receiving written information, each participant had the opportunity to ask further questions and subsequently signed a consent form. Participants who met the eligibility criteria were randomized (at a 1:1 ratio) into the SCP or placebo group using a web-based random number generator [37].

2.6. The Inclusion and Exclusion Criteria

The inclusion criteria were (1) healthy women aged 35 to 55 (2) of phototypes I to IV (on the Fitzpatrick scale) (3) in a state of good physical and mental health and social well-being (4) who provided personal informed consent to participate in the study, (5) were personally present at the institute on the specified days, and (6) were willing and able to comply with the study rules and the fixed schedule.

The exclusion criteria were any deviation from the above inclusion criteria; acute skin diseases (e.g., atopic eczema, atopic dermatitis, psoriasis); food allergies related to the supplemented products; gastrointestinal diseases or digestive disorders; the use of topical medication on the test sites within 6 weeks prior to study entry, systemic medication containing anti-inflammatory agents or antibiotics within 2 weeks prior to study entry, systemic medication containing corticosteroids and/or antihistamines within 4 weeks prior to study entry, or other systemic medication within 4 weeks prior to study entry; systemic disease in the subject at study entry; pregnancy or lactation; immunological diseases; severe disease or severe diabetes; alcohol and drug abuse within 6 months prior to study entry; participation in other studies involving cosmetic products within 2 weeks before study entry or during the study; participation in a study involving a pharmaceutical preparation and/or the intake of dietary supplements within 4 weeks before study entry; changes in lifestyle or dietary habits during this study besides the use of the test products; treatment with leave-on products or oily or moisturizing skin-cleansing products on the test areas or changes to the usual skin care routine; intense sun or artificial UV exposure (through the use of a solarium) of the test areas within 1 week before study entry or during this study; swimming, sauna use, or intense sports activity within 1 day before the measurements; smoking; a lack of compliance; and an intellectual or mental inability to follow the study instructions.

2.7. Subjects

A total of 66 healthy female subjects with an average age of 46.1 $\pm$ 5.6 years were included in this study. All of the participants were randomly assigned into either the treatment group with a daily dose of 2.5 g of the SCPs or the placebo group. In the placebo group, a daily dose of 2.5 g of maltodextrin (Walter GmbH, Olpe, Germany) was supplemented. The samples were packed in individual sachets. Each woman who met the inclusion criteria and was recruited for the trial consumed 2.5 g of a sachet of either the SCPs or the placebo daily over the study period of 8 weeks.

The preparations were taken orally by the subjects at home according to the investigator's instructions. The powders were to be dissolved in water or another cold liquid, with the exception of milk.

A conditioning period of at least 7 days preceded the start of the oral treatment and data collection. During this time, the test subjects were instructed to refrain from using leave-on products on the test areas and not to change their usual skin care routine. In addition, treatment with dermatological therapeutics on the test areas was prohibited for 6 weeks prior to the start of the trial. During this study, changes in lifestyle or dietary habits, the use of additional nutritional or vitamin supplements, treatment of the test areas with cosmetic and dermatological skin care products, and intensive sun or UV light exposure were also forbidden.

2.8. Product Safety

Collagen peptides are characterized by a very high safety profile. No clinical indications of allergies have been observed to date. No incompatibilities with other diets or medications have ever been described in the medical literature. Collagen peptides were awarded a GRAS status by the American Food and Drug Administration. "GRAS" is an official abbreviation for "Generally Recognized As Safe" [36]. In addition, experimental studies have clearly shown that there is no systemic toxicity [38].

During this study, the tolerability of the SCP treatment was assessed through dermatological examinations and interviews with the participants before, during, and after the study.

2.9. Measurements

The wrinkle area around the left eye (the lateral canthus) was defined as the test site for the eye wrinkle measurements (with 1 test site per body region). Skin elasticity and hydration were measured on the inside of the right forearm. The test areas on the forearms were 5 cm × 5 cm. On each measurement day, the subjects had to expose their uncovered test sites to the climate of the room (21.5 °C; 50% relative humidity) for at least 30 min. Two measurement times were set for the primary objective.

The wrinkle measurements were conducted immediately before the start of the product treatment (X_0) and after 8 weeks (X_8) of daily product intake. In addition, an interim analysis was performed after 4 weeks (X_4). For the secondary objectives, skin elasticity and hydration were examined at X_0, X_4, and X_8.

The compliance of the study participants (dosage and type of intake) and tolerability of the products were assessed after 1, 4, and, again, 6 weeks of administration.

The eye wrinkle volume was measured at the outer corner of the eye (lateral canthus) using the PRIMOS® Compact optical 3-dimensional in vivo measuring device (GF Messtechnik GmbH, Teltow, Germany). Three measurements were performed per test site. The size of the measurement area was 30 mm × 40 mm. Post-baseline measurements were performed using the overlay function. The original images of the reference files at baseline and the corresponding measurement files of the post-baseline measurements were matched using the 3-dimensional matching function for each participant. The height of the images was calculated using the standard procedure with mathematical filters. The eye fold volume (in mm^3) of a selected fold was calculated from these height images using the PRIMOS® software. This was carried out for all three images taken per measurement site and at each time. The mean value of the three individual measurements was subsequently calculated for each test site and measurement time point.

Skin elasticity was measured using the Cutometer® MPA 580 (Courage & Khazaka, Cologne, Germany), as described by Segger et al. [39,40]. In brief, the stretching of the skin was recorded in response to negative pressure generated by using a vacuum (350 mbar) over the skin test area. The exposure and non-exposure time was 5 s, with one cycle per measurement. The R5 value (Ur/Ue, instantaneous recovery/elastic deformation) was recorded to analyze the skin's elasticity. This parameter has been shown to be particularly useful in detecting age-related skin changes [41,42]. The measurements were repeated three times at each test site.

The assessment of the skin's surface hydration according to electrical capacitance was performed using the Corneometer® CM 825 (Courage & Khazaka), which measures the reactive capacitance of the skin by using the stratum corneum as a dielectric membrane. The measurements are arbitrarily expressed as hydration indices that rise with increasing skin hydration. A total of 15 individual measurements were taken for each application site and the controls.

2.10. The Statistical Analysis

Statistical analyses were performed using SPSS Statistics (IBM SPSS Statistics for Windows, Version 23.0, Armonk, NY, USA: IBM Corp.). Normality was assessed using the one-sample Kolmogorov–Smirnov (KS) test. The hypothesis of a normal distribution was accepted if the KS test yielded $p > 0.05$. For the cell culture experiments and for the batch-to-batch analysis, the descriptive results were expressed relative to those from the untreated control experiment. For the cell culture experiments, statistical significance was determined based on the KS test results: if normality was confirmed ($p > 0.05$), a parametric one-sample t-test was applied; otherwise, the 1-sample Wilcoxon signed-rank test was used to detect differences between the SCP treatment and the untreated control. For normally distributed data (proofed using the KS test), a one-way ANOVA with Tukey's HSD post hoc test was used to compare batches; otherwise, the Kruskal–Wallis test was used with Dunn–Bonferroni correction for pairwise comparisons. In the clinical trial, baseline and intervention-related group differences were analyzed using either the independent t-test (for normally distributed data) or the Mann–Whitney U test (for non-normally distributed data). Differences between the treatment conditions were considered statistically significant if $p < 0.05$. The data were presented as described in the legends in the figures and tables. The effect size was calculated using Cohen's d for inter-group differences after an 8-week treatment.

3. Results

3.1. The In Vitro Test

The present in vitro study was designed to investigate the possible impact of bovine-derived SCPs on the biosynthesis of ECM macromolecules for the first time. For these cell culture experiments, primary human dermal fibroblasts were used. In a second experimental approach, different batches of the used product were tested regarding their efficacy.

3.1.1. The Bioactivity of the SCPs

The biosynthesis of the most important molecules in the dermal extracellular matrix was statistically significantly increased ($p < 0.05$) after the SCP treatment compared to that with the untreated controls (Figure 1).

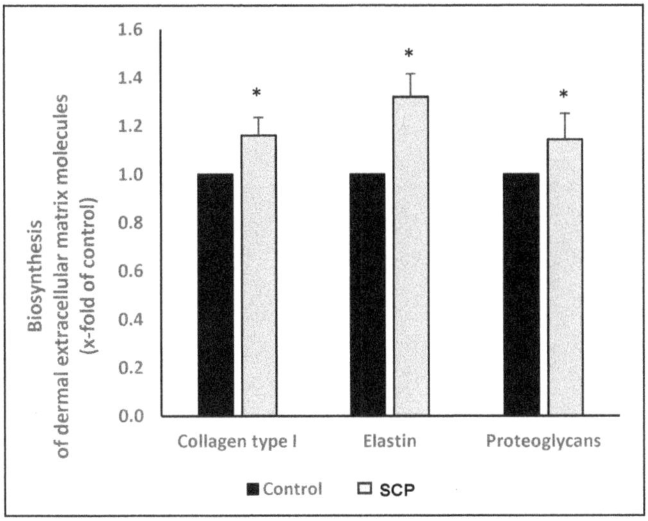

Figure 1. The stimulation of collagen type I, elastin, and proteoglycan biosynthesis after the SCP treatment compared to that in the untreated controls. The data represent the mean ± SD for n = 12; * $p < 0.05$.

The amount of collagen type I, elastin, and proteoglycans was statistically significantly increased ($p < 0.05$) after the SCP supplementation compared to that in the untreated controls (Table 2). The stimulatory effect of SCP treatment on the biosynthesis of ECM molecules was also confirmed at the gene expression level. The mRNA expression levels for type I collagen, elastin, and the proteoglycans decorin and biglycan were also significantly increased ($p < 0.05$) after the SCP supplementation.

Table 2. The stimulatory effect on the mRNA expression of different ECM macromolecules after a 24 h treatment with SCP.

	mRNA Expression	*p*-Value
Collagen type I	2.06 ± 0.17	<0.05
Elastin	1.17 ± 0.12	<0.05
Decorin	1.15 ± 0.21	<0.05
Biglycan	1.30 ± 0.13	<0.05

The data represent the mean ± SD for n = 12 compared to the untreated controls.

3.1.2. The Batch-to-Batch Analysis of the Test Product

A batch-to-batch analysis of four different SCP batches produced between 2018 and 2024 showed no statistically significant differences between the tested products regarding their efficacy in stimulating matrix biosynthesis. All of the tested batches had a similar pronounced stimulatory effect on synthesizing type I collagen, proteoglycans, and elastin (Table 3).

Table 3. The consistent stimulatory effect of the different bovine-derived SCP batches on the synthesis of ECM molecules in the human dermal fibroblasts.

Batch No.	1	2	3	4	*p*-Value
Collagen type I	1.22 ± 0.09	1.19 ± 0.06	1.20 ± 0.15	1.19 ± 0.09	n.s.
Proteoglycans	1.19 ± 0.09	1.17 ± 0.09	1.18 ± 0.09	1.22 ± 0.06	n.s.
Elastin	1.26 ± 0.14	1.34 ± 0.14	1.34 ± 0.30	1.31 ± 0.37	n.s.

The different batches were produced between 2018 and 2024 by GELITA. The data represent the mean ± SD for n = 12; n.s. = not statistically significant.

3.2. The Clinical Trial

The CONSORT (Consolidated Standards of Reporting Trials) flow diagram for this controlled trial showed that 66 Caucasian women between the ages of 35 and 55 were included (Figure 2). The women, with an average age of 46.1 ± 5.6 years (46.7 ± 6.0 for the SCP group; 44.9 ± 5.6 for the placebo), had phototypes I–IV (on the Fitzpatrick scale) and were in generally good condition in terms of their physical and mental health. The participants were randomly assigned into the placebo or SCP group. All measured parameters were compared between the groups to assess the homogeneity of the data at baseline between the SCP and placebo groups. The data showed no statistically significant differences between the two study groups for any of the primary and secondary outcomes at baseline.

The data were evaluated based on the "Per-Protocol" (PP) population. All of the study participants who fulfilled the inclusion criteria and had given informed consent were included after randomization. As all of the recruited women had fulfilled the requirements according to the study protocol and no drop-outs occurred, the PP population was identical to the "intention-to-treat" (ITT) population. The assessment of the tolerability of the product revealed no adverse effects during the evaluation after 1, 4, and 6 weeks. All of the study participants were compliant with the study protocol. There were no discontinuations of the RCT related to product uptake or the study process in general. No inconveniences or adverse reactions were reported.

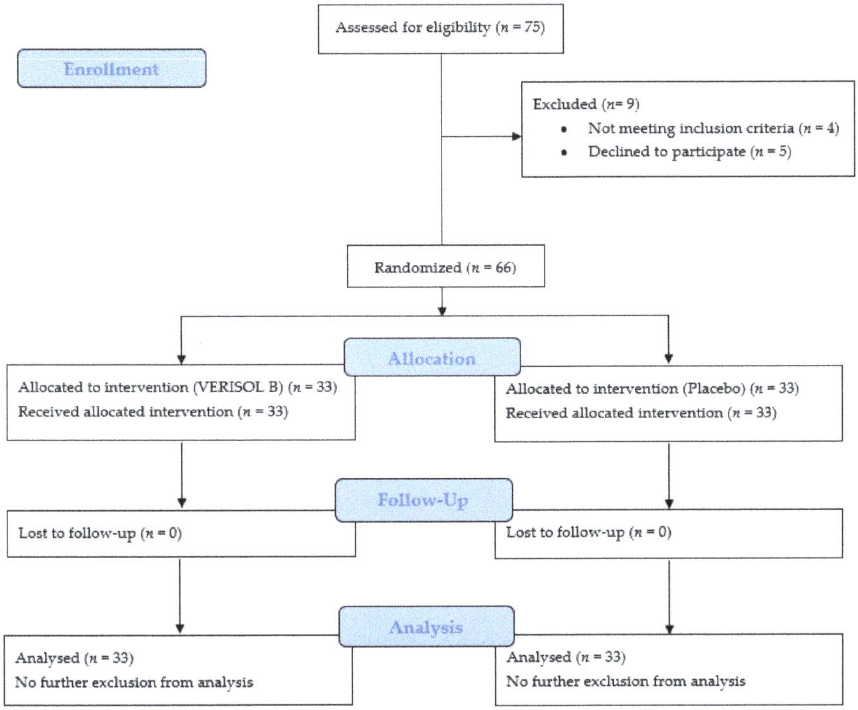

Figure 2. Flow chart of subject recruitment, randomization, and follow-up. VERISOL B (SCPs).

3.2.1. Eye Wrinkle Volume

At baseline, the two treatment groups had no statistically significant difference in their eye wrinkle volume (Table 3). After 4 weeks of the treatment with the SCPs, eye wrinkle volume decreased by approximately 9% ($p < 0.05$; Cohen's ≥ 0.8) compared to that in the placebo group (Figure 3, Table 4). At the end of this study, after 8 weeks, eye wrinkle volume was statistically significantly reduced, by 25%, in the participants who had received SCPs compared to that in the placebo group ($p < 0.05$; Table 4). The analysis of the individual cases revealed a maximum reduction in the eye wrinkle volume of more than 51%.

Table 4. Statistically significant improvement in various skin parameters in healthy women after SCP supplementation for 8 weeks.

	Group	n	$\bar{X}_0 \pm SD$	$\bar{X}_4 \pm SD$	*p*-Value * (After 4 Weeks)	$\bar{X}_8 \pm SD$	*p*-Value * (After 8 Weeks)
Eye wrinkle volume (mm^3)	SCPs	33	1.52 ± 0.24	1.41 ± 0.31	<0.05	1.26 ± 0.11	<0.05
	Placebo	33	1.49 ± 0.35	1.55 ± 0.27		1.58 ± 0.17	
Skin elasticity (Ur/Ue)	SCPs	33	0.53 ± 0.08	0.57 ± 0.05	<0.05	0.61 ± 0.08	<0.05
	Placebo	33	0.54 ± 0.10	0.54 ± 0.02		0.56 ± 0.06	
Skin hydration (AU)	SCPs	33	38 ± 3.9	46 ± 2.9	<0.05	49 ± 4.2	<0.05
	Placebo	33	37 ± 4.4	39 ± 4.7		39 ± 5.3	

The data represent the mean ± SD for 33 study participants. * Statistical significance of group comparison.

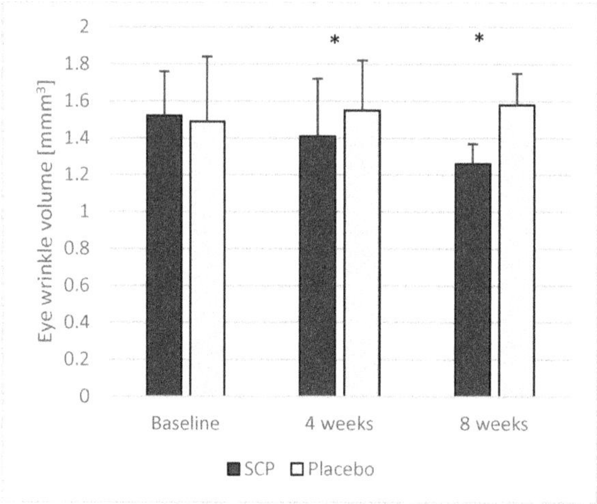

Figure 3. Statistically significant reductions in eye wrinkle volume after 4 and 8 weeks of treatment with SCPs compared to that in untreated controls. Data represent mean ± SD; * $p < 0.05$.

During the course of this study, there was also a statistically significant ($p < 0.05$) decrease in the eye wrinkle volume in the SCP group after 4 and 8 weeks.

3.2.2. Skin Elasticity

At the beginning of this study, there was no statistically significant difference in skin elasticity between the SCP and placebo groups (Table 3). Compared to that in the placebo, a statistically significant ($p < 0.05$) improvement in skin elasticity of 6% was observed in the SCP group after 4 weeks of treatment (Figure 4; Table 4). At the end of this study (Figure 4; Table 4), their skin elasticity had increased by 9% compared to that in the placebo group ($p < 0.05$; Cohen's d ≥ 0.8).

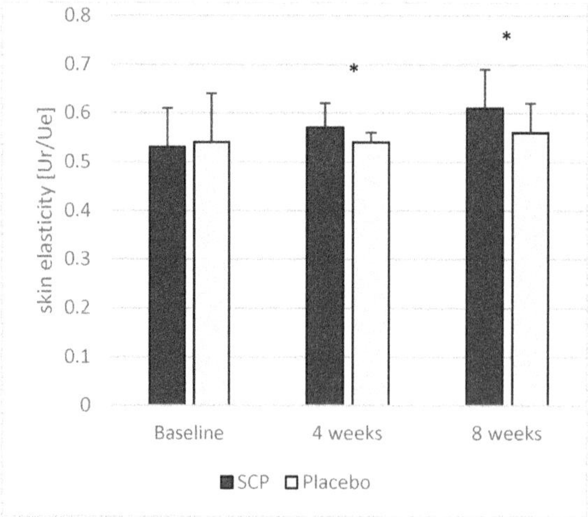

Figure 4. Statistically significant increase in skin elasticity after 4 and 8 weeks of treatment with SCPs compared to that in untreated controls. Data represent mean ± SD; * $p < 0.05$.

A statistically significant ($p < 0.05$) improvement in skin elasticity was observed in the treatment group after 4 and 8 weeks of SCP supplementation. In contrast, no changes in elasticity over time were observed in the placebo group.

3.2.3. Skin Hydration

At the beginning of this study, no statistically significant differences were observed in skin hydration between the SCP and placebo groups (Table 3). After 4 weeks, skin hydration was statistically significantly ($p < 0.05$) increased in the SCP group compared to that in the control group (Figure 5; Table 4). After 8 weeks, at the end of the study period, skin hydration was significantly ($p < 0.05$) increased, by 29%, in the treatment group. Compared to that in the control group, the skin hydration in the treatment group was increased by 26% ($p < 0.05$; Cohen's d ≥ 0.8) at the end of the investigation (Figure 5; Table 4).

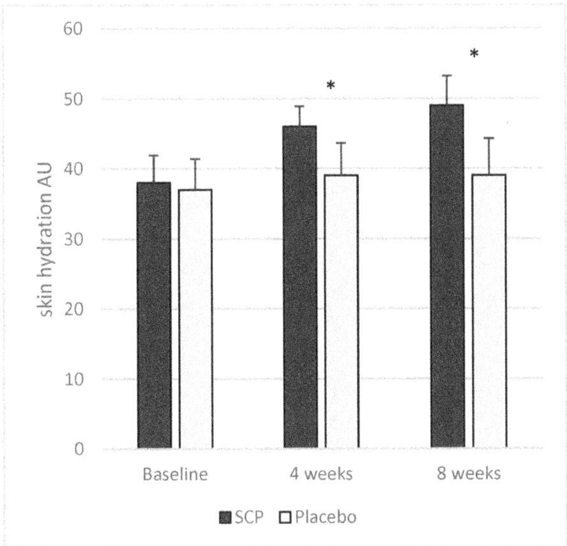

Figure 5. Statistically significant increase in skin hydration after 4 and 8 weeks of treatment with SCPs compared to that in untreated controls. Data represent mean ± SD; * $p < 0.05$.

In contrast to the treatment group, no significant changes in hydration were observed in the placebo group during the study period.

4. Discussion

Human skin undergoes a continuous aging process. This chronological or intrinsic aging is characterized by decreased thickness of the skin and a loss of skin elasticity and hydration [9,43–49]. As the major physiological changes in the skin are known to be located in the dermis, topical products such as creams and lotions may have a limited efficacy and duration. Most of these products do not reach the deeper layer of the skin, explaining their temporary effect on the skin's appearance.

Over the last decade, orally administered supplements for improving skin health have become increasingly popular. The consumption of products derived from collagen in particular has risen significantly. The effects of such collagen hydrolysates or collagen peptides on skin physiology have been studied by several groups worldwide. Experimental studies have shown that supplementation with specific collagen peptides stimulates the synthesis of the dermal extracellular matrix and provides antioxidant protection [9,50–58]. These studies suggest that the intake of collagen peptides can improve the appearance and function of the skin.

Numerous controlled clinical studies have now investigated the effects of orally administered collagen products on various skin parameters [9,10,21,25,32,59–62]. Mainly, collagen peptides derived from fish, pigs, or cattle are marketed. However, the mechanism of action of collagen peptides is not fully understood to date. It is known that, in addition to the molecular weight distribution of peptides, their specific production process plays a crucial role. The special enzymatic degradation of collagen results in specific collagen peptides that can bind to certain cell receptors, ensuring their high bioactivity. The importance of the production process could help to explain why collagen-derived products differ significantly in their efficacy. A systematic review of orally supplied collagen showed a stimulatory effect on the synthesis of matrix molecules [30]. The products studied differ in the recommended daily dosages, the mean molecular weights of the collagen peptides, and the source of collagen.

To ensure the efficacy of the bovine SCPs used in this clinical trial, the same peptides were tested in in vitro studies on primary human dermal fibroblasts. A batch-to-batch analysis was conducted to assess the reproducibility of the measured effects.

The current cell culture experiments clearly demonstrated a pronounced stimulatory effect of the SCPs on the expression of dermal extracellular matrix macromolecules. After supplementation with the SCPs, a significant increase in the biosynthesis of type I collagen, elastin, and proteoglycans was observed.

The current results are consistent with those of several experimental studies in which human fibroblasts demonstrated a stimulatory effect following collagen peptide supplementation on the extracellular matrix molecules [27,28,54]. Although these studies utilized different collagen peptides from various species, their results consistently showed an increase, primarily in collagen production. In the present study, the stimulatory efficacy of the SCPs was confirmed at both the mRNA and protein levels.

The small differences observed in the efficacy of the different batches confirm the stable production process and constant quality of the SCP samples.

This study investigated the efficacy of 2.5 g/day of SCPs of bovine origin in terms of their effect on skin physiology. The results confirmed a statistically significant decrease in eye wrinkle volume and a significant increase in skin elasticity and hydration after 4 weeks of treatment compared to these values int he placebo. After 8 weeks of SCP supplementation, the effects on the measured skin parameters were even more pronounced.

Skin elasticity is a crucial indicator of skin aging. In a clinical study on post-menopausal women, the decrease in skin elasticity per year was compared with its development in pre-menopausal women [47]. Based on the results, a decrease in skin elasticity of 0.55% per year after the menopause was calculated. It is known that elasticity, along with skin hydration [63–65], is mainly influenced by dermal collagen. Marini et al. (2012) described a correlation between skin elasticity and hydration and type I collagen synthesis after oral treatment [66].

In the current study, the skin elasticity was already statistically significantly increased after 4 weeks of treatment compared to that in the placebo ($p < 0.05$). After 8 weeks of treatment, the skin's elasticity was increased by 9% compared to that in the placebo. In one participant, a maximum increase in skin elasticity of 51% was observed after 8 weeks of treatment. These results closely resemble the findings of a study that investigated porcine-derived SCPs [32]. Here, an average increase in skin elasticity of 7% was observed. Another clinical study [67] used a combination of collagen peptides with various vitamins. In this study, skin elasticity was improved by 8% after 12 weeks of supplementation compared to that in the placebo. Additionally, the authors reported an increase in skin hydration of 28% in the treatment group and 9% in the placebo group.

In the current clinical trial, a 26% improvement in skin hydration was observed after 8 weeks of the SCP treatment compared to the controls. After a treatment time of only 4 weeks, skin hydration was increased by 18%, indicating that SCP intake seems to have a fast impact on water-binding structures in the dermal tissue. This observed effect was confirmed in in vitro experiments, where a significant increase in proteoglycans was determined. As proteoglycans have a very high water-binding capacity, this might explain the clinically proven results.

Reducing wrinkles, especially in the face, is key to a healthier and younger appearance. Topically applied creams and lotions are often ineffective, as they do not penetrate the deeper layers of the skin [68,69]. Usually, such treatments have only a very short-term effect, if any.

Collagen and elastin are known to be the main components of the dermis responsible for maintaining the skin's structure, firmness, and elasticity. All of these factors together directly impact wrinkle formation. In particular, the conversion of elastic fibers and the progressive degradation of collagen bundles promote wrinkles.

The current clinical data showed a pronounced reduction in wrinkles in the treatment group within 4 weeks compared to those in the placebo. After 8 weeks, this effect was even more pronounced: wrinkles decreased by an average of 25% after SCP intake compared to those in the untreated controls.

The positive effects on skin moisture and elasticity and the reduction in eye wrinkle volume observed in this study align with comparable clinical studies where similar skin parameters were investigated, although not using bovine-derived collagen peptides. For example, studies performed using a very similar porcine-derived collagen peptide product [25,32] demonstrated comparable positive effects on skin health. In these studies, a daily dose of 2.5 g was also administered. Additionally, a study using fish-derived collagen peptides also showed positive effects [33]. In this study, the daily oral intake of 5 g of the product led to a significant reduction in eye wrinkles, as well as improved skin moisture and elasticity. These data seem to suggest that the clinical effectiveness does not primarily depend on the animal species from which the collagen peptides are derived.

5. Conclusions

The results of the RCT involving healthy women aged 35 to 55 years indicated that a 4-week treatment with bovine-derived SCP led to a statistically significant reduction in eye wrinkle volume, along with an increase in skin elasticity and hydration, compared to these values in the untreated control subjects. After 8 weeks of supplementation, the positive clinical effects were even more pronounced (Cohen's $d \geq 0.8$). Cell culture experiments with human dermal fibroblasts demonstrated that SCP supplementation resulted in a statistically significant increase in collagen, elastin, and proteoglycan biosynthesis compared to that in the controls. Although further studies are needed, it can be speculated that the positive clinical effects on wrinkle volume, skin elasticity, and skin hydration following SCP supplementation observed are due to the stimulatory effect of these bioactive collagen peptides on the synthesis of type I collagen, elastin, and proteoglycans. The analysis of different SCP batches confirmed the product's stimulatory efficacy and demonstrated its consistent quality. No adverse events were reported in this clinical study, and no side effects were observed. The product was found to be safe and well tolerated. Further studies should continue to investigate its mechanism of action and effects on other ECM macromolecules. Moreover, it would be interesting to investigate whether the sources of collagen peptides derived from different species have an impact on the efficacy of the product further.

In summary, orally administered SCP treatment has been shown to significantly improve the health and appearance of the skin. However, the preclinical and clinical results presented here are specific to the SCPs used in this study and may not necessarily apply to other collagen products.

Author Contributions: Conceptualization, E.P. and S.O.; methodology, E.P. and S.O.; validation, E.P.; formal analysis, S.O.; investigation, E.P.; resources, S.O.; data curation, S.O.; writing—original draft preparation, S.O.; writing—review and editing, E.P., D.Z. and S.O.; visualization, D.Z.; supervision, E.P.; project administration, E.P.; funding acquisition, S.O. All authors have read and agreed to the published version of the manuscript.

Funding: The specific collagen peptides (product name VERISOL B®) were provided by GELITA GmbH, Eberbach Germany. It should be emphasized that the sponsor had no influence on this study's design, the conduct of this study, or the analysis of the data.

Institutional Review Board Statement: This study was conducted according to the guidelines of the Declaration of Helsinki and approved by the Institutional Review Board of the Ethics Committee of the Hamburg Medical Association (protocol code: PV5894; date of approval: 24 September 2018).

Informed Consent Statement: Informed consent was obtained from all of the subjects involved in this study.

Data Availability Statement: The datasets used and/or analyzed during the current study are available from the corresponding author on reasonable request.

Acknowledgments: We would like to thank all of the participants, as well as the staff of the study center, who supported us with the examination.

Conflicts of Interest: The authors declare no conflicts of interest. S.O. has given scientific lectures that were in part supported by GELITA AG. He is also a coinventor of patents concerning the use of collagen peptides. All of the procedures performed in the studies involving human participants were performed in accordance with the ethical standards of the institutional and/or national research committee and with the 1964 Declaration of Helsinki and its later amendments or comparable ethical standards. The funders had no role in the design of this study; in the collection, analyses, or interpretation of the data; in the writing of the manuscript; or in the decision to publish the results. The authors responsible for the data's assessment and analysis have neither financial nor competing interests concerning the outcomes of this investigation.

Abbreviations

The following abbreviations are used in this manuscript:

ECM	Extracellular Matrix
EDTA	Ethylenediaminetetraacetic Acid
FCS	Fetal Calf Serum
GAPDH	Glyceraldehyde 3-Phosphate Dehydrogenase
GCP	Good Clinical Practice
GRAS	Generally Recognized As Safe
ITT	Intention-to-Treat
kD	Kilodalton
KS	Kolmogorov–Smirnov
PP	Per-Protocol
RCT	Randomized Controlled Trial
RT-PCR	Real-Time Polymerase Chain Reaction
SCP	Specific Collagen Peptide
SPSS	Statistical Package for the Social Sciences
TM	Melting Temperature
UV	Ultraviolet

References

1. Smalls, L.K.; Randall Wickett, R.; Visscher, M.O. Effect of Dermal Thickness, Tissue Composition, and Body Site on Skin Biomechanical Properties. *Skin Res. Technol.* **2006**, *12*, 43–49. [CrossRef] [PubMed]
2. Calleja-Agius, J.; Muscat-Baron, Y.; Brincat, M.P. Skin Ageing. *Menopause Int.* **2007**, *13*, 60–64. [CrossRef]
3. Takema, Y.; Yorimoto, Y.; Kawai, M.; Imokawa, G. Age-Related Changes in the Elastic Properties and Thickness of Human Facial Skin. *Br. J. Dermatol.* **1994**, *131*, 641–648. [CrossRef] [PubMed]
4. Takema, Y.; Yorimoto, Y.; Kawai, M. The Relationship between Age-Related Changes in the Physical Properties and Development of Wrinkles in Human Facial Skin. *J. Soc. Cosmet. Chem.* **1995**, *46*, 163–173.
5. Boelsma, E.; Hendriks, H.F.; Roza, L. Nutritional Skin Care: Health Effects of Micronutrients and Fatty Acids. *Am. J. Clin. Nutr.* **2001**, *73*, 853–864. [CrossRef]
6. Aguirre-Cruz, G.; León-López, A.; Cruz-Gómez, V.; Jiménez-Alvarado, R.; Aguirre-Álvarez, G. Collagen Hydrolysates for Skin Protection: Oral Administration and Topical Formulation. *Antioxidants* **2020**, *9*, 181. [CrossRef] [PubMed]
7. Cho, S. The Role of Functional Foods in Cutaneous Anti-Aging. *J. Lifestyle Med.* **2014**, *4*, 8–16. [CrossRef]
8. Czajka, A.; Kania, E.M.; Genovese, L.; Corbo, A.; Merone, G.; Luci, C.; Sibilla, S. Daily Oral Supplementation with Collagen Peptides Combined with Vitamins and Other Bioactive Compounds Improves Skin Elasticity and Has a Beneficial Effect on Joint and General Wellbeing. *Nutr. Res.* **2018**, *57*, 97–108. [CrossRef]
9. Liang, J.; Pei, X.; Zhang, Z.; Wang, N.; Wang, J.; Li, Y. The Protective Effects of Long-Term Oral Administration of Marine Collagen Hydrolysate from Chum Salmon on Collagen Matrix Homeostasis in the Chronological Aged Skin of Sprague-Dawley Male Rats. *J. Food Sci.* **2010**, *75*, H230–H238. [CrossRef]
10. Lin, P.; Alexander, R.A.; Liang, C.-H.; Liu, C.; Lin, Y.-H.; Lin, Y.-H.; Chan, L.-P.; Kuan, C.-M. Collagen Formula with Djulis for Improvement of Skin Hydration, Brightness, Texture, Crow's Feet, and Collagen Content: A Double-Blind, Randomized, Placebo-Controlled Trial. *J. Cosmet. Dermatol.* **2021**, *20*, 188–194. [CrossRef]
11. Morgado-Carrasco, D.; Gil-Lianes, J.; Jourdain, E.; Piquero-Casals, J. Oral Supplementation and Systemic Drugs for Skin Aging: A Narrative Review. *Actas Dermosifiliogr.* **2023**, *114*, 114–124. [CrossRef] [PubMed]
12. Pincemail, J.; Meziane, S. On the Potential Role of the Antioxidant Couple Vitamin E/Selenium Taken by the Oral Route in Skin and Hair Health. *Antioxidants* **2022**, *11*, 2270. [CrossRef]
13. Tarshish, E.; Hermoni, K.; Sharoni, Y.; Muizzuddin, N. Effect of Lumenato Oral Supplementation on Plasma Carotenoid Levels and Improvement of Visual and Experiential Skin Attributes. *J. Cosmet. Dermatol.* **2022**, *21*, 4042–4052. [CrossRef]
14. Wertz, P.W. Roles of Lipids in the Permeability Barriers of Skin and Oral Mucosa. *Int. J. Mol. Sci.* **2021**, *22*, 5229. [CrossRef] [PubMed]
15. Zhang, K.; Zhao, C.; Liu, K.; Feng, R.; Zhao, Y.; Zong, Y.; Du, R. Oral Administration of Deer Bone Collagen Peptide Can Enhance the Skin Hydration Ability and Antioxidant Ability of Aging Mice Induced by D-Gal, and Regulate the Synthesis and Degradation of Collagen. *Nutrients* **2024**, *16*, 1548. [CrossRef] [PubMed]
16. Draelos, Z.D. Nutrition and Enhancing Youthful-Appearing Skin. *Clin. Dermatol.* **2010**, *28*, 400–408. [CrossRef]
17. Bagheri Miyab, K.; Alipoor, E.; Vaghardoost, R.; Saberi Isfeedvajani, M.; Yaseri, M.; Djafarian, K.; Hosseinzadeh-Attar, M.J. The Effect of a Hydrolyzed Collagen-Based Supplement on Wound Healing in Patients with Burn: A Randomized Double-Blind Pilot Clinical Trial. *Burns* **2020**, *46*, 156–163. [CrossRef]
18. Venzin, C.; Jacot, V.; Berdichevsky, A.; Karol, A.A.; Seliktar, D.; von Rechenberg, B.; Nuss, K.M. Biocompatibility of Pegylated Fibrinogen and Its Effect on Healing of Full-Thickness Skin Defects: A Preliminary Study in Rats. *J. Biotechnol. Biomater.* **2016**, *6*, online. [CrossRef]
19. Vollmer, D.L.; West, V.A.; Lephart, E.D. Enhancing Skin Health: By Oral Administration of Natural Compounds and Minerals with Implications to the Dermal Microbiome. *Int. J. Mol. Sci.* **2018**, *19*, 3059. [CrossRef]
20. Wang, J.; Xu, M.; Liang, R.; Zhao, M.; Zhang, Z.; Li, Y. Oral Administration of Marine Collagen Peptides Prepared from Chum Salmon (Oncorhynchus Keta) Improves Wound Healing Following Cesarean Section in Rats. *Food Nutr. Res.* **2015**, *59*, 26411. [CrossRef]
21. Zague, V. A New View Concerning the Effects of Collagen Hydrolysate Intake on Skin Properties. *Arch. Dermatol. Res.* **2008**, *300*, 479–483. [CrossRef] [PubMed]
22. Ohara, H.; Matsumoto, H.; Ito, K.; Iwai, K.; Sato, K. Comparison of Quantity and Structures of Hydroxyproline-Containing Peptides in Human Blood after Oral Ingestion of Gelatin Hydrolysates from Different Sources. *J. Agric. Food Chem.* **2007**, *55*, 1532–1535. [CrossRef]
23. Iwai, K.; Hasegawa, T.; Taguchi, Y.; Morimatsu, F.; Sato, K.; Nakamura, Y.; Higashi, A.; Kido, Y.; Nakabo, Y.; Ohtsuki, K. Identification of Food-Derived Collagen Peptides in Human Blood after Oral Ingestion of Gelatin Hydrolysates. *J. Agric. Food Chem.* **2005**, *53*, 6531–6536. [CrossRef]
24. Oesser, S.; Adam, M.; Babel, W.; Seifert, J. Oral Administration of (14)C Labeled Gelatin Hydrolysate Leads to an Accumulation of Radioactivity in Cartilage of Mice (C57/BL). *J. Nutr.* **1999**, *129*, 1891–1895. [CrossRef] [PubMed]

25. Proksch, E.; Schunck, M.; Zague, V.; Segger, D.; Degwert, J.; Oesser, S. Oral Intake of Specific Bioactive Collagen Peptides Reduces Skin Wrinkles and Increases Dermal Matrix Synthesis. *Skin Pharmacol. Physiol.* **2014**, *27*, 113–119. [CrossRef] [PubMed]
26. Dumas, M.; Chaudagne, C.; Bonté, F.; Meybeck, A. In Vitro Biosynthesis of Type I and III Collagens by Human Dermal Fibroblasts from Donors of Increasing Age. *Mech. Ageing Dev.* **1994**, *73*, 179–187. [CrossRef]
27. Sanchez, A.; Blanco, M.; Correa, B.; Perez-Martin, R.I.; Sotelo, C.G. Effect of Fish Collagen Hydrolysates on Type I Collagen mRNA Levels of Human Dermal Fibroblast Culture. *Mar. Drugs* **2018**, *16*, 144. [CrossRef]
28. Zague, V.; do Amaral, J.B.; Rezende Teixeira, P.; de Oliveira Niero, E.L.; Lauand, C.; Machado-Santelli, G.M. Collagen Peptides Modulate the Metabolism of Extracellular Matrix by Human Dermal Fibroblasts Derived from Sun-Protected and Sun-Exposed Body Sites. *Cell Biol. Int.* **2018**, *42*, 95–104. [CrossRef]
29. Chai, H.-J.; Li, J.-H.; Huang, H.-N.; Li, T.-L.; Chan, Y.-L.; Shiau, C.-Y.; Wu, C.-J. Effects of Sizes and Conformations of Fish-Scale Collagen Peptides on Facial Skin Qualities and Transdermal Penetration Efficiency. *J. Biomed. Biotechnol.* **2010**, *2010*, 757301. [CrossRef]
30. Choi, F.D.; Sung, C.T.; Juhasz, M.L.W.; Mesinkovsk, N.A. Oral Collagen Supplementation: A Systematic Review of Dermatological Applications. *J. Drugs Dermatol.* **2019**, *18*, 9–16.
31. Barati, M.; Jabbari, M.; Navekar, R.; Farahmand, F.; Zeinalian, R.; Salehi-Sahlabadi, A.; Abbaszadeh, N.; Mokari-Yamchi, A.; Davoodi, S.H. Collagen Supplementation for Skin Health: A Mechanistic Systematic Review. *J. Cosmet. Dermatol.* **2020**, *19*, 2820–2829. [CrossRef] [PubMed]
32. Proksch, E.; Segger, D.; Degwert, J.; Schunck, M.; Zague, V.; Oesser, S. Oral Supplementation of Specific Collagen Peptides Has Beneficial Effects on Human Skin Physiology: A Double-Blind, Placebo-Controlled Study. *Skin Pharmacol. Physiol.* **2014**, *27*, 47–55. [CrossRef]
33. Oesser, S.; Schunk, M.; Proksch, E. Positive Effect of Fish-Derived Bioactive Collagen Peptides on Skin Health. *Int. J. Nutraceuticals Funct. Foods Nov. Foods* **2020**, *1*, 127–133. [CrossRef]
34. Wang, H. A Review of the Effects of Collagen Treatment in Clinical Studies. *Polymers* **2021**, *13*, 3868. [CrossRef] [PubMed]
35. Pu, S.-Y.; Huang, Y.-L.; Pu, C.-M.; Kang, Y.-N.; Hoang, K.D.; Chen, K.-H.; Chen, C. Effects of Oral Collagen for Skin Anti-Aging: A Systematic Review and Meta-Analysis. *Nutrients* **2023**, *15*, 2080. [CrossRef] [PubMed]
36. *Evaluation of the Health Aspects of Gelatin as a Food Ingredient*; Federation of American Societies for Experimental Biology: Bethesda, MD, USA; Life Sciences Research Office, Food and Drug Administration: Washington, DC, USA, 1975.
37. Urbaniak, G.C.; Plous, S. Research Randomizer (Version 3.0). Available online: https://www.randomizer.org/about/ (accessed on 29 January 2013).
38. 7 Final Report on the Safety Assessment of Hydrolyzed Collagen. *J. Am. Coll. Toxicol.* **1985**, *4*, 199–221. [CrossRef]
39. Segger, D.; Schönlau, F. Supplementation with Evelle Improves Skin Smoothness and Elasticity in a Double-Blind, Placebo-Controlled Study with 62 Women. *J. Dermatol. Treat.* **2004**, *15*, 222–226. [CrossRef]
40. Segger, D.; Matthies, A.; Saldeen, T. Supplementation with Eskimo Skin Care Improves Skin Elasticity in Women. A Pilot Study. *J. Dermatol. Treat.* **2008**, *19*, 279–283. [CrossRef]
41. Krueger, N.; Luebberding, S.; Oltmer, M.; Streker, M.; Kerscher, M. Age-Related Changes in Skin Mechanical Properties: A Quantitative Evaluation of 120 Female Subjects. *Skin Res. Technol.* **2011**, *17*, 141–148. [CrossRef]
42. Ryu, H.S.; Joo, Y.H.; Kim, S.O.; Park, K.C.; Youn, S.W. Influence of Age and Regional Differences on Skin Elasticity as Measured by the Cutometer. *Skin Res. Technol.* **2008**, *14*, 354–358. [CrossRef]
43. Batisse, D.; Bazin, R.; Baldeweck, T.; Querleux, B.; Lévêque, J.-L. Influence of Age on the Wrinkling Capacities of Skin. *Skin Res. Technol.* **2002**, *8*, 148–154. [CrossRef] [PubMed]
44. Frei, V.; Perrier, E.; Orly, I.; Huc, A.; Augustin, C.; Damour, O. Activation of Fibroblast Metabolism in a Dermal and Skin Equivalent Model: A Screening Test for Activity of Peptides. *Int. J. Cosmet. Sci.* **1998**, *20*, 159–173. [CrossRef]
45. Paul, C.; Maumus-Robert, S.; Mazereeuw-Hautier, J.; Guyen, C.N.; Saudez, X.; Schmitt, A.M. Prevalence and Risk Factors for Xerosis in the Elderly: A Cross-Sectional Epidemiological Study in Primary Care. *Dermatology* **2011**, *223*, 260–265. [CrossRef] [PubMed]
46. Nishimori, Y.; Edwards, C.; Pearse, A.; Matsumoto, K.; Kawai, M.; Marks, R. Degenerative Alterations of Dermal Collagen Fiber Bundles in Photodamaged Human Skin and UV-Irradiated Hairless Mouse Skin: Possible Effect on Decreasing Skin Mechanical Properties and Appearance of Wrinkles. *J. Investig. Dermatol.* **2001**, *117*, 1458–1463. [CrossRef]
47. Sumino, H.; Ichikawa, S.; Abe, M.; Endo, Y.; Ishikawa, O.; Kurabayashi, M. Effects of Aging, Menopause, and Hormone Replacement Therapy on Forearm Skin Elasticity in Women. *J. Am. Geriatr. Soc.* **2004**, *52*, 945–949. [CrossRef] [PubMed]
48. Vuillermoz, B.; Wegrowski, Y.; Contet-Audonneau, J.-L.; Danoux, L.; Pauly, G.; Maquart, F.-X. Influence of Aging on Glycosaminoglycans and Small Leucine-Rich Proteoglycans Production by Skin Fibroblasts. *Mol. Cell. Biochem.* **2005**, *277*, 63–72. [CrossRef]
49. Reilly, D.M.; Lozano, J. Skin Collagen through the Lifestages: Importance for Skin Health and Beauty. *Plast. Aesthet. Res.* **2021**, *8*, 2. [CrossRef]

50. Zague, V.; de Freitas, V.; da Costa Rosa, M.; de Castro, G.Á.; Jaeger, R.G.; Machado-Santelli, G.M. Collagen Hydrolysate Intake Increases Skin Collagen Expression and Suppresses Matrix Metalloproteinase 2 Activity. *J. Med. Food* **2011**, *14*, 618–624. [CrossRef]
51. Matsuda, N.; Koyama, Y.; Hosaka, Y.; Ueda, H.; Watanabe, T.; Araya, T.; Irie, S.; Takehana, K. Effects of Ingestion of Collagen Peptide on Collagen Fibrils and Glycosaminoglycans in the Dermis. *J. Nutr. Sci. Vitaminol.* **2006**, *52*, 211–215. [CrossRef]
52. Tanaka, M.; Koyama, Y.; Nomura, Y. Effects of Collagen Peptide Ingestion on UV-B-Induced Skin Damage. *Biosci. Biotechnol. Biochem.* **2009**, *73*, 930–932. [CrossRef]
53. Zhang, Z.; Wang, J.; Ding, Y.; Dai, X.; Li, Y. Oral Administration of Marine Collagen Peptides from Chum Salmon Skin Enhances Cutaneous Wound Healing and Angiogenesis in Rats. *J. Sci. Food Agric.* **2011**, *91*, 2173–2179. [CrossRef]
54. Lin, P.; Hua, N.; Hsu, Y.-C.; Kan, K.-W.; Chen, J.-H.; Lin, Y.-H.; Lin, Y.-H.; Kuan, C.-M. Oral Collagen Drink for Antiaging: Antioxidation, Facilitation of the Increase of Collagen Synthesis, and Improvement of Protein Folding and DNA Repair in Human Skin Fibroblasts. *Oxid. Med. Cell. Longev.* **2020**, *2020*, 8031795. [CrossRef] [PubMed]
55. Prokopová, A.; Pavlačková, J.; Mokrejš, P.; Gál, R. Collagen Hydrolysate Prepared from Chicken By-Product as a Functional Polymer in Cosmetic Formulation. *Molecules* **2021**, *26*, 2021. [CrossRef] [PubMed]
56. Brandao-Rangel, M.A.R.; Oliveira, C.R.; da Silva Olímpio, F.R.; Aimbire, F.; Mateus-Silva, J.R.; Chaluppe, F.A.; Vieira, R.P. Hydrolyzed Collagen Induces an Anti-Inflammatory Response That Induces Proliferation of Skin Fibroblast and Keratinocytes. *Nutrients* **2022**, *14*, 4975. [CrossRef] [PubMed]
57. Lee, M.; Kim, D.; Park, S.-H.; Jung, J.; Cho, W.; Yu, A.R.; Lee, J. Fish Collagen Peptide (Naticol®) Protects the Skin from Dryness, Wrinkle Formation, and Melanogenesis Both In Vitro and In Vivo. *Prev. Nutr. Food Sci.* **2022**, *27*, 423–435. [CrossRef]
58. Wauquier, F.; Boutin-Wittrant, L.; Bouvret, E.; Le Faouder, J.; Roux, V.; Macian, N.; Pickering, G.; Wittrant, Y. Benefits of Circulating Human Metabolites from Fish Cartilage Hydrolysate on Primary Human Dermal Fibroblasts, an Ex Vivo Clinical Investigation for Skin Health Applications. *Nutrients* **2022**, *14*, 5027. [CrossRef]
59. Bauza, E.; Oberto, G.; Berghi, A.; Dal, C.F.; Domloge, N. Collagen-like Peptide Exhibits a Remarkable Antiwrinkle Effect on the Skin When Topically Applied: In Vivo Study. *Int. J. Tissue React.* **2004**, *26*, 105–111.
60. Maia Campos, P.M.B.G.; Melo, M.O.; Siqueira César, F.C. Topical Application and Oral Supplementation of Peptides in the Improvement of Skin Viscoelasticity and Density. *J. Cosmet. Dermatol.* **2019**, *18*, 1693–1699. [CrossRef]
61. Sangsuwan, W.; Asawanonda, P. Four-Weeks Daily Intake of Oral Collagen Hydrolysate Results in Improved Skin Elasticity, Especially in Sun-Exposed Areas: A Randomized, Double-Blind, Placebo-Controlled Trial. *J. Dermatol. Treat.* **2021**, *32*, 991–996. [CrossRef]
62. Evans, M.; Lewis, E.D.; Zakaria, N.; Pelipyagina, T.; Guthrie, N. A Randomized, Triple-Blind, Placebo-Controlled, Parallel Study to Evaluate the Efficacy of a Freshwater Marine Collagen on Skin Wrinkles and Elasticity. *J. Cosmet. Dermatol.* **2021**, *20*, 825–834. [CrossRef]
63. Varani, J.; Warner, R.L.; Gharaee-Kermani, M.; Phan, S.H.; Kang, S.; Chung, J.H.; Wang, Z.Q.; Datta, S.C.; Fisher, G.J.; Voorhees, J.J. Vitamin A Antagonizes Decreased Cell Growth and Elevated Collagen-Degrading Matrix Metalloproteinases and Stimulates Collagen Accumulation in Naturally Aged Human Skin. *J. Investig. Dermatol.* **2000**, *114*, 480–486. [CrossRef] [PubMed]
64. Fisher, G.J.; Wang, Z.Q.; Datta, S.C.; Varani, J.; Kang, S.; Voorhees, J.J. Pathophysiology of Premature Skin Aging Induced by Ultraviolet Light. *N. Engl. J. Med.* **1997**, *337*, 1419–1428. [CrossRef] [PubMed]
65. Schwartz, S.R.; Hammon, K.A.; Gafner, A.; Dahl, A.; Guttman, N.; Fong, M.; Schauss, A.G. Novel Hydrolyzed Chicken Sternal Cartilage Extract Improves Facial Epidermis and Connective Tissue in Healthy Adult Females: A Randomized, Double-Blind, Placebo-Controlled Trial. *Altern. Ther. Health Med.* **2019**, *25*, 12–29.
66. Marini, A.; Grether-Beck, S.; Jaenicke, T.; Weber, M.; Burki, C.; Formann, P.; Brenden, H.; Schönlau, F.; Krutmann, J. Pycnogenol® Effects on Skin Elasticity and Hydration Coincide with Increased Gene Expressions of Collagen Type I and Hyaluronic Acid Synthase in Women. *Skin Pharmacol. Physiol.* **2012**, *25*, 86–92. [CrossRef]
67. Bolke, L.; Schlippe, G.; Gerß, J.; Voss, W. A Collagen Supplement Improves Skin Hydration, Elasticity, Roughness, and Density: Results of a Randomized, Placebo-Controlled, Blind Study. *Nutrients* **2019**, *11*, 2494. [CrossRef] [PubMed]
68. Lodén, M. Barrier Recovery and Influence of Irritant Stimuli in Skin Treated with a Moisturizing Cream. *Contact Dermat.* **1997**, *36*, 256–260. [CrossRef]
69. Draelos, Z.D. Active Agents in Common Skin Care Products. *Plast. Reconstr. Surg.* **2010**, *125*, 719–724. [CrossRef]

cosmetics

MDPI

Article

Photoprotective Effect and Potential Mechanisms of Gardeniae Fructus Extract in UVB-Irradiated HaCaT Cells

Kaile Zong [1], Xiang Li [2], Fangni Zhou [2], Junzi Dong [2], Qing Huang [1,*] and Jianxin Wu [1,*]

[1] Skin Health and Cosmetic Development & Evaluation Laboratory, China Pharmaceutical University, Nanjing 211198, China; 3320021289@stu.cpu.edu.cn

[2] Shanghai Junyu Biotechnology Group Co., Ltd., Shanghai 200050, China; 15150072089@163.com (X.L.); zhoufangni@juny.cn (F.Z.); djz@juny.cn (J.D.)

* Correspondence: huangqing@cpu.edu.cn (Q.H.); wujianxin@cpu.edu.cn (J.W.)

Abstract: Gardeniae Fructus (GF), the desiccative mature fruitage of *Gardenia jasminoides* J. Ellis (*G. jasminoides*), is a traditional herbal medicine in China with potential value against skin photodamage. However, the phytochemical basis and mechanisms underlying GF's anti-photodamage effects remain unclear. In this study, the chemical components in GF extract (GFE) were analyzed using ultra-high-performance liquid chromatography coupled with tandem mass spectrometry (UPLC-MS/MS), and iridoids were identified as the main components. The antioxidant, anti-inflammatory, and barrier-repair effects of GFE in UVB-induced photodamage were assessed through in vitro experiments. Additionally, the potential mechanisms of GFE against skin photodamage were predicted using proteomics and network pharmacology. The results showed that GFE significantly increased the levels of total superoxide dismutase (T-SOD), catalase (CAT), and glutathione peroxidase (GSH-Px) induced by UVB, while decreasing reactive oxygen species (ROS) and malondialdehyde (MDA) contents. GFE also inhibited the secretion of interleukin-6 (IL-6) and interleukin-1β (IL-1β). Additionally, GFE upregulated the expression of filaggrin (FLG), loricrin (LOR), and involucrin (IVL) in 3D epidermal skin models. Proteomic analysis and network pharmacology indicated that the iridoid components identified in GFE ameliorated UVB-induced damage probably by regulating cell cycle-related proteins and signaling pathways, though this requires further experimental confirmation. Overall, the results provide essential evidence to support the development of GFE as a skincare active ingredient.

Keywords: Gardeniae Fructus; plant extract; biological activity; UVB protection; antioxidant

1. Introduction

The skin is the body's protective barrier against environmental hazards, one of which, exposure to solar ultraviolet (UV) radiation, is unavoidable. UV can be categorized into three types: UVA (320 to 400 nm), UVB (290 to 320 nm), and UVC (200 to 290 nm). Among them, UVA and UVB account for 95% and 5%, respectively, of solar radiation at the Earth's surface, while most UVC is absorbed by the ozone layer [1,2]. UV exposure stimulates the production of vitamin D, which is essential in preventing rickets and osteomalacia [3]. However, excessive UV irradiation can cause skin damage. Although UVB accounts for a smaller proportion, it is the primary cause of acute damage to normal human skin [4]. Prolonged exposure to UVB radiation destructs the skin's antioxidant system via the generation of excessive reactive oxygen species (ROS), which results in oxidative stress damage due to alterations in the activity of multiple antioxidant enzymes, including superoxide dismutase (SOD), catalase (CAT), glutathione peroxidase (GSH-Px), and glutathione reductase (GR) [5–7]. Long-term overproduction of ROS can lead to cellular structural and functional damage, triggering the activation of multiple pro-inflammatory cytokines and pathways, which induces skin inflammation [8,9]. Meanwhile, UVB irradiation negatively impacts epidermal morphology and barrier function, causing changes in stratum corneum lipids and keratin structure [10–12]. Furthermore, although the epidermis of the skin absorbs almost all UVB, about 5–10% can still reach and affect the upper layer of the dermis [13]. Repetitive irradiation with UVB leads to significant changes in the three-dimensional structure of the skin's elastic fibers, resulting in decreased elasticity and the formation of wrinkles [14].

In recent years, there has been growing interest in developing functional foods and medicinal products with photoprotective properties for skin applications [15,16]. Gardeniae Fructus (GF), the dried mature fruit of *Gardenia jasminoides* J. Ellis (*G. jasminoides*), which belongs to the family Rubiaceae, has been recognized by the Ministry of Health of China as one of the first traditional Chinese medicinal herbs with dual purposes of food and medicine. In China, *G. jasminoides* is widely distributed in the eastern, central, south-western, and southern provinces. The primary bioactive ingredients in GF are iridoid glycosides, such as geniposide, genipin 1-gentiobioside, geniposidic acid, etc. [17]. Modern pharmacological studies have shown that GF exhibits various beneficial activities, including anti-inflammatory, antioxidant, anti-aging, and neuroprotective effects [18–22]. Additionally, GF is commonly utilized as an active ingredient in cosmetic products with soothing effects. Although GF has been reported to protect cells from UV irradiation, the underlying mechanisms have not been investigated [23,24].

This study aimed to investigate the protective effect of GF extract (GFE) on UVB-damaged HaCaT cells and explore the potential mechanisms of GFE's anti-photodamage properties. In this study, UPLC-MS/MS was employed to analyze the iridoid compounds in GFE. Subsequently, we evaluated the antioxidant, anti-inflammatory, and skin-barrier-repair effects of GFE on UV-irradiated HaCaT cells and 3D epidermal models. In addition, proteomics and network pharmacology were employed to elucidate the potential targets and mechanisms underlying the anti-photodamage effects of the iridoids identified in GFE.

2. Materials and Methods

2.1. Preparation and Analysis of GFE

GFE was provided by Shanghai Junyu Biotechnology Group Co., Ltd. (Shanghai, China) and prepared by the following steps [25]. GF powder was extracted twice with a 20-times volume of water at 80 °C for 2 h. Two filtrates were combined and decolorized via active carbon. Then, the liquid was purified with macroporous adsorption resin and desorbed by ethanol to enrich iridoids. The extract was filtered with a 0.45 μm microporous filter membrane and lyophilized to obtain GFE.

UPLC-MS/MS analyses were performed using a Vanquish UPLC system with an ACQUITY UPLC BEH Amide (2.1 mm × 100 mm, 1.7 μm; Waters, Milford, MA, USA) coupled to an Orbitrap Exploris 120 mass spectrometer. The MS/MS spectra were acquired by acquisition software (Xcalibur; Thermo Fisher Scientific, Waltham, MA, USA) in the information-dependent acquisition (IDA) mode. Identification of peaks was performed using the database provided by Shanghai BIOTREE biomedical technology Co., Ltd. (Shanghai, China) and the corresponding cleavage law matching method.

2.2. Cell Culture and UVB Irradiation

HaCaT cells were obtained from Zhejiang Meisen Cell Technology Co., Ltd. (Jinhua, China) and cultured in Dulbecco's modified Eagle's medium supplemented with 10%

fetal bovine serum, 100 units/mL penicillin, and 100 μg/mL streptomycin in a 5% CO_2 incubator at 37 °C.

UV crosslinker (Luyor, Shanghai, China) UVB lamps (302 nm) were applied for cell modeling at 10 mJ/cm^2 [26]. The HaCaT cells had a thin film of phosphate-buffered saline (PBS) on them and were irradiated by UVB. After irradiation, the medium was replaced with fresh medium and incubated for 24 h.

2.3. Detection of Reactive Oxygen Species (ROS), SOD, CAT, GSH-Px, and MDA

The HaCaT cells were treated with different concentrations of GFE (12.5, 25, and 50 μg/mL) and UVB irradiation and then stained by DCFH-DA (Beyotime, Shanghai, China). ROS contents were measured via a fluorescence microplate reader for the same number of cells and observed by fluorescence microscopy.

The levels of SOD, CAT, GSH-Px, and MDA were measured using a commercially available assay kit (Jiancheng Biological Engineering, Nanjing, China).

2.4. Enzyme-Linked Immunosorbent Assay

Contents of IL-6 and IL-1β were detected by enzyme-linked immunosorbent assay kits (TransGen Biotech, Beijing, China). HaCaT cells were irradiated by UVB, treated with GFE (12.5, 25, and 50 μg/mL), and analyzed for IL-6 and IL-1β at 24 h.

2.5. Immunofluorescence Staining

Epidermal skin models obtained from Regenovo were exposed to UVB (302 nm) at 400 mJ/cm^2 and treated with 50 μg/mL GFE for 24 h. Then, each skin model was frozen and cut. The sections were blocked and incubated with primary antibodies, including filaggrin (FLG), loricrin (LOR), and involucrin (IVL) (Abcam, Cambridge, UK), overnight. After washing with PBS/T three times, the sections were incubated with Alexa: 488-conjugated goat anti-rabbit IgG H&L (Abcam, Cambridge, UK) and DAPI (Beyotime Biotechnology, Shanghai, China). The sections were observed under a fluorescent inverted microscope.

2.6. Proteomic Analysis

The HaCaT cells were treated with 50 μg/mL GFE and irradiated by UVB for 24 h. Then, the HaCaT cells were collected and stored at −80 °C. For each tube containing cell precipitate, the contents were lysed, digested, and desalted. Then, liquid chromatography–mass spectrometry (LC-MS) analysis was performed using a Vanquish Neo UHPLC system coupled to an Orbitrap Astral mass spectrometer (Thermo Scientific, Waltham, MA, USA) for data-independent acquisition (DIA) analysis. Differentially expressed proteins (DEPs) were identified with a p-value less than 0.05 and a fold change greater than 1.5.

2.7. Network Pharmacology

UV photodamage targets were collected from the GeneCards, OMIM, and CTD databases with the keyword "photodamage". Meanwhile, the targets of the iridoids identified in GFE were predicted via SwissTargetPrediction and the CTD database. The intersected genes were then revealed by Venn diagrams.

The intersected targets were imported into STRING, the species was limited to "Homo sapiens", and confidence data > 0.7 were used to construct the PPI network. A visual PPI network and an "ingredient–target" network were constructed using Cytoscape 3.9.0 software.

Protein structures were downloaded from the PDB database, and ingredient structures were obtained from the Pubchem database. Then, molecular docking was plotted and displayed with Auto Dock Vina 1.25 and PyMol 2.3.0 [27].

2.8. Statistical Analysis

Statistical analysis was performed using GraphPad Prism 9 software. The student's *t*-test was performed for the differential analyses, and $p < 0.05$ was considered statistically significant.

3. Results

3.1. Analysis of GFE by UPLC-MS/MS

The total positive and negative ion chromatograms of GFE are shown in Figure 1A,B. Ten iridoids were identified in GFE (Figure 1C and Table S1): (1) genipin 1-gentiobioside, (2) geniposidic acid, (3) shanzhiside methyl ester, (4) deacetylasperulosidic acid, (5) genipin, (6) geniposide, (7) feretoside, (8) gardenoside, (9) asperuloside, and (10) shanzhiside.

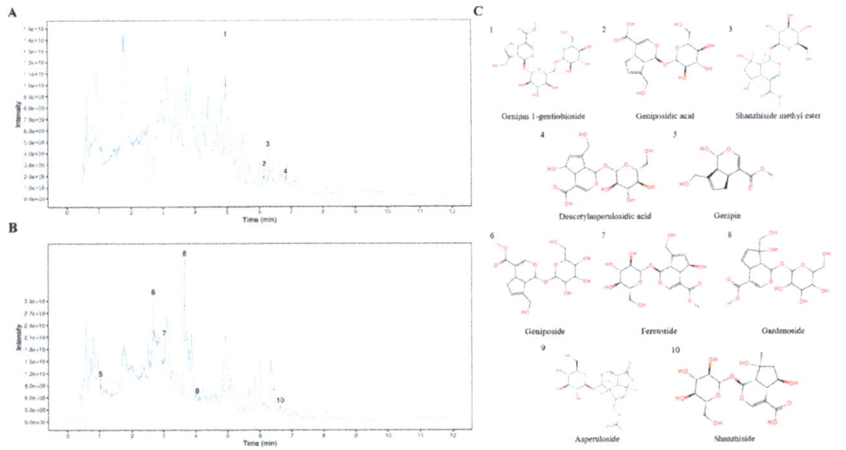

Figure 1. UPLC-MS/MS total ion chromatograms in the positive (**A**) and negative (**B**) ion models of GFE. (**C**) Molecular structure of constituents.

3.2. Effects of GFE on Antioxidants in HaCaT Cells

UVB irradiation induces oxidative damage and promotes intracellular ROS production. In this study, the intracellular ROS levels in HaCaT cells were stained via DCFH-DA. Compared to the UVB exposure group, the production of ROS was significantly inhibited in the GFE groups (Figure 2A,B). Additionally, the levels of T-SOD, CAT, and GSH-Px were decreased under UVB irradiation, while GFE effectively increased the contents of T-SOD, CAT, and GSH-Px (Figure 2C–E). Figure 2F demonstrates that the contents of MDA were decreased in GFE-treated HaCaT cells.

3.3. Effects of GFE on Anti-Inflammation in HaCaT Cells

Skin oxidative stress can further amplify skin inflammation induced by UVB exposure [28]. In the results, the levels of both IL-6 and IL-1β in the supernatants were increased under UVB irradiation, while treatment with GFE inhibited the secretion of IL-6 and IL-1β (Figure 3).

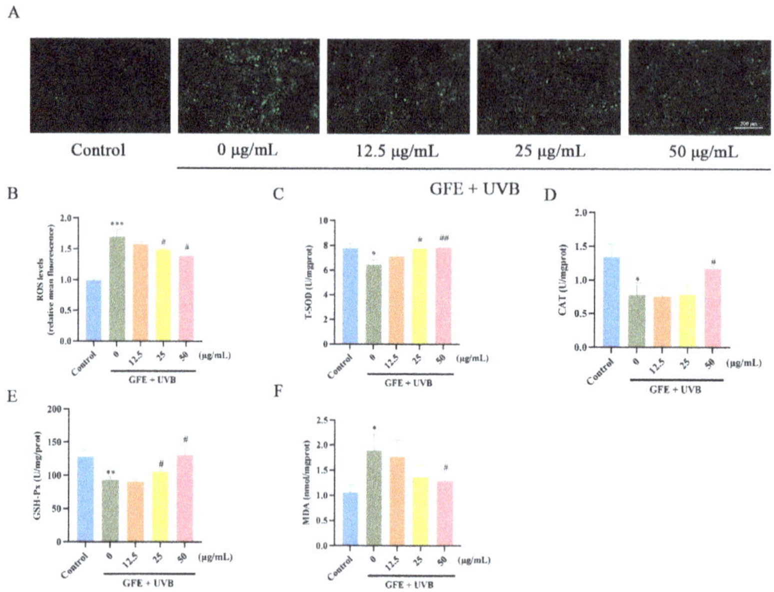

Figure 2. The effects of GFE on the oxidative stress in HaCaT cells. (**A**,**B**) The levels of intercellular ROS were stained with DCFH-DA, photographed by fluorescence microscopy, and measured under a fluorescence microplate reader; scale bar = 50 μm. The levels of (**C**) T-SOD, (**D**) CAT, (**E**) GSH-Px, and (**F**) MDA were determined by specialized test kits (mean ± SD, n = 3). * $p < 0.05$, ** $p < 0.01$, and *** $p < 0.001$ vs. the control; # $p < 0.05$ and ## $p < 0.01$ vs. UVB.

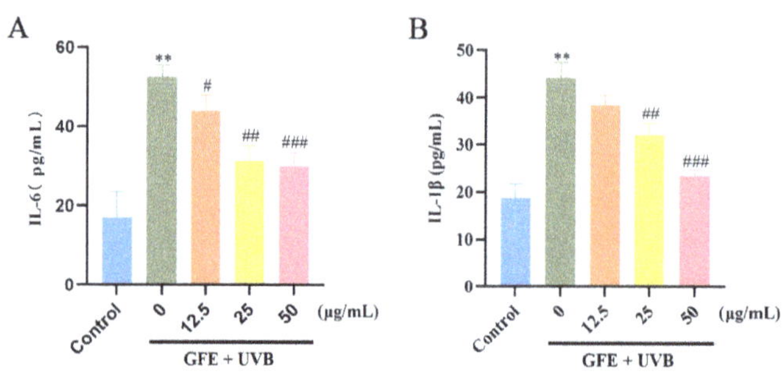

Figure 3. The effects of GFE on inflammation in HaCaT cells. The contents of (**A**) IL-6 and (**B**) IL-1β were quantified by ELISA (mean ± SD, n = 3). ** $p < 0.01$ vs. the control; # $p < 0.05$, ## $p < 0.01$, and ### $p < 0.001$ vs. UVB.

3.4. Effects of GFE on Barrier Repair in Three-Dimensional Epidermal Models

Filaggrin (FLG), loricrin (LOR), and involucrin (IVL) are the important barrier proteins that regulate skin permeability and prevent water evaporation [29]. Our research showed that UVB exposure diminished barrier function by downgrading the expression of FLG, IVL, and LOR in the three-dimensional epidermal models. After GFE treatment, the barrier functions of the skin models were restored and the expression of barrier-related proteins (FLG, IVL, and LOR) was increased (Figure 4).

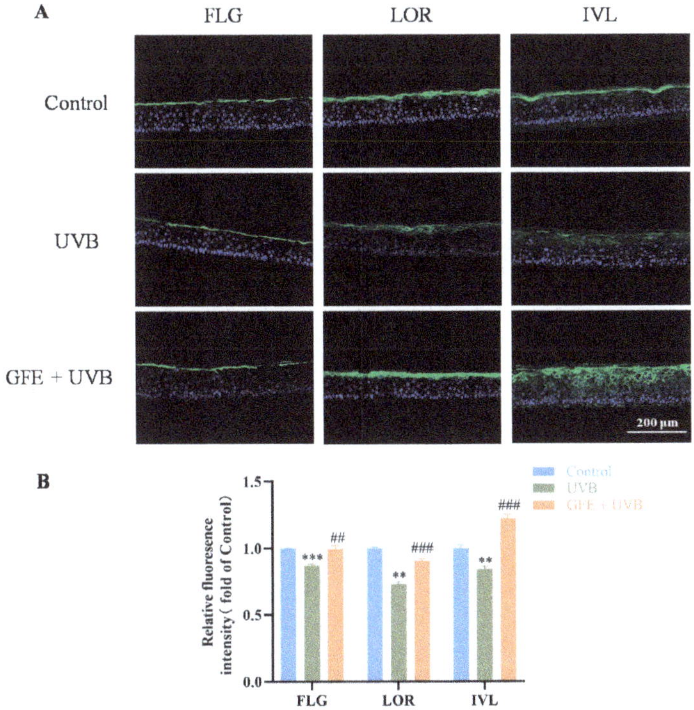

Figure 4. The effects of GFE on the barrier functions in three-dimensional epidermal models. (**A**) The levels of FLG, LOR, and IVL were stained by immunofluorescence and (**B**) measured via Imagj J 1.53q. Bar = 50 μm (mean ± SD, n = 3). ** $p < 0.01$ and *** $p < 0.001$ vs. the control; ## $p < 0.01$ and ### $p < 0.001$ vs. UVB.

3.5. Proteomic Analysis

DEPs among the control, UV, and UV-GFE groups were analyzed by proteomics. The results revealed that 234 and 186 DEPs increased and decreased between the control group and UV group, respectively (Figure 5A). Subsequently, cluster analysis of these DEPs was performed, and the results were visualized as a heatmap (Figure 5B). The control group and UV group were separated distinctly, suggesting that the identified DEPs were representative. Kyoto Encyclopedia of Genes and Genomes (KEGG) results revealed that UVB irradiation induced cellular photodamage mainly by interfering with the cell cycle (Figure 5C). Additionally, Gene Set Enrichment Analysis (GSEA) showed that UVB treatment could damage epidermal cells by activating the G2/M checkpoint (Figure 5D).

Between the UV-GFE and UV groups, 38 DEPs were upregulated and 36 DEPs were downregulated (Figure 6A). Gene Ontology (GO) analysis (Figure 6B) showed that GFE treatment exerted a significant impact on meiotic nuclear division, meiotic spindle assembly, and the meiotic cell cycle in the biological process (BP) category. Cell components (CCs) that were influenced mostly included axon hillocks, mitotic spindle poles, and spindle poles. In the molecular function (MF) category, the functions of retinal dehydrogenase activity, oxidoreductase activity, and aldehyde dehydrogenase (NAD+) activity were mainly altered. Next, reactome enrichment analysis revealed that the DEPs were enriched in the cell cycle (Figure 6C). Interestingly, the GESA results indicated that GFE treatment significantly downregulated the G2/M checkpoint, which was upregulated by UVB treatment (Figure 6D).

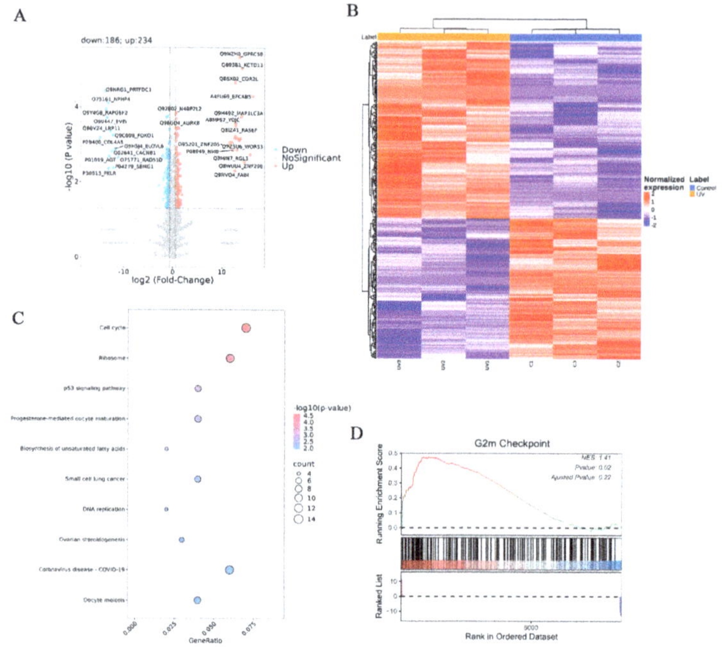

Figure 5. Proteomic analysis of the control and UV groups. (**A**) DEPs between the control group and UV group shown in a volcano plot; (**B**) cluster analysis of DEPs between the control group and UV group shown in a heatmap; (**C**) KEGG analysis between the control group and UV group; (**D**) GSEA between the control group and UV group.

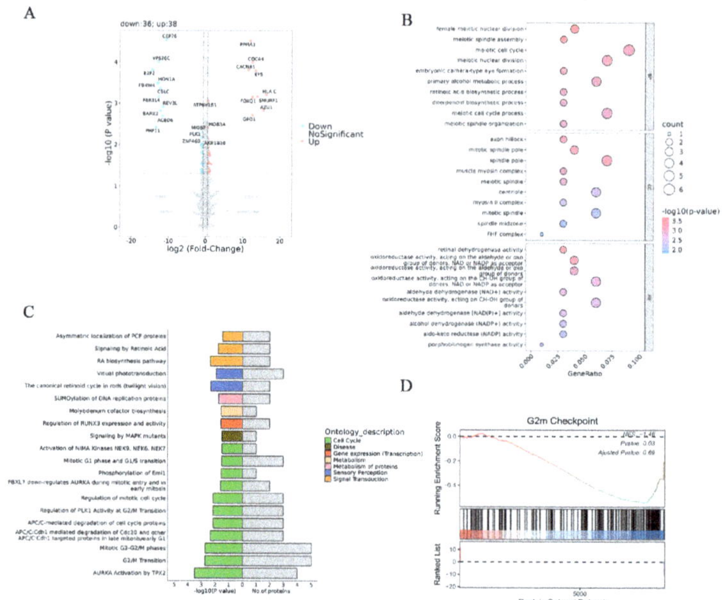

Figure 6. Proteomic analysis of UV and UV-GFE groups. (**A**) DEPs between the UV-GFE group and UV group shown in a volcano plot; (**B**) GO enrichment analysis of DEPs between the UV-GFE group and UV group; (**C**) reactome analysis of DEPs between the UV-GFE group and UV group; (**D**) GSEA between the UV-GFE group and UV group.

3.6. Network Pharmacology

After removing duplicates, a total of 201 genes associated with photodamage were collected from the GeneCards, OMIM, and CTD databases. Additionally, 203 genes related to iridoids in GFE were predicted using SwissTargetPrediction and the CTD database. A Venn diagram showed that 24 targets were intersected between GFE and photodamage (Figure 7A).

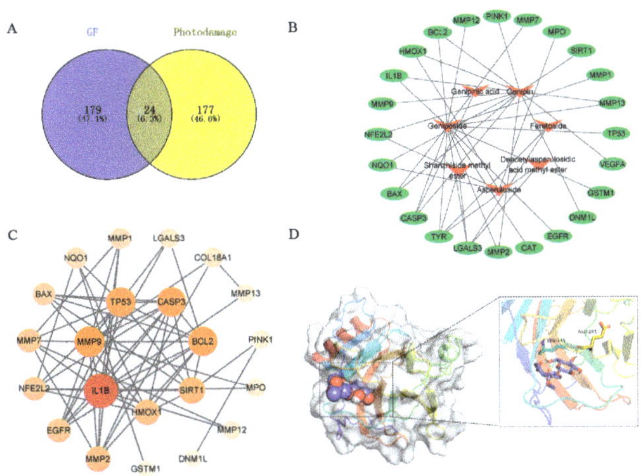

Figure 7. Network pharmacology of GF and photodamage. (**A**) Venn diagram; (**B**) "component–target" network; (**C**) PPI network; (**D**) molecular docking of geniposide–p53.

Subsequently, a "component–target" network was constructed and visualized using Cytoscape 3.9.0 (Figure 7B). The red polygon nodes represent the compounds, the green ellipsis nodes represent the genes, and each edge indicates an interaction between them. The network consisted of 31 nodes and 40 edges, with geniposide being associated with the most targets.

In order to predict the potential targets of GFE anti-photodamage, PPI network analysis was performed. The Centiscape analysis of the key targets is shown in Table 1. PPI network results revealed the genes with values above the average for degree centrality (5.64), betweenness centrality (0.055), and closeness centrality (0.497) obtained from the Centiscape analysis, including IL1B, MMP9, TP53, BCL2, and SIRT1.

Table 1. Centiscape analysis of the key targets in the PPI network.

No.	Target	Degree Centrality	Betweenness Centrality	Closeness Centrality
1	IL1B	14	0.302057308	0.7
2	MMP9	11	0.124122346	0.6
3	TP53	10	0.138510101	0.636363636
4	BCL2	10	0.079234179	0.617647059
5	SIRT1	8	0.187934962	0.583333333

Based on network pharmacology and proteomics, geniposide and TP53 (p53) were selected for molecular docking. Typically, a binding energy less than −5 kcal/mol signifies an excellent binding affinity. The results indicated that geniposide–p53 had two hydrogen bonds with Leu145 and Glu221 and exhibited a favorable binding activity, with a binding energy of −6.0 kcal/mol.

4. Discussion

Repeated or long-term exposure to UVB can induce skin photodamage due to its penetration of the epidermis and upper dermis, manifesting as sunburn, tanning, immunosuppression, and the release of pro-inflammatory mediators [30]. Keratinocytes are the most predominant cells in the epidermis of the skin. Researchers have revealed that UVB radiation could lead to DNA damage and accumulation of ROS in keratinocytes, resulting in cell death [31]. Additionally, apoptosis of keratinocytes is accompanied by inflammation and disruption of the epidermal barrier [32]. Therefore, it is essential to find a safe and effective treatment to prevent skin photodamage. In recent years, researchers have shown an increasing interest in the application of natural products in the cosmetic industry [33]. For example, ferulic acid, caffeine, and rutin have shown excellent efficacy in skin photoprotection and sunscreen cosmetic applications [34–37]. GF, with a positive effect on anti-photoaging, has garnered our attention.

Iridoids are a special class of monoterpenoids found in GFE, serving as some of the main chemical components with numerous pharmacological activities, such as antioxidant, anti-inflammatory, and hypoglycemic properties [38–41]. In this study, ten iridoids in GFE were identified via LC-MS/MS. Among them, geniposide is the main active ingredient in GFE and has the largest content [25]. To further evaluate the effects of GFE against UVB-induced skin damage, a UVB-irradiated HaCaT cell model was established. Oxidative stress and inflammation are the main negative effects caused by UVB. UVB irradiation induced oxidative stress damage in HaCaT cells, which presented as a decrease in antioxidant enzyme (T-SOD, CAT, and GSH-Px) levels and an increase in ROS and lipid peroxides (MDA). GFE treatment effectively combated UVB-induced cellular oxidative damage. Meanwhile, GFE inhibited the IL-6 and IL-1β levels caused by UVB exposure. UVB-induced skin photodamage is accompanied by impairment of the skin barrier, and skin barrier function is closely associated with the terminal differentiation-associated proteins, such as FLG, LOR, and IVL [42]. In this study, UVB-irradiated epidermal models exhibited significant downregulation of FLG, LOR, and IVL expression, while GFE effectively ameliorated the UVB-induced epidermal barrier damage by upregulating the expression of these barrier-related proteins.

Recently, proteomics has served as a crucial tool for uncovering the potential targets in TCM. In our studies, proteomic analysis indicated that UVB radiation affected various cell cycle-related proteins in HaCaT cells. Cell cycle checkpoints are typically regulated by a variety of cell cycle regulatory proteins [43]. The activation of targets in the G2/M checkpoint also confirmed the damage of keratinocyte DNA by UVB radiation, which is one of the main factors affecting DNA synthesis in keratinocytes [44]. Between the UV and UV-GFE groups, reactome enrichment analysis revealed that the UV-GFE group mitigated UVB-induced skin photodamage mainly by affecting the cell cycle. In addition, GFE treatment improved UVB-induced cell cycle arrest at the G2/M phase in HaCaT cells. Combined with the results of pharmacological experiments, GFE might act on multiple targets and cause complex pharmacological changes, resulting in a comprehensive effect on UVB-induced skin damage.

The network pharmacology method is widely utilized to explore the complex relationship between TCM and targets. The results of topological analyses indicated that geniposide was associated with more targets. Geniposide is the most abundant iridoid in GFE, and it can exert anti-photoaging effects by inhibiting oxidative stress in human skin cells [24]. Additionally, it can regulate the G2/M phase and inhibit inflammatory responses to improve apoptosis, suggesting that it might be the main active component in GFE against photodamage [45]. PPI network results revealed that apoptosis- and cell cycle-related targets, such as TP53, CASP3, and BCL2, were essential targets in GFE-mediated anti-photodamage, which corresponded to the proteomic results. TP53 (p53), the key target affecting the cell cycle and apoptosis, might play an essential role in regulating photodamage [46]. In order to demonstrate the binding modes between geniposide and p53, molecular docking was performed. Our research found that the geniposide–p53 complex showed stable binding, with a binding energy of -6.0 kcal/mol. The findings of this study lay a foundation for the further development of GFE in cosmetic raw materials and products capable of repairing damage induced by UVB. The anti-photodamage components and the mechanisms of GFE still need to be further verified.

5. Conclusions

Our study reveals the effects and potential mechanisms of GFE in ameliorating UVB-induced photodamage in HaCaT cells based on pharmacological experiments, proteomics, and network pharmacology. GFE treatment effectively ameliorates UVB-induced skin photodamage by regulating the cell cycle, including the G2/M checkpoint and p53 pathways, indicating its potential as an active ingredient against skin photoaging.

Supplementary Materials: The following supporting information can be downloaded at https://www.mdpi.com/article/10.3390/cosmetics12020072/s1: Table S1: Ten iridoids identified in GFE.

Author Contributions: Conceptualization, J.W., Q.H., and K.Z.; methodology, software, validation, and visualization, K.Z., X.L., and F.Z.; investigation and data curation, J.D.; writing—original draft preparation, K.Z. and J.D.; writing—review and editing, Q.H. and J.W.; project administration, funding acquisition, J.W., Q.H., and J.D. All authors have read and agreed to the published version of the manuscript.

Funding: This research was supported by the 2019 high-level talent introduction scientific research start-up fund of China Pharmaceutical University (3150020056).

Institutional Review Board Statement: Not applicable.

Informed Consent Statement: Not applicable.

Data Availability Statement: The data that support the findings of this study are available from the corresponding author upon reasonable request.

Acknowledgments: The authors are grateful for the research material provided by Shanghai Junyu Biotechnology Group Co., Ltd., located in Shanghai, China, and wish to thank Hangzhou CIRS KEXIN Biotech Co., Ltd. for its participation in the conduction of the studies.

Conflicts of Interest: X.L., F.Z., and J.D. were employed by the Shanghai Junyu Biotechnology Group Co., Ltd. The remaining authors declare that the research was conducted in the absence of any commercial or financial relationships that could be construed as a potential conflict of interest.

References

1. Wong, Q.Y.A.; Chew, F.T. Defining skin aging and its risk factors: A systematic review and meta-analysis. *Sci. Rep.* **2021**, *11*, 22075. [CrossRef] [PubMed]
2. Gan, C.; Liu, T.; Jia, X.; Huang, X.; Qin, X.; Wang, X.; Sheng, J.; Xu, H. ATP supplementation suppresses UVB-induced photoaging in HaCaT cells via upregulation of expression of SIRT3 and SOD2. *Ski. Res. Technol.* **2023**, *29*, e13303. [CrossRef] [PubMed]
3. Passeron, T.; Bouillon, R.; Callender, V.; Cestari, T.; Diepgen, T.L.; Green, A.C.; van der Pols, J.C.; Bernard, B.A.; Ly, F.; Bernerd, F.; et al. Sunscreen photoprotection and vitamin D status. *Br. J. Dermatol.* **2019**, *181*, 916–931. [CrossRef]
4. de la Coba, F.; Aguilera, J.; Korbee, N.; de Gálvez, M.V.; Herrera-Ceballos, E.; Álvarez-Gómez, F.; Figueroa, F.L. UVA and UVB Photoprotective Capabilities of Topical Formulations Containing Mycosporine-like Amino Acids (MAAs) through Different Biological Effective Protection Factors (BEPFs). *Mar. Drugs* **2019**, *17*, 55. [CrossRef]
5. Solano, F. Photoprotection and Skin Pigmentation: Melanin-Related Molecules and Some Other New Agents Obtained from Natural Sources. *Molecules* **2020**, *25*, 1537. [CrossRef]
6. Park, J.; Kim, D.; Lee, M.; Han, S.; Jun, W.; Jung, H.M.; Koo, Y.K.; Na, G.H.; Han, S.H.; Han, J.; et al. Enzyme-Treated Caviar Prevents UVB Irradiation-Induced Skin Photoaging. *Mar. Drugs* **2022**, *20*, 685. [CrossRef]

7. Md Jaffri, J. Reactive Oxygen Species and Antioxidant System in Selected Skin Disorders. *Malays. J. Med. Sci.* **2023**, *30*, 7–20. [CrossRef]
8. Prasanth, M.I.; Sivamaruthi, B.S.; Chaiyasut, C.; Tencomnao, T. A Review of the Role of Green Tea (*Camellia sinensis*) in Antiphotoaging, Stress Resistance, Neuroprotection, and Autophagy. *Nutrients* **2019**, *11*, 474. [CrossRef]
9. Hussain, T.; Tan, B.; Yin, Y.; Blachier, F.; Tossou, M.C.; Rahu, N. Oxidative Stress and Inflammation: What Polyphenols Can Do for Us? *Oxidative Med. Cell. Longev.* **2016**, *2016*, 7432797. [CrossRef]
10. Ansary, T.M.; Hossain, M.R.; Kamiya, K.; Komine, M.; Ohtsuki, M. Inflammatory Molecules Associated with Ultraviolet Radiation-Mediated Skin Aging. *Int. J. Mol. Sci.* **2021**, *22*, 3947. [CrossRef]
11. Diao, P.; He, H.; Tang, J.; Xiong, L.; Li, L. Natural compounds protect the skin from airborne particulate matter by attenuating oxidative stress. *Biomed. Pharmacother.* **2021**, *138*, 111534. [CrossRef] [PubMed]
12. Biniek, K.; Levi, K.; Dauskardt, R.H. Solar UV radiation reduces the barrier function of human skin. *Proc. Natl. Acad. Sci. USA* **2012**, *109*, 17111–17116. [CrossRef]
13. Mavrogonatou, E.; Angelopoulou, M.; Rizou, S.V.; Pratsinis, H.; Gorgoulis, V.G.; Kletsas, D. Activation of the JNKs/ATM-p53 axis is indispensable for the cytoprotection of dermal fibroblasts exposed to UVB radiation. *Cell Death Dis.* **2022**, *13*, 647. [CrossRef]
14. Mayangsari, E.; Mustika, A.; Nurdiana, N.; Samad, N.A. Comparison of UVA vs UVB Photoaging Rat Models in Short-term Exposure. *Med. Arch.* **2024**, *78*, 88–91. [CrossRef]
15. Cavinato, M.; Waltenberger, B.; Baraldo, G.; Grade, C.V.C.; Stuppner, H.; Jansen-Dürr, P. Plant extracts and natural compounds used against UVB-induced photoaging. *Biogerontology* **2017**, *18*, 499–516. [CrossRef]
16. Tan, H.; Ren, H.; Chai, J.; Zhai, C.; Li, T.; Zhou, X.; Lee, J.; Li, X.; Zhao, Y. Protective effect of ginseng berry saponin conversion products on skin photodamage caused by UVB in vitro and in vivo. *Food Res. Int.* **2024**, *198*, 115379. [CrossRef]
17. Fu, H.; Zhang, Y.; An, Q.; Wang, D.; You, S.; Zhao, D.; Zhang, J.; Wang, C.; Li, M. Anti-Photoaging Effect of Rhodiola rosea Fermented by Lactobacillus plantarum on UVA-Damaged Fibroblasts. *Nutrients* **2022**, *14*, 2324. [CrossRef]
18. Guo, S.; Bao, L.; Li, C.; Sun, J.; Zhao, R.; Cui, X. Antiviral activity of iridoid glycosides extracted from Fructus Gardeniae against influenza A virus by PACT-dependent suppression of viral RNA replication. *Sci. Rep.* **2020**, *10*, 1897. [CrossRef]
19. Kim, S.H.; Shin, M.R.; Lee, A.R.; Seo, B.I.; Park, H.J.; Roh, S.S. Improvement of Inflammation through Antioxidant Pathway of Gardeniae Fructus 50% EtOH Extract (GE) from Acute Reflux Esophagitis Rats. *BioMed Res. Int.* **2020**, *2020*, 4826176. [CrossRef]
20. Nguyen, L.T.H.; Ahn, S.H.; Nguyen, U.T.; Yang, I.J.; Shin, H.M. Geniposide, a Principal Component of Gardeniae Fructus, Protects Skin from Diesel Exhaust Particulate Matter-Induced Oxidative Damage. *Evid. -Based Complement. Altern. Med.* **2021**, *2021*, 8847358. [CrossRef]
21. De Tollenaere, M.; Chapuis, E.; Martinez, J.; Paulus, C.; Dupont, J.; Don Simoni, E.; Robe, P.; Sennelier-Portet, B.; Auriol, D.; Scandolera, A.; et al. Gardenia jasminoides Extract, with a Melatonin-like Activity, Protects against Digital Stress and Reverses Signs of Aging. *Int. J. Mol. Sci.* **2023**, *24*, 4948. [CrossRef] [PubMed]
22. Zhang, H.; Lai, Q.; Li, Y.; Liu, Y.; Yang, M. Learning and memory improvement and neuroprotection of Gardenia jasminoides (*Fructus gardenia*) extract on ischemic brain injury rats. *J. Ethnopharmacol.* **2017**, *196*, 225–235. [CrossRef] [PubMed]
23. Kato, T.; Hino, S.; Horie, N.; Shimoyama, T.; Kaneko, T.; Kusama, K.; Sakagami, H. Anti-UV activity of Kampo medicines and constituent plant extracts: Re-evaluation with skin keratinocyte system. *Vivo* **2014**, *28*, 571–578.
24. Shin, D.; Lee, S.; Huang, Y.H.; Lim, H.W.; Lee, Y.; Jang, K.; Cho, Y.; Park, S.J.; Kim, D.D.; Lim, C.J. Protective properties of geniposide against UV-B-induced photooxidative stress in human dermal fibroblasts. *Pharm. Biol.* **2018**, *56*, 176–182. [CrossRef]
25. Dong, J.; Li, B.; Xu, P.; Kong, L.; Zhang, W.; Wu, D.; Tian, Y.; Wang, X. Skin barrier repair efficacy and safety evaluation of an extract from Gardenia jasminoides J. Ellis. *Nat. Prod. Res.* **2024**, 1–7. [CrossRef]
26. Ke, H.; Zhang, X.; Liang, S.; Zhou, C.; Hu, Y.; Huang, Q.; Wu, J. Study on the anti-skin aging effect and mechanism of Sijunzi Tang based on network pharmacology and experimental validation. *J. Ethnopharmacol.* **2024**, *333*, 118421. [CrossRef]
27. Zong, K.; Xu, K.; Zhang, X.; Wang, P.; Wang, Z.; Yang, S.; Li, H.; Ke, H.; He, S.; Hu, Y.; et al. Exploring the mechanism of Huangqin Tang against skin lipid accumulation through network pharmacology and experimental validation. *J. Ethnopharmacol.* **2023**, *313*, 116581. [CrossRef]
28. Baek, J.; Kim, J.H.; Park, J.; Kim, D.H.; Sa, S.; Han, J.S.; Kim, W. 1-Kestose Blocks UVB-Induced Skin Inflammation and Promotes Type I Procollagen Synthesis via Regulating MAPK/AP-1, NF-κB and TGF-β/Smad Pathway. *J. Microbiol. Biotechnol.* **2024**, *34*, 911–919. [CrossRef]
29. Jang, A.Y.; Choi, J.; Rod-In, W.; Choi, K.Y.; Lee, D.H.; Park, W.J. In Vitro Anti-Inflammatory and Skin Protective Effects of Codium fragile Extract on Macrophages and Human Keratinocytes in Atopic Dermatitis. *J. Microbiol. Biotechnol.* **2024**, *34*, 940–948. [CrossRef]
30. He, H.; Huang, W.; Xiong, L.; Ma, C.; Wang, Y.; Sun, P.; Shi, D.; Li, L.; Yan, H.; Wu, Y. FUNDC1-mediated mitophagy regulates photodamage independently of the PINK1/Parkin-dependent pathway. *Free Radic. Biol. Med.* **2024**, *225*, 630–640. [CrossRef]
31. Tang, Z.; Tong, X.; Huang, J.; Liu, L.; Wang, D.; Yang, S. Research progress of keratinocyte-programmed cell death in UV-induced Skin photodamage. *Photodermatol. Photoimmunol. Photomed.* **2021**, *37*, 442–448. [CrossRef] [PubMed]

32. Cheng, W.; Di, F.; Li, L.; Pu, C.; Wang, C.; Zhang, J. Anti-Photodamage Effect of Agaricus blazei Murill Polysaccharide on UVB-Damaged HaCaT Cells. *Int. J. Mol. Sci.* **2024**, *25*, 4676. [CrossRef]
33. Milutinov, J.; Pavlović, N.; Ćirin, D.; Atanacković Krstonošić, M.; Krstonošić, V. The Potential of Natural Compounds in UV Protection Products. *Molecules* **2024**, *29*, 5409. [CrossRef]
34. Morocho-Jácome, A.L.; Freire, T.B.; de Oliveira, A.C.; de Almeida, T.S.; Rosado, C.; Velasco, M.V.R.; Baby, A.R. In vivo SPF from multifunctional sunscreen systems developed with natural compounds-A review. *J. Cosmet. Dermatol.* **2021**, *20*, 729–737. [CrossRef]
35. Peres, D.D.; Sarruf, F.D.; de Oliveira, C.A.; Velasco, M.V.R.; Baby, A.R. Ferulic acid photoprotective properties in association with UV filters: Multifunctional sunscreen with improved SPF and UVA-PF. *J. Photochem. Photobiol. B Biol.* **2018**, *185*, 46–49. [CrossRef]
36. Rosado, C.; Tokunaga, V.K.; Sauce, R.; de Oliveira, C.A.; Sarruf, F.D.; Parise-Filho, R.; Maurício, E.; de Almeida, T.S.; Velasco, M.V.R.; Baby, A.R. Another Reason for Using Caffeine in Dermocosmetics: Sunscreen Adjuvant. *Front. Physiol.* **2019**, *10*, 519. [CrossRef]
37. Tomazelli, L.C.; de Assis Ramos, M.M.; Sauce, R.; Cândido, T.M.; Sarruf, F.D.; de Oliveira Pinto, C.A.S.; de Oliveira, C.A.; Rosado, C.; Velasco, M.V.R.; Baby, A.R. SPF enhancement provided by rutin in a multifunctional sunscreen. *Int. J. Pharm.* **2018**, *552*, 401–406. [CrossRef]
38. Wang, L.; Chen, S.; Liu, S.; Biu, A.M.; Han, Y.; Jin, X.; Liang, C.; Liu, Y.; Li, J.; Fang, S.; et al. A comprehensive review of ethnopharmacology, chemical constituents, pharmacological effects, pharmacokinetics, toxicology, and quality control of gardeniae fructus. *J. Ethnopharmacol.* **2024**, *320*, 117397. [CrossRef]
39. Sun, K.X.; Chen, Y.Y.; Li, Z.; Zheng, S.J.; Wan, W.J.; Ji, Y.; Hu, K. Genipin relieves diabetic retinopathy by down-regulation of advanced glycation end products via the mitochondrial metabolism related signaling pathway. *World J. Diabetes* **2023**, *14*, 1349–1368. [CrossRef]
40. Li, H.; Yang, D.H.; Zhang, Y.; Zheng, F.; Gao, F.; Sun, J.; Shi, G. Geniposide suppresses NLRP3 inflammasome-mediated pyroptosis via the AMPK signaling pathway to mitigate myocardial ischemia/reperfusion injury. *Chin. Med.* **2022**, *17*, 73. [CrossRef]
41. Chen, Y.I.; Cheng, Y.W.; Tzeng, C.Y.; Lee, Y.C.; Chang, Y.N.; Lee, S.C.; Tsai, C.C.; Chen, J.C.; Tzen, J.T.; Chang, S.L. Peroxisome proliferator-activated receptor activating hypoglycemic effect of Gardenia jasminoides Ellis aqueous extract and improvement of insulin sensitivity in steroid induced insulin resistant rats. *BMC Complement. Altern. Med.* **2014**, *14*, 30. [CrossRef] [PubMed]
42. Jing, C.; Guo, J.; Li, Z.; Xu, X.; Wang, J.; Zhai, L.; Liu, J.; Sun, G.; Wang, F.; Xu, Y.; et al. Screening and Research on Skin Barrier Damage Protective Efficacy of Different Mannosylerythritol Lipids. *Molecules* **2022**, *27*, 4648. [CrossRef] [PubMed]
43. Faurschou, A.; Gniadecki, R.; Calay, D.; Wulf, H.C. TNF-alpha impairs the S-G2/M cell cycle checkpoint and cyclobutane pyrimidine dimer repair in premalignant skin cells: Role of the PI3K-Akt pathway. *J. Investig. Dermatol.* **2008**, *128*, 2069–2077. [CrossRef]
44. Chilampalli, C.; Guillermo, R.; Zhang, X.; Kaushik, R.S.; Young, A.; Zeman, D.; Hildreth, M.B.; Fahmy, H.; Dwivedi, C. Effects of magnolol on UVB-induced skin cancer development in mice and its possible mechanism of action. *BMC Cancer* **2011**, *11*, 456. [CrossRef]
45. Chen, W.J.; Bao, T.Z.; Chen, K.; Zhu, C.M.; Wan, F.; Tan, Y.L.; Yan, F. [Effects of geniposide on SNP-induced apoptosis of chondrocyte and cell cycle]. *China J. Orthop. Traumatol.* **2013**, *26*, 232–235.
46. Liu, F.J.; Chen, X.Y.; Yang, J.; Zhao, Z.; Wang, Q.L.; Li, P.; Jiang, Y.; Li, H.J. Revealing active components and action mechanism of Fritillariae Bulbus against non-small cell lung cancer through spectrum-effect relationship and proteomics. *Phytomedicine* **2023**, *110*, 154635. [CrossRef]

cosmetics

MDPI

Article

Exploring Skin Longevity Pathways: *Rosa hybrid* Extract-Mediated AMP-Activated Protein Kinase Activation, Antioxidant, and Autophagic Mechanisms in Human Keratinocytes

Olivier Gouin [1], Andrea Cavagnino [1], Gayané Azadiguian [2], Sibylle Jäger [3], Gilles Comte [4], Mohammed Bendahmane [5], Lionel Breton [6], Martin A. Baraibar [1,*,†] and Annie F. Black [2,†]

1 Oxiproteomics, 2 Rue Antoine Etex, 94000 Créteil, France
2 Lancôme, 62 Quai Charles Pasqua, 92300 Levallois Perret, France; gayane.azadiguian@loreal.com (G.A.)
3 L'Oréal Recherche & Innovation, 1 Av. Eugène Schueller, 93600 Aulnay-sous-Bois, France; sibylle.jager@loreal.com
4 UMR Ecologie Microbienne, CNRS, INRAE, VetAgro Sup, UCBL, Université de Lyon, 69007 Lyon, France
5 Laboratoire Reproduction et Développement des Plantes, INRAE-CNRS-Lyon, Ecole Normale Supérieure de Lyon, 69007 Lyon, France
6 CILIA Consulting, 18 Av de Paris, 78000 Versailles, France; btwocg@gmail.com
* Correspondence: martin.baraibar@oxiproteomics.fr
† These authors contributed equally to this work.

Abstract: The skin is continuously exposed to environmental changes, rendering it vulnerable to damage from external stressors that contribute to premature skin aging. This study aims to explore skin longevity pathways stimulated by a rose extract (RE) derived from petals. Human keratinocytes treated with RE exhibited a significant increase in NRF2 (NF-E2-related factor 2; ≈2–4% of induction) and LAMP2A (Lysosome-Associated Membrane Protein 2A; ≈6–12% of induction) levels. The presence of RE significantly mitigated the increase in carbonylation levels (≈34–37% of protection) and the number of labeled $P16^{INK4A}$ cells (≈60–72% of protection), associated with proliferation arrest, both induced by exposure to BAP (Benzo[*a*]pyrene) coupled with UV-A (Ultraviolet A) irradiation. The beneficial effects mediated by RE were inhibited by Compound C, a specific AMPK inhibitor (AMP-activated protein kinase). The involvement of the AMPK pathway in mediating the beneficial effects of RE has been confirmed by assessing its activation through the evaluation of its phosphorylation state which was significantly elevated in the presence of RE compared to the stress condition. In conclusion, the activation of the AMPK pathway enhances antioxidant defenses and promotes autophagy. This dual action, mediated by RE, helps protect skin cells from oxidative damage and senescence while maintaining proteostasis, skin integrity, and cellular proliferation under pollution-induced stress (BAP + UV-A). These findings highlight the potential in mitigating age-related skin changes through the modulation of longevity pathways.

Keywords: rose extract; autophagy; senescence; AMPK; carbonylation; proteostasis; skin; longevity

1. Introduction

The skin is not only a vital protective barrier but also functions as a dynamic sensory organ, forming the primary line of defense against environmental aggressors. This complex organ is perpetually exposed to a myriad of environmental changes, making it susceptible

https://doi.org/10.3390/cosmetics12020057

to frequent damage from various external stressors. As a responsive shield, the skin actively counters and safeguards the body from these continual challenges, playing a critical role in maintaining overall health and well-being. To maintain its biological functions, the skin possesses the remarkable ability to repair and regenerate its components when damaged, involving mechanisms such as detoxification and the elimination of damaged proteins. Skin aging is a complex biological process influenced by both intrinsic factors, predominantly genetic, and extrinsic factors that affect an individual from conception to death, contributing to the biological and clinical signs of skin aging [1]. Environmental factors are key actors in premature aging, impairing skin homeostasis, which results in the accumulation of damages and a reduced ability to repair them. This decline is further contrasted by a cell's internal energy sensor, the AMPK (AMP-activated protein kinase) pathway, which can trigger cellular processes that promote longevity. However, with age, AMPK function can become impaired, potentially contributing to the diminished ability of skin cells to maintain themselves [2]. Among environmental factors, Ultraviolet Radiation (UVR) is the main actor in skin dysfunction, inducing oxidative damage and harmful genetic modifications. The urban environment, in which the skin is exposed to pollutants (particulate matters, volatile organic compounds (VOCs), and polycyclic aromatic hydrocarbons (PAHs)), contributes to premature skin aging and reduces the functionality of the skin [3–5]. Other exogenous aggressors, such as dysbiosis (*Cutibacterium acnes* pathogenic subtypes or *Staphylococcus aureus*) [6] and brutal changes in environmental temperature are also known to be involved in oxidative stress induction within the skin. Oxidative stress results from an imbalance between the production and accumulation of oxygen reactive species (ROS) in cells and tissues and impaired ability of skin to detoxify these reactive products due to the accumulation of oxidative modification of macromolecules [7]. An accumulation of damaged proteins (proteostasis collapse) is associated with several age-related morbidities, such as amyotrophic lateral sclerosis (ALS), Alzheimer's disease, Parkinson's disease, and cataract [8].

Among these modifications, protein carbonylation is a harmful irreversible modification identified as a major hallmark of severe oxidative damage related to oxidative disorders [9–12]. This modification is associated with protein dysfunction and misfolding and must be cleared to maintain skin homeostasis. Carbonylation can negatively affect, or totally abrogate, protein catalytic functions and may trigger the formation of potentially cytotoxic protein aggregates.

Antioxidant defenses, including enzymes, such as superoxide dismutases (SODs), glutathione peroxidases (GPXs), peroxiredoxins (PRDXs), and catalase (CAT), as well as non-enzymatic molecules, such as glutathione (GSH), protect skin proteins from oxidative stress. Among these antioxidant defenses, the NF-E2-related factor 2 (NRF2) is the master cytoprotective and antioxidant regulating the expression of ~250 genes by the antioxidant response element (ARE). Products of these genes are involved in antioxidant response, redox homeostasis, the detoxification of toxic compounds, mitochondrial biogenesis, and many other processes [9]. Once damaged by oxidative stress, the intracellular cellular components' degradation can be mediated by autophagy.

Autophagy, essential for cell survival, is one of the clearance mechanisms involved in the degradation of damaged cellular components. The NRF2 target LAMP2A (Lysosome-Associated Membrane Protein 2A) [13] plays a significant role in chaperone-mediated autophagy (CMA) by facilitating the recognition, binding, translocation, and degradation of specific proteins targeted for lysosomal degradation. Its presence on the lysosomal membrane is essential for the efficient functioning of CMA, ensuring the removal of damaged proteins from the cell. The autophagy pathway is finely regulated and requires the activation of the AMP-activated protein kinase (AMPK) signaling pathway, which coordinates cell growth, autophagy, and metabolism.

Furthermore, a functional association between the AMPK and NRF2 pathways has been described to underlie the crosstalk between energy homeostasis and oxidative clearance upon AMPK activation upstream of NRF2 [3,14]. NRF2 and LAMP2A also play critical roles in mitochondrial function, ensuring both structural integrity and functional effectiveness [15,16].

In addition, previous work has shown that rose petal extracts contain large quantities of phenolic compounds, which have been described as stimulating antioxidant enzymes such as CAT and GPXs [17], but research on the detailed compositions of these extracts is lacking, and their characterization could lead us to obtain a better understanding of their overall activity.

In this study, rose extract was subjected to a chromatographic analysis coupled with mass spectrometry in order to annotate its major volatile and non-volatile compounds. The main objective was to identify the presence of compounds previously reported in the literature as being able to interfere with skin aging processes.

We evaluated the effects of a rose extract, obtained using a supercritical CO_2 extraction process, on primary human keratinocytes, focusing on cellular resilience via antioxidant defenses, autophagy stimulation, and the modulation of the AMPK pathway. We show the effect of rose extract on the AMPK signaling pathway and that targeting this pathway may be a crucial strategy for promoting lifelong healthy skin.

2. Materials and Methods

2.1. Extraction from Rosa hybrid Flowers

Rosa hybrid "DELFLOBLA" is a variety of floribunda garden rose with the selection code number N° 09.8649.1 and is the result of breeding between the varieties "MEICHIBON" and "DELGRAMAUE" performed in 2009, which was first observed in Delbard nurseries in Malicorne, France, in 2010. The extraction of active compounds from rosa hybrid flowers was performed using a supercritical fluid extraction (SFE) method. Supercritical carbon dioxide (CO_2) was used as the primary extraction fluid modified with an ethanol co-solvent. This co-solvent/active compound mixture was then used for GC-MS analysis tests and on cell cultures.

2.2. Volatile Molecules Analysis by GC-MS

Volatile chemical composition was determined by injecting 1 μL of the rose extract into GC-MS equipment using an HP 6890 GC system connected to an HP 5973 Mass Selective Detector equipped with a DB-5 MS capillary column (60 m × 0.25 mm i.d., 0.25 μm film thickness; Agilent Technologies, Palo Alto, CA, USA). The column temperature was held at 60 °C for 3 min, programmed to increase at 5 °C/min to 310 °C, and then held at this temperature for 15 min. Helium was used as carrier gas with a flow rate of 2.3 mL/min and a split ratio of 20:1. The temperature of the injector was maintained at 290 °C. The MS was set to scan in the range of m/z 35–500 amu with an ionization energy of 70 eV. The chemical components were identified by a comparison of their retention indices and MS spectra with those reported in the literature and using Masshunter Qualitative Analysis software (B07.00, Agilent Technologies, USA) matching with standard reference databases (NIST23, Wiley275 and CNRS libraries). The retention index of each component was calculated relative to a standard mix of n-alkanes (C7–C40, Sigma-Aldrich, St. Louis, MO, USA), analyzed under identical experimental conditions. The concentration of the components was computed based on the percentage of their relative peak area (%).

2.3. Non-Volatile Molecules Analysis by UHPLC-MS

Non-volatile molecules were analyzed using Agilent Technologies Accurate-Mass ESI-Q-TOF LCMS 6530 with an LC 1290 Infinity system equipped with a DAD detector. Separation was carried out at 40 °C using a 120 EC-C18 column (3.0 × 100 mm, 2.7 µm; Agilent Technologies, Palo Alto, CA, USA). The CO_2 extract was diluted in ethanol (1:10), and 1.5 µL was injected at the head of the column. The column was eluted at 0.7 mL/min with a solvent gradient using solvent A (water with formic acid 0.1% (*v*/*v*)) and solvent B (methanol with formic acid 0.1% (*v*/*v*)). After 1 min at 2%, the proportion of solvent B increased from 2% to 100% over 23 min, followed by a 3 min isocratic phase with 100% solvent B; the proportion decreased from 100% back to 2% in 0.5 min, and equilibration occurred with 2% of solvent B over 2.5 min until the end of the run at 30 min. Mass analyses were conducted in positive and negative mode in the range of 100–1500 m/z, with nebulization gas (nitrogen) at a flow of 10 L/min and 40 psg pressure. The capillary tension and the fragmentor were 3500 V and 100 V, respectively. Complementary tandem mass spectrometry analyses (MS/MS) were performed with a collision energy of 20 eV to elucidate molecular structures. The chemical components were identified using their structural data (UV and mass) and the SIRIUS software version 5.8.5 [18–20].

2.4. Primary Cultured Keratinocytes

Keratinocytes (Biopredic International, Saint-Grégoire, France) were cultured in OxiProteomics® medium at 37 °C in 5% CO_2 humidified air. Cytotoxicity MTS assay (G542, Promega; Charbonnières-les-Bains, France) was carried out according to the manufacturer's guidelines, and specific absorbance was recorded (Varioskan, Thermofisher™, Asnières-sur-Seine, France). The rose extract was diluted (*w*/*v*) in the medium at the defined doses for 48 h. 5-Aminoimidazole-4-carboxamide riboside (AICAR) at an amount of 0.5 mM was used as the positive inducer of the AMPK pathway. To confirm AMPK pathway activation, Compound C, a specific inhibitor of the AMPK pathway, was used at an amount of 10 µM. For "protective" property evaluation, cells were then exposed to Benzo[*a*]pyrene (BAP; CRM40071; 20 µM; Sigma-Aldrich-Merck KGaA, Darmstadt, Germany), diluted using Hank's Balanced Salt Solution (HBSS), and irradiated with UV-A (LED source, emission peak at λ = 365 nm; 2,4 J/cm^2) using the OxiProteomics® irradiation system after 48 h of contact with the rose extract. Just after the irradiation process, cells were transferred into a fresh culturing medium and incubated in optimal condition of culture for 2 h before sampling.

2.5. Protein Extraction and Western Blot

Keratinocytes were subjected to protein extraction in an aqueous buffer using the optimized OxiProteomics® buffer for Western blotting. Accurate measurement of protein concentration was carried out using the Bradford Protein Assay Dye Reagent (Bio-Rad™, Marnes-la-Coquette, France) according to the manufacturer's guidelines. Extracted proteins were separated by high-resolution electrophoresis (SDS-PAGE—gradient 4–20%; Thermo Scientific™). After migration, proteins were transferred from gel to a 0.2 µm nitrocellulose membrane (Bio-Rad™). The membrane was washed in a Tris-buffered saline (TBS) solution at pH 7.6 with 0.1% of Tween (TBS-T) before being incubated for 30 min in TBS with 3% of Bovine Serum Albumin (TBS-BSA) for saturation. Just after the saturation step, the membrane was incubated at 4 °C overnight in a fresh TBS-BSA solution containing the primary antibodies (Table 1). The day after, the excess primary antibody was eliminated by washing with TBS-T solution, and then the membrane was incubated for 1 h with the secondary antibody coupled to a fluorophore (Table 1). Three (3) additional washes with TBS-T were performed to remove the excess secondary antibodies. The digital acquisition of images of P_AMPK, AMPK, P_ACC, and Actin was performed using the "iBright" system (Thermofisher™).

Table 1. Antibodies used.

Host	Target	Provider/Ref.	Application
Rabbit	P_ACCS79	Abcam (Paris, France) ab222774	Western blot
Rabbit	P_AMPKT183/T172	Abcam ab23875	Western blot
Mouse	AMPK	Abcam ab110036	Western blot
Rabbit	Actin	Abcam ab179467	Western blot
Rabbit	P16	Invitrogen (Les Ulis, France) MA5–27905	In situ visualization
Rabbit	LAMP2A	Invitrogen 10332473	In situ visualization
Rabbit	Ki67	Abcam ab92742	In situ visualization
Rabbit	NRF2	Abcam ab137550	In situ visualization
Goat	Anti-mouse Alexafluor 488	Invitrogen A11001	Secondary antibody
Goat	Anti-Rabbit Alexafluor 467	Invitrogen A21244	Secondary antibody

2.6. Detection, Visualization, and Quantification of Biomarkers

Keratinocytes were fixed with a solution containing 95% ethanol and 5% acetic acid. Oxidatively damaged (carbonylated) proteins were labeled using an OxiProteomics® fluorescent probe (Ex = 647 nm/Em = 650 nm) functionalized to specifically bind to carbonyl moieties and DAPI (4′,6-diamidino-2-phenylindole) for nuclear labeling in Phosphate-Buffered Saline, pH 7.4 (PBS). For immunodetection, a saturating step was carried out on the non-specific sites with a solution of PBS containing 3% BSA (PBS-BSA). Keratinocytes were incubated with diluted primary antibodies in a PBS-BSA solution (Table 1). The excess primary antibodies were eliminated with washing using a solution of PBS with 0.1% Tween (PBS-T), and then cells were then incubated for 1 h with the secondary antibody coupled to a fluorophore (Table 1) in PBS-BSA. The cellular nuclei were labeled using DAPI. Finally, the antibody and excess DAPI were removed with a sequence of washing steps with PBS-T. Fluorescent images were collected with an epi-fluorescent microscope (Thermofisher™, EVOS M5000 Imaging System) and analyzed with ImageJ software version 1.53 (U. S. National Institutes of Health, Bethesda, MD, USA).

A % of induction (compared to the experimental group of reference, namely the control or stress group) was obtained by using the following equation:

$$\text{Induction (\% group X vs. Reference group)} = ((\text{Biomarker\_Levels\_Group\_X}/\text{Biomarker\_Levels\_Ref\_Group}) - 1) \times 100$$

Also, a protective value (%) was obtained for the experimental groups using, as references, the control group, considered at maximum efficiency (100%), and the stress group, considered at minimum efficiency (0%).

$$\text{Protection (\% group X)} = ((\text{Biomarker\_Levels\_Group\_Stress} - \text{Biomarker\_Levels\_Group\_X})/(\text{Biomarker\_Levels\_Group\_Stress} - \text{Biomarker\_Levels\_Group\_Control}) \times 100$$

2.7. Histogram Illustration and Statistical Analysis

The quantification of biomarkers was normalized in relation to the control (considered at 100%), finally obtaining a mean value and a standard deviation. Statistical analyses were carried out using the "GraphPad" software version 10.0.3 (La Jolla, CA, USA) by using a one-way ANOVA and Dunnett's post hoc test for multi-comparison analyses (vs. the respective stress group) as well as Student's binary t-test or Mann–Whitney comparisons between groups (confidence interval of 95%).

3. Results

3.1. Rose Extract Induces No Cytotoxic Effect on Primary Human Keratinocytes and Stimulates Antioxidant Defense and Autophagy

Initially, the cytotoxic impact of rose extract was assessed across concentrations ranging from 0.1% to 0.00005% (w/v) by exposing primary human keratinocytes for 48 h. Then, the effects of rose extract on primary keratinocytes after 48 h of contact, at basal levels, were explored by the evaluation of NRF2 (Figure 1B,C) and LAMP2A levels (Figure 1D,E), detected in situ (on cells) via epifluorescence microscopy.

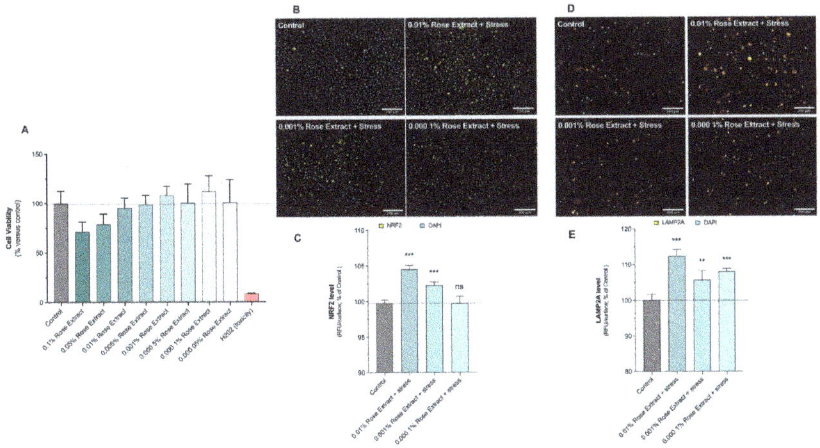

Figure 1. Rose extract boosted antioxidative defense. (**A**) Cell viability upon rose extract contact was carried out using MTS assay. Keratinocytes were exposed for 48 h to rose extract, ranging from 0.00005% to 0.1%. Data are shown as mean +/− S.D. from 6 replicates per condition. Representative images of (**B**) NRF2 levels and (**D**) LAMP2A levels in keratinocytes upon UV-A irradiation and Benzopyrene (BAP) at 20 µM. Nuclei were stained in cyan (DAPI coloration). Scale bar, 200 µm. Quantification of (**C**) NRF2 levels and (**E**) LAMP2A levels of each experimental group is expressed as relative values (% vs. control) and shown as mean +/− S.D. from 4 replicates per condition. ***, $p < 0.001$; **, $p < 0.01$; ns, not significant. One-way ANOVA and Dunnett's post hoc test were used for multi-comparisons vs. control (alpha = 0.05).

The presence of rose extract at concentrations of 0.01% (112.4 ± 1.8, 12% induction), 0.001% (105.7 ± 2.5, 6% induction), and 0.0001% (108.1 ± 0.8, 8% induction) showed stimulatory effects on detoxification mechanisms through increased levels of LAMP2A under basal conditions (100 ± 1.7). LAMP2A is a lysosomal protein involved in cell detoxification by hydrolyzing damaged proteins playing a crucial role in the process of chaperone-mediated autophagy (CMA), a selective form of autophagy that targets specific proteins for degradation within lysosomes. NRF2 is a transcriptional factor which is activated following cell damage to produce downstream target genes involved in cell defense and detoxification. Rose extract at concentrations of 0.01% (104.5 ± 0.6, 4% induction) and 0.001% (102.3 ± 0.5, 2% induction) significantly increased NRF2 levels under basal conditions (100 ± 0.5). Thus, the antioxidant and detoxification effects of rose extract were confirmed by the significant increase in NRF2 levels and the stimulation of detoxification mechanisms through elevated LAMP2A levels under basal conditions.

3.2. Treatment with Rose Extract Counteracts Pollution (BAP + UV-A)-Induced Oxidative Damage

The effects of rose extract on human primary keratinocytes exposed to oxidative damages such as those caused by pollution (BAP) and UV-A were assessed through in situ (on cell) detection, visualization, and the quantification of carbonylation levels. A significant increase in oxidative damage (185.9 ± 9.7) (carbonylation levels) was observed upon stress (BAP + UV-A) exposure when compared to the control group (100 ± 5.1) (Figure 2). Interestingly, rose extract at concentrations of 0.001% (154.2 ± 14.8; 37% protection) and 0.0001% (156.9 ± 16.7; 34% protection) counteracted stress-induced increases in the levels of carbonylation. The maintenance of proteostasis and the regulation of carbonylation levels in cells are pivotal for cellular health and functionality. Effective proteostasis ensures the balance and integrity of the proteome, preventing the accumulation of misfolded or damaged proteins. The beneficial effects of these processes also include enhanced cell viability, improved cellular responses to stress, and the overall promotion of longevity and skin health, spotlighting their vital role in cellular homeostasis.

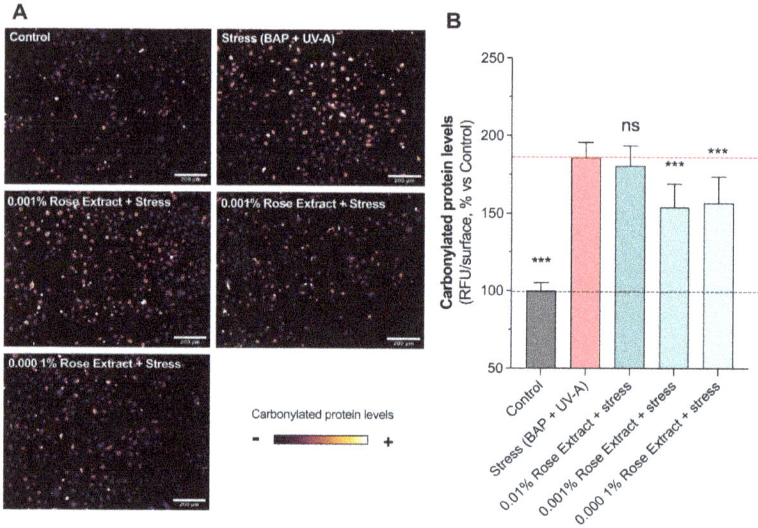

Figure 2. Induced oxidative protein damage caused by Benzopyrene (BAP) and UV-A was counteracted by rose extract. Representative images of (**A**) oxidative damage (carbonyl levels, visualized in orange range) in keratinocytes upon BAP and UV-A irradiation at 20 µM. Scale bar, 200 µm. Quantification of (**B**) carbonyl levels of each experimental group are expressed as relative values (% vs. control) and shown as mean +/− S.D from at least 3 independent experiments. ***, $p < 0.001$; ns, not significant. One-way ANOVA and Dunnett's post hoc test were used for multi-comparisons vs. stress group (alpha = 0.05).

3.3. Treatment with Rose Extract Counteracts Pollution (BAP + UV-A)-Mediated Cellular Senescence and Cell Growth Arrest

$P16^{INK4A}$, a cell cycle regulator, is considered as a marker of cellular senescence as its expression was found to be significantly increased in senescent cells during natural aging or in age-related pathologies [21,22]. A significant increase in labeled $P16^{INK4A}$ cells (332.4 ± 18.61) was observed in keratinocytes exposed to BAP and UV-A when compared to the control (100 ± 10.7) (non-stressed group). Interestingly, rose extract at concentrations of 0.01% (186.0 ± 31.9 63% protection), 0.001% (165.6 ± 26.2; 72% protection), and 0.0001% (193.3 ± 25.7; 60% protection) significantly counteracted the increased number of labeled $P16^{INK4A}$ cells mediated by UV-A and pollution, suggesting a beneficial effect on stress-induced cellular senescence (Figure 3A,B). The effects of rose extract were also observed on the proliferation marker KI67 (Figure 3C,D). P16^{INK4A} is a nuclear protein that is present during all active phases of the cell cycle (G1, S, G2, and mitosis) but absent in resting cells (G0 phase) [23]. Exposure to stress conditions (BAP + UV-A) significantly decreased the levels of KI67 (39.2 ± 13.1) in comparison to the control group (100 ± 4.7). The rose extract at concentrations of 0.01% (80.4 ± 10.7; 68% protection), 0.001% (82.1 ± 13.0; 70% protection), and 0.0001% (97.1 ± 15.1; 92% protection) preserved keratinocytes from undergoing cell growth arrest and maintained the proliferative (KI67-positive) bulk of cells, confirming its beneficial effects on cellular homeostasis under stress.

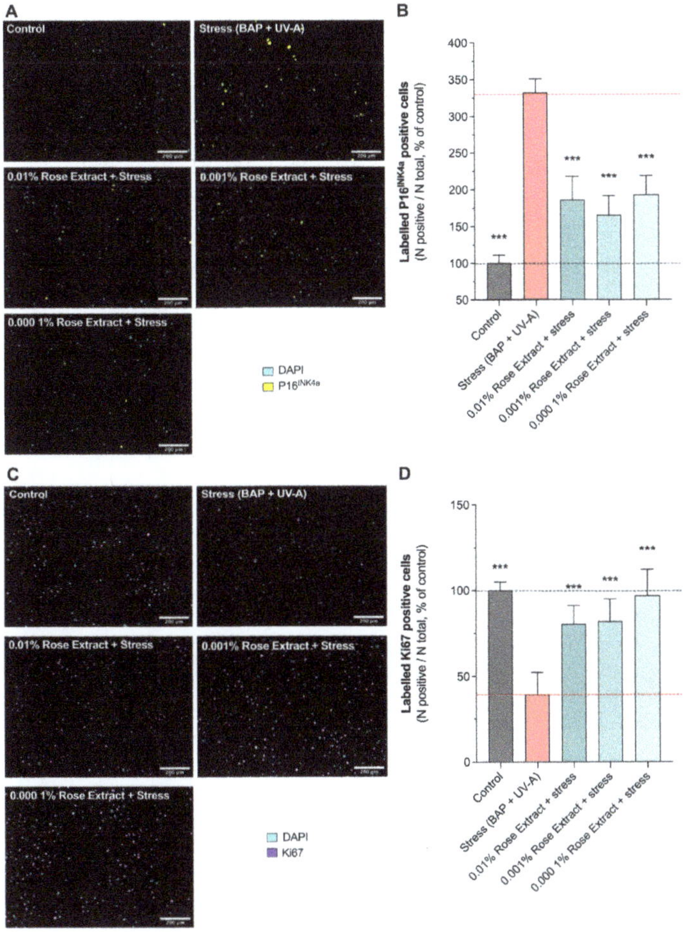

Figure 3. Induced cellular senescence and cell growth arrest caused by Benzopyrene (BAP) and UV-A were counteracted by rose extract. Representative images of (**A**) labeled P16^{INK4A} cells (yellow), and (**C**) labeled Ki67 cells (purple) in keratinocytes upon Benzopyrene (BAP) and UV-A irradiation at 20 μM. Nuclei were stained in cyan (DAPI coloration). Scale bar, 200 μm. Quantification of (**B**) labeled P16^{INK4A} cells, and (**D**) labeled Ki67 cells of each experimental group are expressed as relative values (% vs. control) and shown as mean +/− S.D from at least 3 independent experiments. ***, $p < 0.001$. One-way ANOVA and Dunnett's post hoc test for multi-comparisons vs. stress group (alpha = 0.05).

3.4. The Beneficial Antioxidant Effect of Rose Extract on Pollution (BAP + UV-A) Exposure Is Mediated by an AMPK-Dependent Mechanism

The AMPK pathway's roles extend to influencing proteostasis, which encompasses the control of protein biogenesis, folding, trafficking, and degradation by maintaining cellular homeostasis, promoting autophagy, and preventing senescence [24,25]. Increased levels of carbonylated proteins (185.9 ± 9.7) mediated by stress were counteracted by rose extract at a concentration of 0.001% (154.2 ± 14.8) (Figure 4). The presence of Compound C, the specific inhibitor of AMPK, at 10 μM totally reversed the beneficial effects of 0.001% of rose extract (178.2 ± 23.9), suggesting that its positive impact is mediated via the AMPK pathway. The group treated with 0.5 mM of AICAR (an AMPK activator) displayed similar modulation (148.2 ± 7.8; 44% of protection) as that treated with rose extract, confirming the beneficial effects of AMPK pathway activation in preventing the oxidative stress-mediated increase in carbonyl levels. The inhibition of the AMPK pathway using 10 μM of Compound C fully blocked the beneficial effect mediated by AICAR (173.7 ± 19.3).

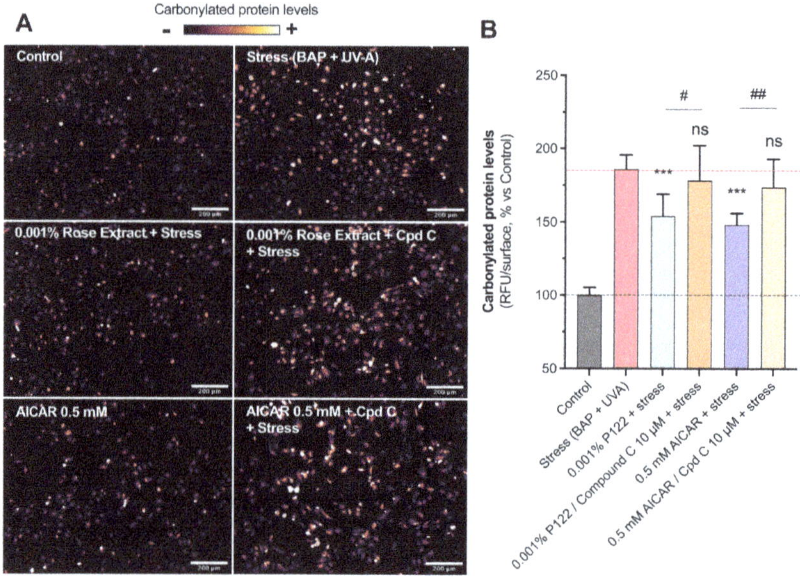

Figure 4. Rose extract counteracted oxidative stress via AMPK-dependent mechanism: Representative images of (**A**) oxidative damage (carbonyl levels, visualized in orange range) in keratinocytes upon Benzopyrene (BAP) and UV-A irradiation at 20 μM. Scale bar, 200 μm. Quantification of (**B**) carbonyl levels of each experimental group are expressed as relative values (% vs. control) and shown as mean +/− S.D from at least 3 independent experiments. ***, $p < 0.001$; ns, not significant. One-way ANOVA and Dunnett's post hoc test were used for multi-comparisons vs. stress group (alpha = 0.05). ##, $p < 0.01$; #, $p < 0.05$; ns, not significant. Student's t-test was used for binary-comparisons (alpha = 0.05).

3.5. The Anti-Cell Senescence and Cell Growth Arrest Effect of Rose Extract Are Mediated by the AMPK Pathway

The involvement of the AMPK pathway in rose extract-mediated anti-senescence and cell proliferation effects was investigated (Figure 5). Rose extract at a concentration of 0.001% significantly reduced (165.6 ± 26.2) the number of labeled $P16^{INK4A}$ cells (332.4 ± 18.6), indicating its anti-senescent properties. This beneficial effect was blocked by Compound C (267.0 ± 48.3), a specific AMPK inhibitor, suggesting that the anti-senescent effects of rose extract are mediated through the AMPK pathway. Similarly, AICAR at

0.5 mM also mitigated the stress-induced increase in labeled P16^{INK4A} cells (113.9 ± 36.6), and this effect was reversed by Compound C (314.0 ± 51.0), further confirming the involvement of the AMPK pathway (Figure 5A,B).

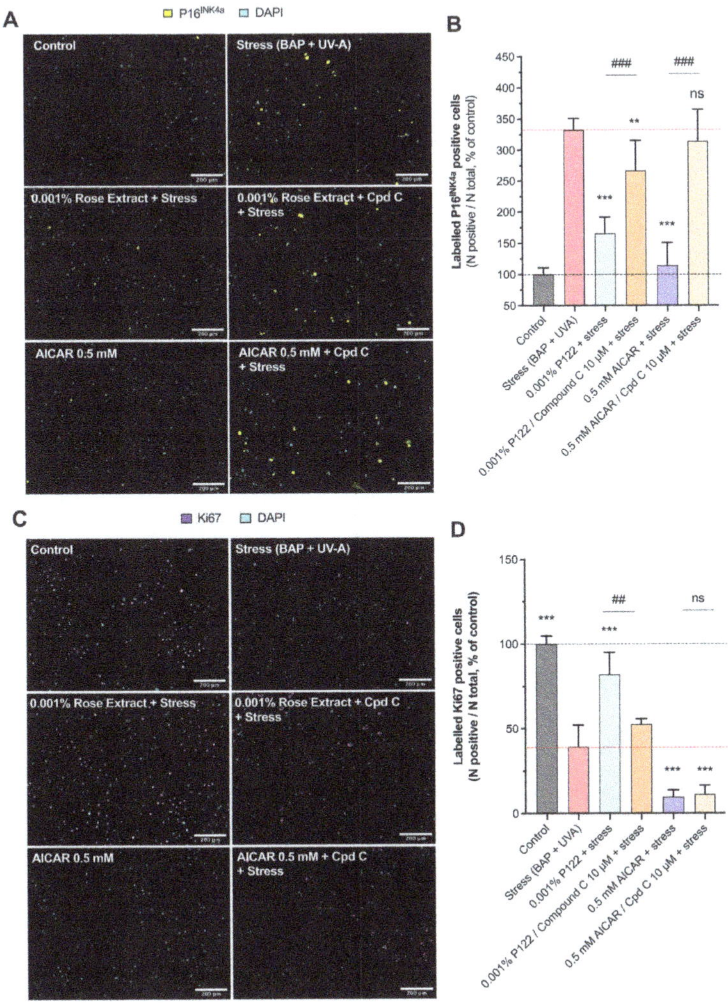

Figure 5. Rose extract counteracts UVA- and pollution-mediated senescence, and cell growth arrest is dependent on AMPK pathway. Representative images of (**A**) labeled P16^{INK4A} cells (yellow) and (**C**) labeled Ki67 cells (purple) in keratinocytes upon Benzopyrene (BAP) and UV-A irradiation at 20 µM. Nuclei were stained in cyan (DAPI coloration). Scale bar, 200 µm. Quantification of (**B**) labeled P16^{INK4A} cells and (**D**) labeled Ki67 cells of each experimental group is expressed as relative values (% vs. control) and shown as mean +/− S.D from at least 3 independent experiments. ***, $p < 0.001$; **, $p < 0.01$; ns, not significant. One-way ANOVA and Dunnett's post hoc test were used for multi-comparisons vs. stress group (alpha = 0.05). ###, $p < 0.001$; ##, $p < 0.01$; ns, not significant. Student's *t*-test was used for binary comparisons (alpha = 0.05).

In addition, rose extract at a concentration of 0.001% prevented the stress-induced decrease in the number of KI67-positive cells (82.1 ± 13.1), indicating its role in promoting cell proliferation. This effect was also counteracted by Compound C (52.7 ± 3.4), reinforcing the role of AMPK activation in this process. However, AICAR at 0.5 mM exhibited antiproliferative properties (9.6 ± 4.1) that were not affected by Compound C (11.2 ± 5.2), suggesting that AICAR's anti-proliferative effects operate through an AMPK-independent mechanism (Figure 5C,D). This aligns with previous findings that AICAR has AMPK-independent anti-proliferative properties [26,27].

3.6. Rose Extract Stimulates the AMPK Pathway, Causing a Beneficial Effect

To confirm the activation of the AMPK pathway by rose extract, we assessed the ratio of P-AMPK to total AMPK (P-AMPK/AMPK) and the levels of P-ACC (a downstream target of AMPK) via a Western blot analysis (Figure 6). Quantification revealed a significant increase in the P-AMPK/AMPK ratio in the stress group (13.4 ± 0.5) compared to the control (1 ± 0.5). Notably, treatment with 0.001% rose extract further elevated this ratio (15.4 ± 0.4, 15% induction), indicating enhanced AMPK pathway activation. Similar results were observed with AICAR at 0.5 mM (14.7 ± 0.6, 10% induction), supporting the activation of the AMPK pathway. The stimulatory effects on the AMPK pathway were fully negated by Compound C, a specific AMPK inhibitor, confirming the specificity of rose extract (10.6 ± 0.5) and AICAR (6.1 ± 1.1) in inducing AMPK activation. Additionally, both rose extract (4.63 ± 0.7) and AICAR (5.7 ± 1.4) significantly increased the P-ACC levels, an effect that was also totally counteracted by Compound C (2.2 ± 0.1 and 2.4 ± 0.4, respectively). These findings provide strong evidence that rose extract activates the AMPK pathway and influences downstream phosphorylation events, demonstrating its potential role in cellular defense mechanisms.

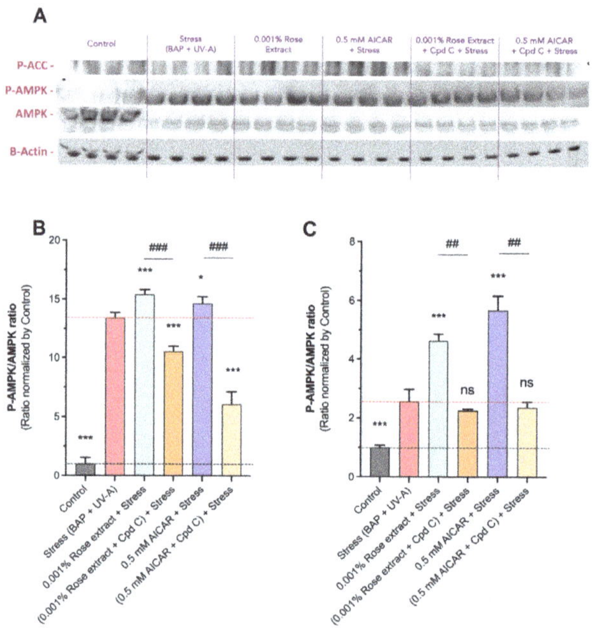

Figure 6. Rose extract mediates beneficial effect in AMPK-dependent manner. (**A**) Representative images of Western blots of P-ACC, P-AMPK, AMPK, and Actin in keratinocytes upon Benzopyrene (BAP) and UV-A irradiation at 20 µM. Semi quantification of (**B**) P-AMPK/AMPK ratio and (**C**) à P_ACC/Actin ratio of each experimental group is expressed as relative values and shown as mean +/− S.D from at least 3 independent experiments. ***, $p < 0.001$; *, $p < 0.05$ ns, not significant. One-way ANOVA and Dunnett's post hoc test were used for multi-comparisons vs. stress group (alpha = 0.05). ###, $p < 0.001$; ##, $p < 0.01$; ns, not significant. Student's *t*-test was used for binary comparisons (alpha = 0.05).

3.7. Volatile and Non-Volatile Compositions of Rose Extract

The characterization of rose extract using hyphenated techniques enabled us to detect 166 different compounds (22 in the LC/MS experiment and 144 in the GC/MS experiment). Transforming GC/MS and LC-MS data into an annotated metabolite profile is a major challenge. Despite the intricacy and complexity of the sample, several key metabolites were successfully annotated with clarity: 16 non-volatile and 107 volatile compounds. These compounds belong to different primary and secondary metabolites, such as tannins, flavonoids, and organic acids for non-volatile compounds and monoterpenes, sesquiterpenes, and fatty acid derivatives for volatile compounds (tables available in Supplementary Materials).

4. Discussion

Twelve hallmarks of aging have been characterized, including DNA instability, telomere attrition, epigenetic alterations, loss of proteostasis, deregulated nutrient sensing, mitochondrial dysfunction, cellular senescence, stem cell exhaustion, altered intercellular communication, disabled macroautophagy, chronic inflammation, and dysbiosis [1,8]. Loss of proteostasis and proteome integrity is a key element of aging. Rosa species, in addition to their global growth, contain elements such as phenolic acid, flavonols, and anthocyanins that have been identified to exhibit beneficial effects on skin biofunction, addressing inflammation and aging [28].

Despite there being limited knowledge about the effects of rose extracts on skin aging, studies have shown that rose extract can stimulate the activity and levels of catalase (CAT) and glutathione peroxidases (GPXs), thereby confirming its antioxidant properties [17]. However, further investigation is needed to clarify additional biological properties of rose extract, particularly in relation to the key hallmarks of aging such as antioxidant defense and autophagy. First, NRF2, a cytoprotective and antioxidant transcription factor regulating ~250 antioxidant downstream target genes, was investigated following 48 h of contact with rose extract at concentrations of 0.01%, 0.001%, and 0.0001% in human keratinocytes. The presence of rose extract significantly increased the levels of NRF2, suggesting its stimulatory effect on antioxidant defense. Premature aging and loss of longevity result from the accumulation of misfolded, oxidized, glycated, or ubiquitinylated proteins upon exposure to exogenous stressors [10,12,29]. To overcome these accumulations of damaged proteins, the degradation of intracellular components by the lysosome can be achieved through lysosome vesicles via chaperone-mediated autophagy and LAMP2A, which facilitates the translocation of damaged proteins into the lysosome lumen [30,31]. The impact of rose extract on autophagy has not been established and needs to be addressed. The presence of rose extract at concentrations of 0.01%, 0.001%, and 0.0001% for 48 h significantly increased the levels of LAMP2A in human keratinocytes, suggesting that rose extract could boost autophagy pathways favoring detoxification. Interestingly, enhanced autophagic activity is associated with improvement in the skin's healthy lifespan (longevity) [32]. Thus, rose extract was characterized for the first time as a key stimulatory factor of antioxidant defense and the detoxification process (autophagy) associated with skin longevity. Taken together, these cellular processes are part of the hallmarks of aging [8,29], and rose extract is addressed as an antioxidant and detoxifying ingredient. To dive deeper into the biological properties of rose extract, the effects on proteostasis were examined. Both UV-A irradiation and BAP are stressors related as oxidative inducers [3,10,33] associated with skin dysfunction, inducing oxidative damage and harmful genetic modification, contributing to premature skin aging [3–5]. One early, direct, and irreversible modification induced by ROS, RNS, RXS, and reactive aldehydes is the carbonylation of proteins that consists of the formation of reactive aldehyde or ketone residues on proteins [10,34]. This harmful irreversible modification process has been identified as a major hallmark of severe oxidative damage associated with protein dysfunction and misfolding related to premature skin aging [10,35]. To assess the antioxidant properties of rose extract, keratinocytes were pre-exposed to the rose extract for 48 h before being subjected to stress induced by BAP and UV-A irradiation. Rose extract at various concentrations significantly protected proteins from oxidative damage (carbonyl levels) induced by BAP and UV-A irradiation, confirming its antioxidant and proteostasis-restoring properties.

Senescence is the primary aging process at the cell level that is elicited by acute or chronic damage [8]. Senescent cells are characterized by the loss of proliferative capacity, mitochondrial dysfunction, significantly altered patterns of expression, and the secretion of bioactive molecules [22,36]. $P16^{INK4A}$ expression is strongly increased in aged tissues of rodents and underscores the significance of senescence as a hallmark of aging and a driver of age-related diseases. The removal of $P16^{INK4A}$ -positive senescent cells has been shown to delay the progression of age-related pathologies, extending both premature and natural aging lifespans [21,37]. Rose extract has the ability to mitigate cellular senescence assessed by quantifying labeled $P16^{INK4A}$ cells in cultured human keratinocytes exposed to BAP and UVA irradiation. As expected, stress UV-A and pollution significantly increased the number of labeled $P16^{INK4A}$ cells, which was contrasted by the rose extract at various concentrations. In addition to being directly related to aging, $P16^{INK4A}$ is a tumor suppressor and cell cycle regulator. $P16^{INK4A}$ inhibits the cyclin-dependent kinases CDK4/6 and CDK2 that directly control the cell cycle, which implicate them in cellular proliferation-dependent mechanisms such as tissue regeneration (renewal), aging, and cellular senescence. Therefore, the ability of rose extract to counteract stress-induced reductions in cell proliferation was investigated. Stress induced by BAP + UV-A significantly decreased the number of labeled KI67 cells (proliferative cells), an effect mitigated by rose extract at different concentrations. This study sheds light on the anti-senescent effects and tissue regeneration properties of rose extract, expanding its potential beyond antioxidant effects and its beneficial effect on on proteostasis.

The AMPK signaling pathway is crucial for maintaining proteome integrity (proteostasis), enhancing protein stability, and removing damaged proteins through proteasomal degradation or autophagy. Additionally, AMPK is involved in preventing cell senescence, promoting cell growth, and regulating mitochondrial biogenesis to control mitochondrial redox homeostasis [2,38]. In senescent cells, AMPK is typically inactivated. The activation of AMPK, such as by metformin, significantly reduces the development of senescent cells, while inhibition of AMPK by Compound C accelerates it [24]. This study aimed to elucidate the pathway through which rose extract exerts its beneficial effects, focusing on the AMPK pathway.

To investigate this, Compound C, a selective AMPK inhibitor, was used to counteract the effects of rose extract at a concentration of 0.001%. Additionally, AICAR, an AMPK activator, was used alone or in combination with Compound C to confirm AMPK activation in cultured human keratinocytes exposed to BAP and UV-A irradiation. The beneficial effects of rose extract at a concentration of 0.001% on reducing carbonyls were counteracted by Compound C, indicating that rose extract mediates its antioxidant effects via an AMPK-dependent mechanism. AICAR exhibited similar effects to rose extract, and these effects were also counteracted by Compound C, further confirming the involvement of the AMPK pathway. Therefore, rose extract appears to act through an AMPK-dependent mechanism to address early oxidative events and the loss of proteostasis associated with oxidative protein modifications like carbonyls. Interestingly, while AMPK activation has been associated with mitigating oxidative stress, these findings also reveal a direct link between AMPK activation and the prevention of or reduction in protein carbonylation.

To further explore the role of the AMPK pathway in the mechanism of action of rose extract, its effects on the number of labeled $P16^{INK4A}$ cells and proliferative (Ki67) cells induced by stress were assessed in the presence or absence of the AMPK inhibitor Compound C. Interestingly, Compound C significantly blocked the rose extract-mediated reduction in the number of labeled $P16^{INK4A}$ cells and the restoration of Ki67 cells. Additionally, AICAR displayed a similar pattern of action to rose extract for the $P16^{INK4A}$ biomarker, which was also counteracted by Compound C. However, AICAR completely blocked the proliferation of keratinocytes, and this antiproliferative effect was not reversed by Compound C, suggesting an AMPK-independent mechanism [26,27]. This suggests that the anti-senescent and tissue regeneration properties of rose extract are mediated by an AMPK-dependent mechanism. To ultimately confirm the activation of the AMPK pathway by rose extract, the phosphorylation of AMPK at T183/T172 (normalized by total AMPK) and a downstream target of the AMPK pathway, phospho-ACC at S79 (normalized by Actin), were analyzed. The presence of rose extract at a concentration of 0.001% and AICAR significantly increased the P-AMPK/AMPK ratio and the P-ACC/ACC ratio compared to the stressed condition. These elevations in P-AMPK and P-ACC, indicative of AMPK pathway activation, were nullified in the presence of Compound C. In summary, these findings confirm that rose extract mediates its antioxidant, anti-senescent, and tissue regeneration properties through an AMPK-dependent mechanism.

The analysis of the rose extract using coupled methods revealed a wide range of phenolic compounds and the presence of other fatty acid derivatives and terpene compounds, which contribute to their characteristic aroma, flavor, and biological activity. Rose petal extract, rich in phenolic compounds, has been previously described to boost antioxidant enzymes like CAT and GPXs [14]; moreover, sesquiterpene compounds have not yet been identified to interfere with the AMPK pathway [39]. Rose petal extract contains terpenes, such as limonene, β-myrcene, linalool, camphor, and γ-terpineol, which are known to display potential biological effects, including anti-inflammatory, antioxidant, and antimicrobial properties. These effects may be linked to the activation of AMP-activated protein kinase (AMPK) [39,40]. Alcohols including 1-hexanol, 2-hexenol, 3-hexenol, 2-ethyl-1-hexanol, benzyl alcohol, terpinen-4-ol, and linalool are well known for their therapeutic properties [17,41]. For instance, linalool, a monoterpene alcohol, has shown anti-inflammatory effects, which could be linked to AMPK modulation to reduce inflammation [42]. Sesquiterpenes such as ocimene and α-murolene have anti-inflammatory and antioxidant properties and play a role in regulating energy metabolism and potentially influencing the AMPK pathway. These compounds reduce markers of oxidative stress and modulate inflammatory processes, which are linked to AMPK activation.

Fatty acids such as linoleic acid and oleic acid have well-established effects on lipid metabolism regulation and inflammation. These polyunsaturated fatty acids can influence the activation of the AMPK pathway, playing a role in managing energy metabolism and inflammation. For example, linoleic acid has been shown to activate AMPK, thus modifying cellular energy balance [43,44].

In addition, resveratrol, another plant compound, is known to activate the AMPK pathway, counteracting stress-mediated increases in the number of senescent cells [45,46].

The effects of various plant extracts on the AMPK pathway in cosmetics have been well characterized. The extract of *Lycium barbarum* (goji berries) is a potent modulator of lipid synthesis in keratinocytes through AMPK activation and SIRT1 signaling. This leads to an increase in the expression of key lipid synthesis genes, strengthening epidermal barrier function [47]. Flavonoid-rich plant extracts have been shown to possess anti-aging effects on the skin. These compounds exhibit antioxidant, anti-inflammatory, and antimicrobial properties, helping to protect skin from the effects of aging [48]. Tea extract (*Camellia sinensis*), particularly the saponins derived from tea seeds, has been shown to suppress sebum production by regulating the AMPK/mTOR pathway, favoring antiphotoaging, stress resistance, neuroprotection, autophagy, and lipid production [49,50]. Like rose extract, other plant extracts that activate the AMPK pathway hold significant potential in the development of skincare products aiming to improve skin health. Their incorporation into formulations can help reinforce the barrier function, reduce inflammation, and prevent skin aging.

In terms of AMPK activation by rose petal extract, while many of these compounds are known to exhibit biological activities, some may influence AMPK activation indirectly through their effects on oxidative stress, inflammation, or metabolic processes. However, further detailed studies are required to clarify the specific role of these compounds in AMPK activation as their effects may vary depending on concentration, combinations, and the biological systems involved.

5. Conclusions

This study provides new insights into the protective mechanisms mediated by AMPK activation in the context of oxidative stress. Interestingly, these findings reveal a novel link between AMPK activation and the preservation of protein carbonylation, particularly under treatment with rose extract. This suggests that AMPK activation may play a critical role in maintaining protein integrity by reducing oxidative protein modifications.

Moreover, this study highlights the beneficial effects of rose extract on antioxidant defense and autophagy, which protect skin cells from oxidative damage, senescence, and the loss of proliferation induced by pollution (BAP and UV-A). Notably, this is the first time that the activation of the AMPK pathway by rose extract has been observed, underscoring its crucial role in mediating these beneficial effects.

Overall, these results confirm that rose extract plays a significant role in maintaining proteome integrity and stability (cellular proteostasis) and in promoting the removal of damaged proteins through the stimulation of autophagy (Figure 7), suggesting its potential to promote healthy skin aging by targeting longevity pathways.

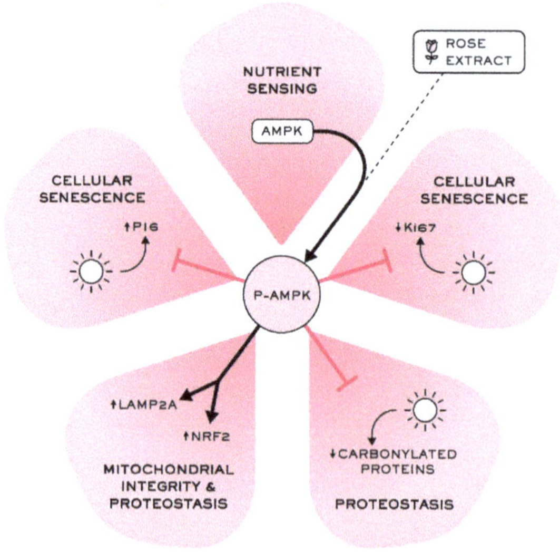

Figure 7. Rose extract, a stimulatory AMPK pathway ingredient, counteracts skin damage caused by UVA irradiation and pollution-mediated oxidative damages, counteracts cellular senescence, and maintains epidermal integrity.

Supplementary Materials: The following supporting information can be downloaded at https://www.mdpi.com/article/10.3390/cosmetics12020057/s1, Table S1. Chromatographic and mass spectral characteristics of compounds annotated by LC-ESI-MS in CO_2 extract; Table S2. Volatile chemical composition of CO_2 extract (abundance: %area).

Author Contributions: O.G. conducted the biological analyses, interpreted the results, and wrote the paper. A.C. and L.B. co-guided the research, interpreted the results, and contributed to writing and reviewing the manuscript. G.A. and S.J. co-guided the project and reviewed the manuscript. G.C. and M.B. performed the extract content analyses and contributed to manuscript writing. A.F.B. and M.A.B. co-guided the study, supervised the research, and reviewed the manuscript. All authors have read and agreed to the published version of the manuscript.

Funding: This research received no external funding.

Institutional Review Board Statement: This study involved the use of biological samples derived from human subjects, in accordance with regulations and ethical standards, approved by the French Minister of Higher Education and Research (AC-2022-5147, last approval date 3 January 2023). Biopredic International holds all regulatory permits for the collection, processing, transfer and export of human tissues and cell-based products to be used exclusively for research (AC-2023-5431, last approval date 19 July 2023).

Informed Consent Statement: Confirmation of written informed consent from donor has been collected and conserved by Biopredic International complying strictly with the ethical rules for donation and collection of human tissues, in view of research use only.

Data Availability Statement: The data presented in this study are available on request from the corresponding author. The data are not publicly available due to confidentiality agreements.

Conflicts of Interest: Olivier Gouin, Andrea Cavagnino and Martin A. Baraibar are employed by Oxiproteomics. Gayané Azadiguian, Sibylle Jäger, and Annie F. Black are employed by L'Oréal. Lionel Breton is employed by CILIA Consulting. The remaining authors declare that the research was conducted in the absence of any commercial or financial relationships that could be construed as a potential conflict of interest.

References

1. Krutmann, J.; Bouloc, A.; Sore, G.; Bernard, B.A.; Passeron, T. The Skin Aging Exposome. *J. Dermatol. Sci.* **2017**, *85*, 152–161. [CrossRef]
2. Xu, W.; Luo, Y.; Yin, J.; Huang, M.; Luo, F. Targeting AMPK Signaling by Polyphenols: A Novel Strategy for Tackling Aging. *Food Funct.* **2023**, *14*, 56–73. [CrossRef] [PubMed]
3. Hamon, M.-P.; Ahmed, E.K.; Baraibar, M.A.; Friguet, B. Proteome Oxidative Modifications and Impairment of Specific Metabolic Pathways During Cellular Senescence and Aging. *Proteomics* **2020**, *20*, e1800421. [CrossRef] [PubMed]
4. Celebi Sozener, Z.; Ozdel Ozturk, B.; Cerci, P.; Turk, M.; Gorgulu Akin, B.; Akdis, M.; Altiner, S.; Ozbey, U.; Ogulur, I.; Mitamura, Y.; et al. Epithelial Barrier Hypothesis: Effect of the External Exposome on the Microbiome and Epithelial Barriers in Allergic Disease. *Allergy* **2022**, *77*, 1418–1449. [CrossRef] [PubMed]
5. Vineis, P.; Robinson, O.; Chadeau-Hyam, M.; Dehghan, A.; Mudway, I.; Dagnino, S. What Is New in the Exposome? *Environ. Int.* **2020**, *143*, 105887. [CrossRef]
6. Azzimonti, B.; Ballacchino, C.; Zanetta, P.; Cucci, M.A.; Monge, C.; Grattarola, M.; Dianzani, C.; Barrera, G.; Pizzimenti, S. Microbiota, Oxidative Stress, and Skin Cancer: An Unexpected Triangle. *Antioxidants* **2023**, *12*, 546. [CrossRef]
7. Md Jaffri, J. Reactive Oxygen Species and Antioxidant System in Selected Skin Disorders. *Malays. J. Med. Sci.* **2023**, *30*, 7–20. [CrossRef]
8. López-Otín, C.; Blasco, M.A.; Partridge, L.; Serrano, M.; Kroemer, G. Hallmarks of Aging: An Expanding Universe. *Cell* **2023**, *186*, 243–278. [CrossRef]
9. Tonelli, C.; Chio, I.I.C.; Tuveson, D.A. Transcriptional Regulation by Nrf2. *Antioxid. Redox Signal.* **2018**, *29*, 1727–1745. [CrossRef]
10. Baraibar, M.A.; Friguet, B. Oxidative Proteome Modifications Target Specific Cellular Pathways during Oxidative Stress, Cellular Senescence and Aging. *Exp. Gerontol.* **2013**, *48*, 620–625. [CrossRef]
11. Baraibar, M.A.; Ladouce, R.; Friguet, B. Proteomic Quantification and Identification of Carbonylated Proteins upon Oxidative Stress and during Cellular Aging. *J. Proteom.* **2013**, *92*, 63–70. [CrossRef]

12. Le Boulch, M.; Ahmed, E.K.; Rogowska-Wrzesinska, A.; Baraibar, M.A.; Friguet, B. Proteome Oxidative Carbonylation during Oxidative Stress-Induced Premature Senescence of WI-38 Human Fibroblasts. *Mech. Ageing Dev.* **2018**, *170*, 59–71. [CrossRef] [PubMed]
13. Pajares, M.; Rojo, A.I.; Arias, E.; Díaz-Carretero, A.; Cuervo, A.M.; Cuadrado, A. Transcription Factor NFE2L2/NRF2 Modulates Chaperone-Mediated Autophagy through the Regulation of LAMP2A. *Autophagy* **2018**, *14*, 1310–1322. [CrossRef] [PubMed]
14. Mo, C.; Wang, L.; Zhang, J.; Numazawa, S.; Tang, H.; Tang, X.; Han, X.; Li, J.; Yang, M.; Wang, Z.; et al. The Crosstalk between Nrf2 and AMPK Signal Pathways Is Important for the Anti-Inflammatory Effect of Berberine in LPS-Stimulated Macrophages and Endotoxin-Shocked Mice. *Antioxid. Redox Signal* **2014**, *20*, 574–588. [CrossRef]
15. Dinkova-Kostova, A.T.; Abramov, A.Y. The Emerging Role of Nrf2 in Mitochondrial Function. *Free Radic. Biol. Med.* **2015**, *88*, 179–188. [CrossRef]
16. Liu, X.; Liao, X.; Rao, X.; Wang, B.; Zhang, J.; Xu, G.; Jiang, X.; Qin, X.; Chen, C.; Zou, Z. The Lysosomal Membrane Protein LAMP-2 Is Dispensable for PINK1/Parkin-Mediated Mitophagy. *FEBS Lett.* **2020**, *594*, 823–840. [CrossRef]
17. Ng, T.B.; Gao, W.; Li, L.; Niu, S.M.; Zhao, L.; Liu, J.; Shi, L.S.; Fu, M.; Liu, F. Rose (*Rosa rugosa*)-Flower Extract Increases the Activities of Antioxidant Enzymes and Their Gene Expression and Reduces Lipid Peroxidation. *Biochem. Cell Biol.* **2005**, *83*, 78–85. [CrossRef]
18. Sparkman, O.D. Identification of Essential Oil Components by Gas Chromatography/Quadrupole Mass Spectroscopy Robert P. Adams. *J. Am. Soc. Mass. Spectrom.* **2005**, *16*, 1902–1903. [CrossRef]
19. Dührkop, K.; Shen, H.; Meusel, M.; Rousu, J.; Böcker, S. Searching Molecular Structure Databases with Tandem Mass Spectra Using CSI:FingerID. *Proc. Natl. Acad. Sci. USA* **2015**, *112*, 12580–12585. [CrossRef]
20. Dührkop, K.; Fleischauer, M.; Ludwig, M.; Aksenov, A.A.; Melnik, A.V.; Meusel, M.; Dorrestein, P.C.; Rousu, J.; Böcker, S. SIRIUS 4: A Rapid Tool for Turning Tandem Mass Spectra into Metabolite Structure Information. *Nat. Methods* **2019**, *16*, 299–302. [CrossRef]
21. Baker, D.J.; Wijshake, T.; Tchkonia, T.; LeBrasseur, N.K.; Childs, B.G.; van de Sluis, B.; Kirkland, J.L.; van Deursen, J.M. Clearance of p16Ink4a-Positive Senescent Cells Delays Ageing-Associated Disorders. *Nature* **2011**, *479*, 232–236. [CrossRef]
22. Safwan-Zaiter, H.; Wagner, N.; Wagner, K.-D. P16INK4A—More Than a Senescence Marker. *Life* **2022**, *12*, 1332. [CrossRef]
23. Alessio, N.; Aprile, D.; Cappabianca, S.; Peluso, G.; Di Bernardo, G.; Galderisi, U. Different Stages of Quiescence, Senescence, and Cell Stress Identified by Molecular Algorithm Based on the Expression of Ki67, RPS6, and Beta-Galactosidase Activity. *Int. J. Mol. Sci.* **2021**, *22*, 3102. [CrossRef]
24. Han, X.; Tai, H.; Wang, X.; Wang, Z.; Zhou, J.; Wei, X.; Ding, Y.; Gong, H.; Mo, C.; Zhang, J.; et al. AMPK Activation Protects Cells from Oxidative Stress-Induced Senescence via Autophagic Flux Restoration and Intracellular NAD^+ Elevation. *Aging Cell* **2016**, *15*, 416–427. [CrossRef]
25. Wang, S.; Li, H.; Yuan, M.; Fan, H.; Cai, Z. Role of AMPK in Autophagy. *Front. Physiol.* **2023**, *13*, 2994.
26. Philippe, C.; Pinson, B.; Dompierre, J.; Pantesco, V.; Viollet, B.; Daignan-Fornier, B.; Moenner, M. AICAR Antiproliferative Properties Involve the AMPK-Independent Activation of the Tumor Suppressors LATS 1 and 2. *Neoplasia* **2018**, *20*, 555–562. [CrossRef]
27. Albrecht, D.; Ceschin, J.; Dompierre, J.; Gueniot, F.; Pinson, B.; Daignan-Fornier, B. Chemo-Genetic Interactions Between Histone Modification and the Antiproliferation Drug AICAR Are Conserved in Yeast and Humans. *Genetics* **2016**, *204*, 1447–1460. [CrossRef]
28. Shin, E.J.; Han, A.; Lee, M.; Song, Y.-R.; Lee, K.M.; Nam, T.-G.; Lee, P.; Lee, S.-Y.; Lim, T.-G. Extraction Conditions for Rosa Gallica Petal Extracts with Anti-Skin Aging Activities. *Food Sci. Biotechnol.* **2019**, *28*, 1439–1446. [CrossRef]
29. Hipp, M.S.; Kasturi, P.; Hartl, F.U. The Proteostasis Network and Its Decline in Ageing. *Nat. Rev. Mol. Cell Biol.* **2019**, *20*, 421–435. [CrossRef]
30. Jeong, D.; Qomaladewi, N.P.; Lee, J.; Park, S.H.; Cho, J.Y. The Role of Autophagy in Skin Fibroblasts, Keratinocytes, Melanocytes, and Epidermal Stem Cells. *J. Investig. Dermatol.* **2020**, *140*, 1691–1697. [CrossRef]
31. Kaushik, S.; Cuervo, A.M. The Coming of Age of Chaperone-Mediated Autophagy. *Nat. Rev. Mol. Cell Biol.* **2018**, *19*, 365–381. [CrossRef] [PubMed]
32. Raz, Y.; Guerrero-Ros, I.; Maier, A.; Slagboom, P.E.; Atzmon, G.; Barzilai, N.; Macian, F. Activation-Induced Autophagy Is Preserved in CD4+ T-Cells in Familial Longevity. *J. Gerontol. Ser. A* **2017**, *72*, 1201–1206. [CrossRef]
33. Ansary, T.M.; Hossain, M.R.; Kamiya, K.; Komine, M.; Ohtsuki, M. Inflammatory Molecules Associated with Ultraviolet Radiation-Mediated Skin Aging. *Int. J. Mol. Sci.* **2021**, *22*, 3974. [CrossRef]
34. Gonos, E.S.; Kapetanou, M.; Sereikaite, J.; Bartosz, G.; Naparło, K.; Grzesik, M.; Sadowska-Bartosz, I. Origin and Pathophysiology of Protein Carbonylation, Nitration and Chlorination in Age-Related Brain Diseases and Aging. *Aging* **2018**, *10*, 868–901. [CrossRef]
35. Nyström, T. Role of Oxidative Carbonylation in Protein Quality Control and Senescence. *EMBO J.* **2005**, *24*, 1311–1317. [CrossRef]

36. Low, E.; Alimohammadiha, G.; Smith, L.A.; Costello, L.F.; Przyborski, S.A.; von Zglinicki, T.; Miwa, S. How Good Is the Evidence That Cellular Senescence Causes Skin Ageing? *Ageing Res. Rev.* **2021**, *71*, 101456. [CrossRef]
37. Baker, D.J.; Childs, B.G.; Durik, M.; Wijers, M.E.; Sieben, C.J.; Zhong, J.; Saltness, R.A.; Jeganathan, K.B.; Verzosa, G.C.; Pezeshki, A.; et al. Naturally Occurring p16Ink4a-Positive Cells Shorten Healthy Lifespan. *Nature* **2016**, *530*, 184–189. [CrossRef]
38. Herzig, S.; Shaw, R.J. AMPK: Guardian of Metabolism and Mitochondrial Homeostasis. *Nat. Rev. Mol. Cell Biol.* **2018**, *19*, 121–135. [CrossRef]
39. Smitha Grace, S.R.; Chandran, G.; Chauhan, J.B. Terpenoids: An Activator of "Fuel-Sensing Enzyme AMPK" with Special Emphasis on Antidiabetic Activity. In *Plant and Human Health, Volume 2: Phytochemistry and Molecular Aspects*; Ozturk, M., Hakeem, K.R., Eds.; Springer International Publishing: Cham, Switzerland, 2019; pp. 227–244. ISBN 978-3-030-03344-6.
40. The Potential Therapeutic Value of Terpenes. Available online: https://accscience.com/journal/ITPS/7/3/10.36922/itps.0332 (accessed on 11 February 2025).
41. Izcara, S.; Perestrelo, R.; Morante-Zarcero, S.; Sierra, I.; Câmara, J.S. Volatilomic Fingerprinting from Edible Flowers. Unravelling Some Impact Compounds Behind Its Attractiveness. *Food Biosci.* **2022**, *50*, 102188. [CrossRef]
42. Tamilmani, P.; Sathibabu Uddandrao, V.V.; Chandrasekaran, P.; Saravanan, G.; Brahma Naidu, P.; Sengottuvelu, S.; Vadivukkarasi, S. Linalool Attenuates Lipid Accumulation and Oxidative Stress in Metabolic Dysfunction-Associated Steatotic Liver Disease via Sirt1/Akt/PPRA-α/AMPK and Nrf-2/HO-1 Signaling Pathways. *Clin. Res. Hepatol. Gastroenterol.* **2023**, *47*, 102231. [CrossRef]
43. Jiang, S.; Wang, Z.; Riethoven, J.-J.; Xia, Y.; Miner, J.; Fromm, M. Conjugated Linoleic Acid Activates AMP-Activated Protein Kinase and Reduces Adiposity More Effectively When Used with Metformin in Mice. *J. Nutr.* **2009**, *139*, 2244–2251. [CrossRef] [PubMed]
44. Watt, M.J.; Steinberg, G.R.; Chen, Z.-P.; Kemp, B.E.; Febbraio, M.A. Fatty Acids Stimulate AMP-Activated Protein Kinase and Enhance Fatty Acid Oxidation in L6 Myotubes. *J. Physiol.* **2006**, *574*, 139–147. [CrossRef] [PubMed]
45. Ido, Y.; Duranton, A.; Lan, F.; Weikel, K.A.; Breton, L.; Ruderman, N.B. Resveratrol Prevents Oxidative Stress-Induced Senescence and Proliferative Dysfunction by Activating the AMPK-FOXO3 Cascade in Cultured Primary Human Keratinocytes. *PLoS ONE* **2015**, *10*, e0115341. [CrossRef] [PubMed]
46. Lan, F.; Weikel, K.A.; Cacicedo, J.M.; Ido, Y. Resveratrol-Induced AMP-Activated Protein Kinase Activation Is Cell-Type Dependent: Lessons from Basic Research for Clinical Application. *Nutrients* **2017**, *9*, 751. [CrossRef]
47. Lee, S.-H.; Seo, H.-S.; Seo, S.J.; Kim, C.-D.; Hong, S.-P. Screening of Plant-Derived Natural Extracts to Identify a Candidate Extract Capable of Enhancing Lipid Synthesis in Keratinocytes. *Ann. Dermatol.* **2022**, *34*, 331–339. [CrossRef]
48. Čižmárová, B.; Hubková, B.; Tomečková, V.; Birková, A. Flavonoids as Promising Natural Compounds in the Prevention and Treatment of Selected Skin Diseases. *Int. J. Mol. Sci.* **2023**, *24*, 6324. [CrossRef]
49. Li, J.; Huang, Y.-C.; Deng, J.-M.; Yu, M.; Zouboulis, C.C.; Wang, G.-L.; Wang, J. Tea (*Camellia sinensis*) Seed Saponins Act as Sebosuppression Agents via the AMPK/mTOR Pathway. *J. Cosmet. Dermatol.* **2025**, *24*, e16793. [CrossRef]
50. Prasanth, M.I.; Sivamaruthi, B.S.; Chaiyasut, C.; Tencomnao, T. A Review of the Role of Green Tea (*Camellia sinensis*) in Antiphotoaging, Stress Resistance, Neuroprotection, and Autophagy. *Nutrients* **2019**, *11*, 474. [CrossRef]

cosmetics

MDPI

Article

In Vivo Application of the Effects of Red-to-Near-Infrared Light Spectroscopy on Skin-Brightening and Anti-Aging Properties via LED Facial Masks

Sunghoon Lee [1,*], Mohammad M. Afandi [2], Jungah Lee [1] and Jongsu Kim [2,*]

[1] Cell Bio Korea Co., Ltd., Bucheon-si 14445, Republic of Korea
[2] Department of Display Science and Engineering, Pukyong National University, Busan 48513, Republic of Korea; andiafandi@pukyong.ac.kr
* Correspondence: nvcrom@naver.com (S.L.); jsukim@pknu.ac.kr (J.K.)

Abstract: Most people want effective anti-aging and skin-brightening products. Although red-to-near-infrared (R/NIR) spectroscopy has recently been used in cosmetology, its practical use with high efficacy for anti-aging and skin brightening remains challenging. Herein, we aimed to determine the efficacy and improvement effects of a newly developed anti-aging and skin-brightening facial mask. A face study was conducted to assess efficacy and improvement effectiveness, with 21 female volunteers with oily, dry, and normal skin conditions applying the product under study (CF Magic Mask) to their face for 4-week periods. The dermatologist investigator evaluated the skin brightness, skin elasticity, eye wrinkles, dead skin cells on the scalp, dermal density, face lifting, scalp sebum, and global appearance. The mean skin-brightening and anti-aging parameters were improved ($p < 0.05$) after the use of the newly developed CF Magic Masks for 4-week periods. Significantly, the scalp sebum and dead skin cells on the scalp showed the greatest improvement, being reduced by about 26.71% and 21.96%, respectively. The global assessment by the volunteers showed moderate efficacy and preference, with no adverse effects or skin irritation indicated after the use of the test product.

Keywords: facial mask; skin brightening; anti-aging properties; red-to-near-infrared light; cosmetics

1. Introduction

In the dynamic realm of cosmetology, there is a growing recognition of the importance of personalized skincare tailored to individual needs and preferences. In this context, cosmetology is evolving, leveraging cutting-edge technologies to meet the growing demand for facial skin treatments [1]. One observable trend accompanying this evolution is the increasing adoption of non-invasive treatments for facial skin enhancement, aligning with the pursuit of an ideal appearance. This trend is reflected in the utilization of cutting-edge technologies such as rhodamine-intense pulsed light [2], black gold delicate pulsed light [3], line-field confocal optical coherence tomography [4], near-infrared diffuse reflectance spectroscopy [5], and nonsurgical modalities [6]. Among the innovative approaches being explored, near-infrared (NIR) spectroscopy has been applied in the field of cosmetology as a non-invasive method for analyzing human skin [7]. The cosmetic industry continues to face a significant challenge in effectively addressing skin brightening and aging concerns. Skin aging is characterized by several visible signs, including the appearance of wrinkles, an uneven skin tone, and dryness [8]. These manifestations are the result of a combination of internal factors inherent in an individual's biology and external factors, such as exposure to oxidative radicals induced by ultraviolet light, which contribute to the overall aging process [9]. Furthermore, hyperpigmentation can arise from prolonged exposure to UV (ultraviolet) radiation, hormonal imbalances, and the natural aging of the skin, as well as inflammation, injuries, and the accumulation of hemosiderin [10]. Brightening approaches are considered an efficient alternative to commonly used brightening agents.

Recent advancements in esthetic dermatology incorporate a growing comprehension of skin physiology to stimulate response mechanisms at both the tissue and molecular levels, aiming to achieve skin rejuvenation and brightening. Previous research provides robust evidence for the superiority of delicate pulsed light therapy over intense pulsed light treatment for the management of facial pigmented dermatoses [3]. Other technologies for facial skin treatment with rhodamine-intense pulsed light are the result of improvements in photodamaged facial skin, including reductions in vascular and pigment irregularities and improvements in skin texture [2]. NIR spectroscopy has been widely embraced for clinical purposes in several fields, due to its general effects on the body [11], especially the skin [12,13]. These supplementary approaches provide insights into skin hydration [14], an essential facet of cosmetology, thereby enriching our broad comprehension of skin physiology. Furthermore, NIR spectroscopy, known for its swiftness and non-intrusiveness, has been complemented by the development of advanced tools such as electrical conductance and capacitance analysis for assessing skin hydration [15,16]. By integrating NIR spectroscopy into these advanced tools, researchers and practitioners are better equipped to assess skin health, develop targeted interventions, and tailor personalized skincare regimens to meet individual needs effectively.

This study investigated the development and composition of a novel non-invasive facial mask integrated into R/NIR spectroscopy. The mask incorporates a set of R/NIR LEDs emitting light at wavelengths of 630, 655, and 842 nm, strategically positioned to ensure safety. These LEDs are utilized to stimulate cell renewal and boost collagen synthesis, aiming to improve skin brightness and reduce signs of aging. Clinical trials involving female participants were conducted to assess the efficacy of the product, with the findings showcasing its positive effects on skin appearance.

2. Methods

2.1. Product

Figure 1a is a photograph of the CF Magic Mask test product that is equipped with 480 red-to-near-infrared (R/NIR) LEDs. To eliminate the risk of direct light radiation through the eyes, a safety guard is implemented with a thick molding mask with a depth of 15 mm and a distance to the nearest LEDs of 20 mm, which completely blocks the light radiation to the eyes, as illustrated in Figure 1b. In detail, the LED groups consist of 144ea 655 nm LEDs, 112ea 630 nm LEDs, and 224ea 842 LEDs with a total rated output of 14.5 W, and the optical spectrum is depicted in Figure 2 and was recorded using a UV–Vis–NIR fiber optic spectroradiometer (USB 4000, Ocean Optics, Orlando, FL, USA).

The mask is provided with a strap for securing it in a circular fashion around the cranial region, coupled with a remote activation mechanism aimed at simplifying the operational procedure. Moreover, a ventilation system incorporating eight linear slits and a circular aperture is established at the oral region to enhance respiratory function for the patients. It has dimensions of 215 $\times$ 309 $\times$ 161 mm, with a light weight of 670 g.

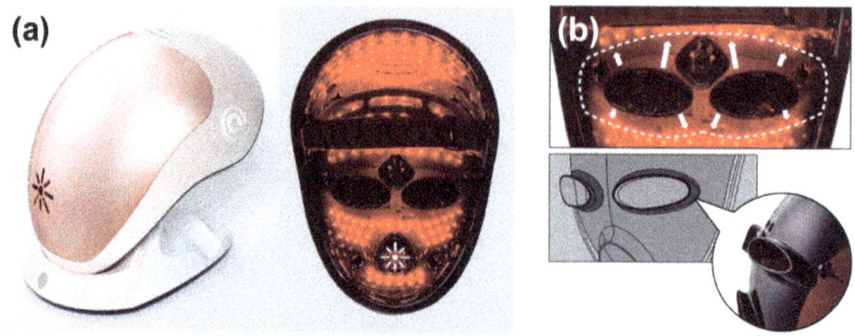

Figure 1. (**a**) The photograph of the CF Magic Mask equipped with 480 red-to-near-infrared (R/NIR) lights and (**b**) safety placement and eye-shield design to eliminate direct radiation from the lights to the eyes, with the arrow showing the 3D-rendering image.

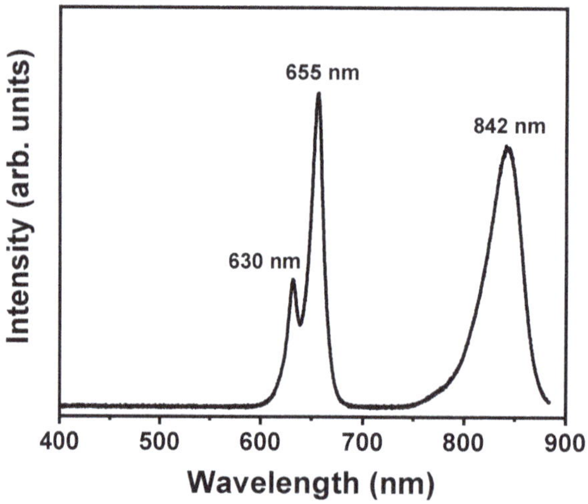

Figure 2. The optical spectrum of the CF Magic Mask consists of three main peaks in the R/NIR region.

2.2. Subjects

Twenty-one female volunteers (aged 29~59) without chronic disease involving the skin were enrolled after being screened and approved by the Review Board of P&K Skin Research Center and Chung-Ang University Hospital Dermatology Department. Written informed consent was obtained from all participants after the risks and benefits of the procedure were explained in detail. Enrolled participants could freely terminate their participation at any time. Participants were excluded if they reported a history of active steroid treatment in the last month and/or had undergone immune-suppressive treatment and scaling therapy in the last three months.

2.3. Treatment Protocol

Subjects were provided with the study product (CF Magic Mask) for 15 min of daily use during the study period of 4 weeks. Before utilization, the participants were required to cleanse their facial skin with water and subsequently pat it dry using a towel. Following this preparatory step, the CF Magic Mask was affixed and securely fastened utilizing the accompanying strap prior to the activation of the LED system. The LED system was subsequently activated for a duration of 15 min for the treatment procedure. The test area was assessed before, 2 weeks after, and 4 weeks after the test product was applied. After the test was completed, a global assessment of efficacy and product preference was conducted through a survey among the subjects. A compliance diary was provided to record the use of the study product. For the safety evaluation, on each visit, researchers comprehensively took into consideration the results of the visual evaluation and safety questionnaire.

2.4. Assessments

For the evaluation, the test subjects were stabilized for 30 min in an atmosphere room under constant temperature and humidity conditions, with an indoor temperature of 20–25 °C and a humidity of 40 to 60%, allowing the skin surface temperature and humidity to adapt to the environment of the measurement space, and while resting, water intake was restricted. For objective measurement, one researcher performed all the measurements, and the same area was measured at each measurement. The investigator evaluated the quality of skin hydration (Corneometer CM825, Courage-Khazaka GmbH, Cologne, Germany), skin elasticity (Cutometer CM580, Courage-Khazaka GmbH, Cologne, Germany), eye wrinkles (PRIMOS premium, GFMesstechnik GmbH, Berlin, Germany), and dermal density using a skin scanner (TPM taberna pro medicum, Lueneburg, Germany). Face lifting and skin brightness measurements were conducted using F-ray and Janus (Image pro plus v7.0), respectively. Dead skin cell measurements on the scalp was recorded using Visioscan VC98. The measurement was performed by extracting keratin using a special film (Coneofix, Medelink, Brossard, QC, Canada), and the image was photographed and measured as the *D.I.* (Desquamation Index) from Equation (1):

$$D.I. = \frac{2A + \sum_{n=1}^{5} Tn \times (n-1)}{6} \tag{1}$$

where *D.I.* is in %, A is the percentage of the area covered by corneocytes, Tn is the percentage of corneocytes in relation to the thickness, and n is the thickness level (1–5). Finally, the scalp sebum measurement was scanned using a Sebumeter SM815 (Courage-Khazaka GmbH, Cologne, Germany). All criteria were evaluated at baseline, week 2, and week 4. The global assessment of efficacy and product preference survey were evaluated on a 5-point ordinal scale (0 = very bad, 1 = bad, 2 = moderate, 3 = good, 4 = very good). Statistical significance was defined as a p-value less than or equal to 0.05. The ordinal nonparametric data obtained were evaluated as a change from baseline using the paired samples t-test and Wilcoxon signed-rank test.

3. Results

All 21 subjects successfully completed the 4-week assessment of the test product with no adverse effects, no skin abnormalities, and no medical or drug history that affected the study. The average age of the 21 participants was 46.67 years, with one in their 20s, two in their 30s, ten in their 40s, and eight in their 50s. To evaluate the skin quality of the test product, the skin parameters were measured before and after the use of the test product. After two weeks of the use of the test product, the skin brightness (L*) had changed from 68.23 ± 3.16 to 69.25 ± 2.96, the value for dead skin cells on the scalp (*D.I.*) had improved from 12.69 ± 5.89 to 7.74 ± 3.24, and scalp sebum ($\mu g/cm^2$) was reduced from 59.95 ± 75.58 to 19.81 ± 23.31. A significant improvement was seen in probability at $p = 0.0005$, $p = 0.001$, and $p = 0.001$ for skin brightness, dead skin cells on the scalp, and scalp sebum, respectively, after the product was used. A population comparison can be seen in the graph in Figure 3. Photographs of participants who applied for the test product after two weeks are depicted in Figure 4. A noticeable increase in skin brightness is observed, indicating the improvement after the use of the test product.

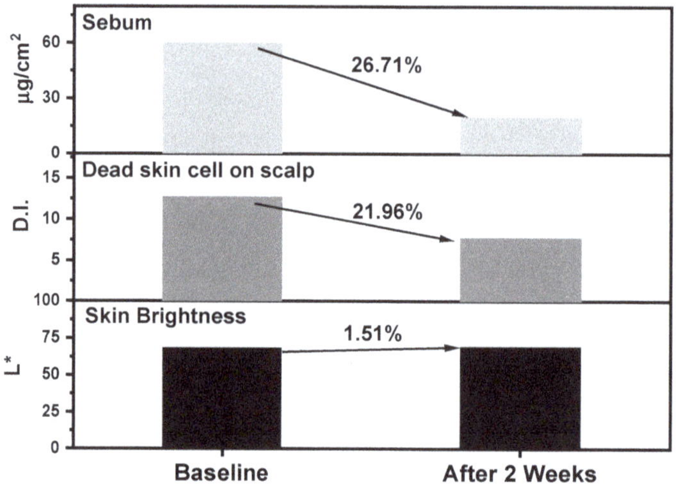

Figure 3. Improvements in skin brightness (L*), dead skin cells on the scalp (*D.I.*), and skin sebum (μg/cm^2) after 2 weeks of the use of the test product.

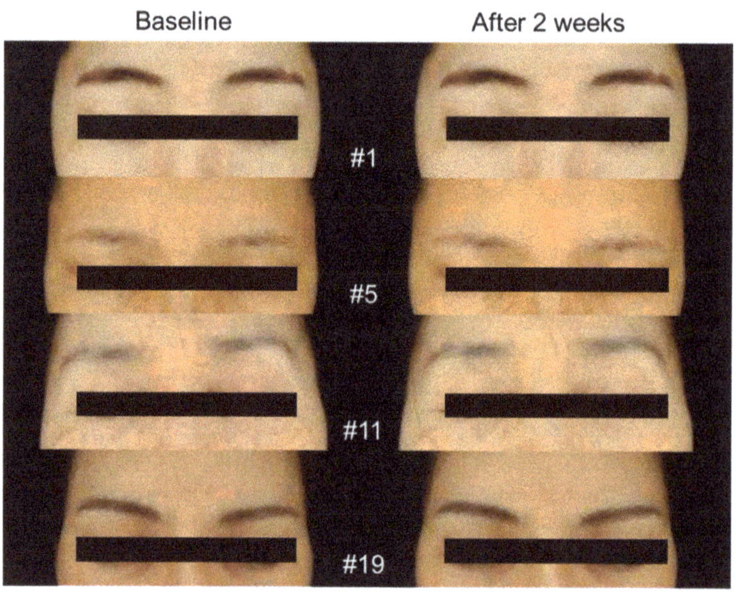

Figure 4. Clinical photographs comparing the skin brightness of the participants at baseline and after 2 weeks of treatment, black shape is used to protect patient privacy.

To evaluate the anti-aging parameters of the test product, skin hydration, elasticity, eye wrinkles, dermal density, and face lifting were investigated by the dermatologist investigator. After the use of the test product for 4 weeks, there was a statistically significant improvement in subject-reported skin hydration of 6.36% ($p = 0.003$) and in skin elasticity of 4.81% ($p = 0.002$), a decrease in eye wrinkles of 4.13% ($p = 0.001$), an improvement in dermal density of 5.52% ($p = 0.0001$), and a decrease in face lifting of 6.37% ($p = 0.0002$). The clinical photographs in Figure 5 show major improvements in the area with a red circle, with fewer eye wrinkles observed.

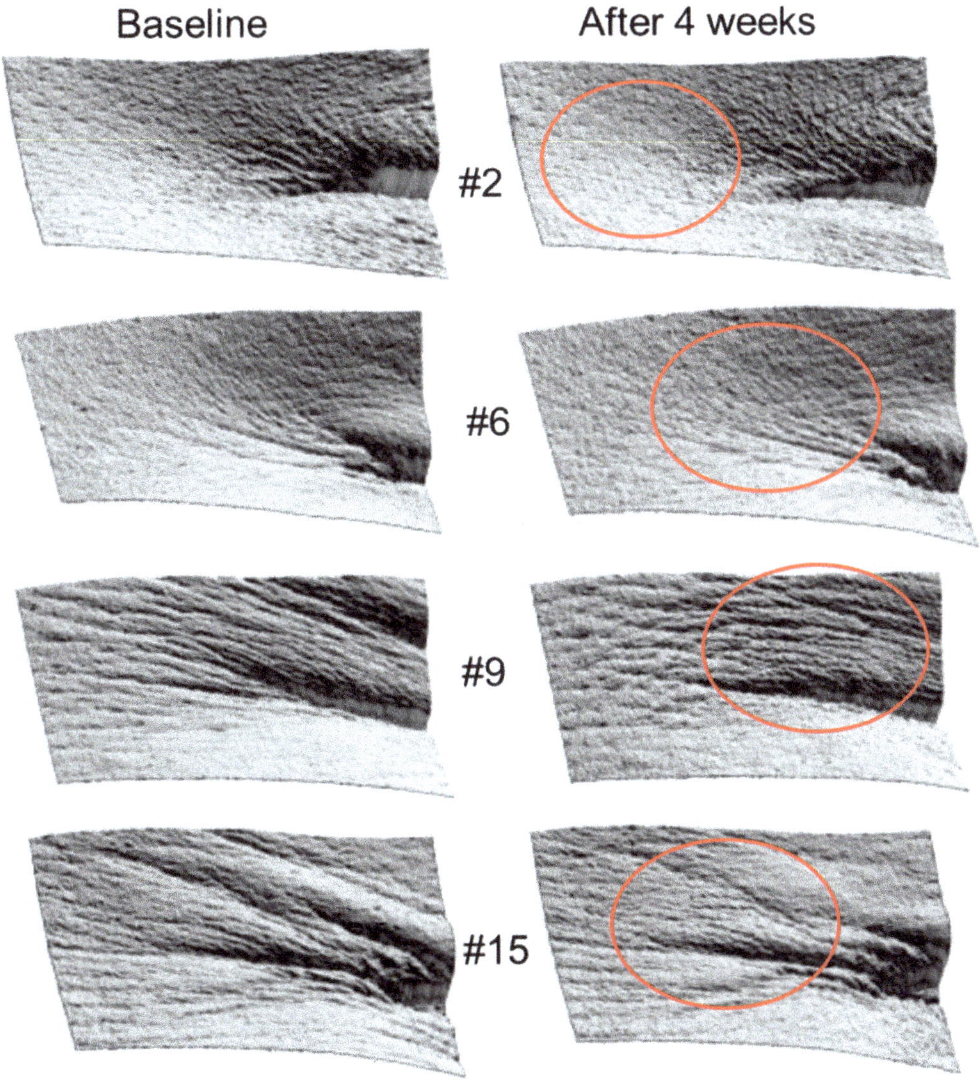

Figure 5. Photograph of the subject showing eye wrinkles at baseline and after 4 weeks of treatment, the red circle highlighting the skin improvement.

The improvement in skin quality after the use of the product was evaluated with global assessment surveys. The results are depicted in Table 1, with 100% of the volunteers giving scores higher than "moderate" for the skin hydration, skin elasticity, face lifting, and skin brightness aspects. Regarding the aspects of eye wrinkles, dermal density, dead skin on the scalp, and scalp sebum, the volunteers were 95.2% in agreement with higher than "moderate" satisfaction.

Table 1. Global assessment of the CF Magic Mask according to 21 volunteers.

Improvement Effect	Number of Subjects (Percentage, %)					Average	Standard Deviation
	4*	3*	2*	1*	0*		
Skin hydration	4 (19.0)	12 (57.2)	5 (23.8)	0 (0)	0 (0)	2.95	0.67
Skin elasticity	3 (14.3)	10 (47.6)	8 (38.1)	0 (0)	0 (0)	2.76	0.70
Eye wrinkles	2 (9.5)	10 (47.6)	8 (38.1)	1 (4.8)	0 (0)	2.62	0.74
Dermal density	3 (14.3)	9 (42.8)	8 (38.1)	1 (4.8)	0 (0)	2.67	0.80
Face lifting	3 (14.3)	9 (42.8)	9 (42.8)	0 (0)	0 (0)	2.71	0.72
Skin brightness	2 (9.5)	13 (62.0)	6 (28.5)	0 (0)	0 (0)	2.81	0.60
Dead skin on the scalp	3 (14.3)	7 (33.3)	10 (47.6)	1 (4.8)	0 (0)	2.57	0.81
Scalp sebum	3 (14.3)	10 (47.6)	7 (33.3)	1 (4.8)	0 (0)	2.71	0.78

Key: 4*, very good; 3*, good; 2*, moderate; 1*, bad; 0*, very bad.

4. Discussion

We initiated the evaluation by addressing the safety aspect of the CF Magic Mask product. Given that exposure to high doses of near-infrared (NIR) light spectroscopy directly targeted at the eyes can cause injury [17,18], implementing a protective shield is imperative in ensuring eye safety. As illustrated in Figure 1b, the CF Magic Mask design features a safety guard crafted from molded silicone to shield against direct radiation from red-to-near-infrared (R/NIR) LEDs. Additionally, there is a deliberate exclusion zone of approximately 1.5 cm around the eyes that is not covered by the LEDs, contributing to the smart and secure design of the device. Consequently, there were no adverse effects observed among the subjects after undergoing four weeks of treatment with the product.

Cosmetology in facial skin treatment has been evolving over the last decade. The trend in this technology is inclined toward non-invasive and reusable technology. Two of the hot topics in cosmetology are skin brightening and anti-aging techniques. Since skin brightening is multifactorial and encompasses abundant light reflection from evenly pigmented skin surfaces, the formulation was designed to brighten the skin tone with innovative treatment. Meanwhile, anti-aging treatment relies on skin hydration to improve skin elasticity. Previous works mentioned that exposure to 630 nm R/NIR LEDs in human skin promotes acute and chronic dermal wound repair and generation [19,20]. The process involves deeper penetration by R/NIR lights, which are absorbed by a set of cellular chromophores to regenerate the cells in the dermal tissue. The 830 nm continuous R/NIR has been successfully demonstrated to heal wounds in mice due to increased collagen production [21]. Accordingly, we used this formulation to design a facial mask that was covered with R/NIR LEDs to promote skin-brightening and anti-aging treatment.

Since collagen is fundamental in providing structure and skin strength and helping new cells grow, i.e., supporting skin health, increasing collagen production is critical in anti-aging treatment. The newly developed CF Magic Mask showed a major improvement in anti-aging parameters by reducing eye wrinkles in the subjects who used the product for four weeks, as illustrated in the clinical photograph in Figure 5. In detail, the average roughness (R_a) on the skin of the subjects near the eyes decreased by about 4.13%, from 23.74 to 22.68. The skin health of the subjects was also increased, indicated by the improvement in dermal density from 17.17% to 18.62%. Furthermore, Figure 6 illustrates the decrease in dead skin cells on the scalp of the volunteers after two weeks of CF Magic Mask treatment. The measurement was performed by extracting the skin keratin using a special film. The results indicate that exposure to R/NIR LEDs promoted skin regeneration by reducing dead skin cells.

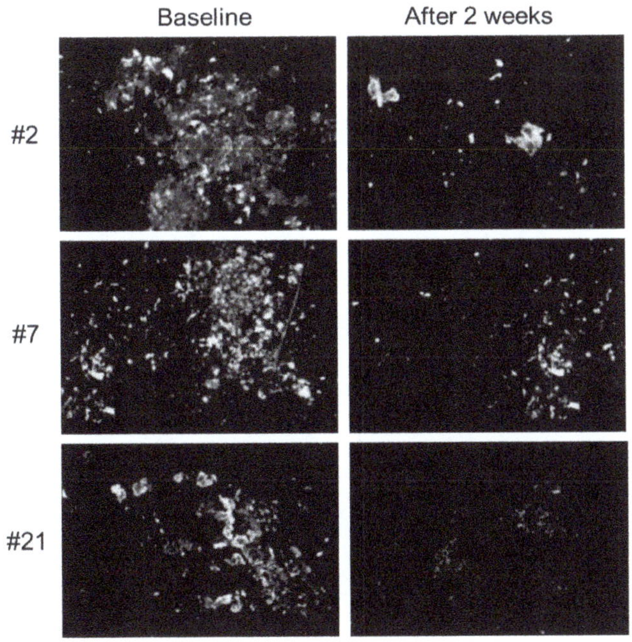

Figure 6. Clinical photograph of the dead skin cells on the scalp at baseline and after 4 weeks of treatment, obtained from Coneofixx.

Regarding the aspect of skin elasticity, after four weeks, R/NIR LED treatment successfully enhanced it, increasing the *R*2 value by about 4.81%, from 0.70 to 0.74. This is further confirmed for the face-lifting aspect, as shown in Figure 7. The lifting measurement is expressed by the skin with contour lines. The contour was then evaluated by drawing a straight line from the center of the circle near the cheekbones and measuring the angle with the straight line spanning to the mouth. The angle of subject #12 was reduced from 31.34 degrees to 28.25 degrees, as illustrated in the clinical photograph in Figure 7. Overall, the average angle among 21 volunteers decreased by 6.37% after four weeks of treatment with the newly developed CF Magic Mask. Thus, 97.6% of volunteers were satisfied with the product, regardless of the skin-brightening and anti-aging parameters, and the product preference survey resulted in 98.8% of volunteers having "moderate" satisfaction, as depicted in Tables 1 and 2, respectively.

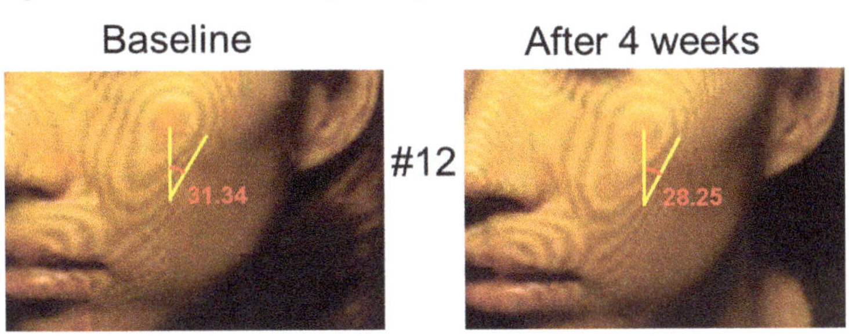

Figure 7. The skin contour angle of subject #12 at baseline and after 4 weeks of the use of the CF Magic Mask.

Table 2. Product preference survey results.

	Number of Subjects (Percentage, %)					Average	Standard Deviation
	4*	3*	2*	1*	0*		
Hydration	5 (23.8)	14 (66.7)	2 (9.5)	0	0	3.14	0.57
Smoothness	4 (19.0)	13 (61.9)	3 (14.3)	1 (4.8)	0	2.95	0.74
Applicability	4 (19.0)	12 (57.1)	5 (23.8)	0	0	2.95	0.67
Absorptiveness	3 (14.3)	14 (66.7)	4 (19.0)	0	0	2.95	0.59
Fragrance	0	9 (42.9)	11 (52.4)	0	1 (4.8)	2.33	0.73
Overall usability	6 (28.6)	11 (52.4)	4 (19.0)	0	0	3.10	0.70

Key: 4*, very good; 3*, good; 2*, moderate; 1*, bad; 0*, very bad.

The safety and efficacy of cosmetic products are essential because of their long-term and continued use by consumers. Although the newly developed CF Magic Mask has a shield that prevents direct R/NIR light reaching the eyes, proper use is essential, and the instructions must be followed and understood prior to using it. A limitation of this study is the assumption that all the volunteers followed the instruction to not take other supplements for skin treatment during the study period; there are also external factors like weather changes. However, with increasing usage, a significant change in skin tone and health is expected. This study was performed in Seoul, Republic of Korea, during the summer period from 1 August to 31 August 2016, when the UV index was at its maximum level.

5. Conclusions

Facial skin conditions treated with the newly developed CF Magic Mask equipped with red-to-near-infrared (R/NIR) LEDs for four weeks demonstrated improvements in skin-brightening and anti-aging parameters. The use of red LEDs promotes skin regeneration, while the utilization of NIR LEDs increases the production of collagen for healthy skin. In total, 97.6% of the volunteers were 97.6% in agreement with "moderate" satisfaction, and 98.8% preferred the product. Finally, there were no adverse effects or skin irritation after the use of the test product during the four weeks of the study period, indicating excellence in safety aspects and continuity, which is important since cosmetic products are used over a long time period.

Author Contributions: Conceptualization, S.L.; Methodology, S.L. and J.L.; Validation, S.L. and M.M.A.; Visualization, M.M.A. and J.L.; Writing—original draft, M.M.A.; Writing—review and editing, M.M.A., S.L. and J.K.; Supervision, S.L. and J.K. All authors participated in the research and the writing of the manuscript. All authors have read and agreed to the published version of the manuscript.

Funding: This research received no external funding.

Institutional Review Board Statement: All volunteers who participated in this research signed a consent form approved by the Review Board of P&K Skin Research Center and Chung-Ang University Hospital Dermatology Department, and clinical practice guidelines were followed (PNK-16801-K1R, 13 September 2016).

Informed Consent Statement: Informed consent was obtained from all subjects involved in the study.

Data Availability Statement: Data will be made available on request.

Conflicts of Interest: Sunghoon Lee is the CEO of Cell Bio Korea Co., Ltd. and J. Lee has been involved as a consultant and expert in Cell Bio Korea Co., Ltd. The authors declare that they have no known competing financial interests or personal relationships that could have appeared to influence the work reported in this research paper.

References

1. Tamrakar, D.G.; Takhur, S.S. Nanotechnology's application in cosmetics: Dermatology and skin care items. *Migr. Lett.* **2023**, *20*, 1–18. [CrossRef]
2. Cannarozzo, G.; Bonciani, D.; Tamburi, F.; Mazzilli, S.; Garofalo, V.; Del Duca, E.; Sannino, M.; Zingoni, T.; Nisticò, S.P. New insight in noninvasive rejuvenation: The role of a rhodamine-intense pulsed light system. *Photobiomodulation Photomed. Laser Surg.* **2019**, *37*, 539–543. [CrossRef] [PubMed]
3. Zhou, M.; Xu, N.; Lan, S.; Liu, W. Efficacy of black gold, delicate pulse light, super photon skin rejuvenation for pigmented dermatoses. *Altern. Ther. Health Med.* **2024**, *31*, 232–237.
4. Jdid, R.; Pedrazzani, M.; Lejeune, F.; Fischman, S.; Cazorla, G.; Forestier, S.; Khalifa, Y.B. Skin dark spot mapping and evaluation of brightening product efficacy using line-field confocal optical coherence tomography (LC-OCT). *Skin Res. Technol.* **2024**, *30*, 1363. [CrossRef]
5. Miyamae, M.; Kawabata, M.; Yamakawa, Y.; Tsuchiya, J.; Ozaki, Y. Non-invasive estimation of skin thickness by near infrared diffuse reflection spectroscopy-separate determination of epidermis and dermis thickness. *J. Near Infrared Spectrosc.* **2022**, *20*, 617–622. [CrossRef]
6. Li, K.; Meng, F.; Li, Y.R.; Tian, Y.; Chen, H.; Jia, Q.; Cai, H.; Jiang, H.B. Application of nonsurgical modalities in improving facial aging. *Int. J. Dent.* **2022**, *2022*, 8332631. [CrossRef] [PubMed]
7. Miyamae, Y.; Yamakawa, Y.; Kawabata, M.; Ozaki, Y. A combined near-infrared diffuse reflectance spectroscopy and principal component analysis method of assessment for the degree of photoaging and physiological aging of human skin. *Anal. Sci.* **2012**, *28*, 1159–1164. [CrossRef] [PubMed]
8. Bruce, S.; Karnik, J.; Dryer, L.; Burkholder, D. Anti-aging proof of concept study: Results and summary. *J. Drugs Dermatol.* **2014**, *13*, 1074–1081. [PubMed]
9. Rattanawiwatpong, P.; Wanitphakdeedecha, R.; Bumrungpert, A.; Maiprasert, M. Anti-aging and brightening effects of a topical treatment containing vitamin C, vitamin E, and raspberry leaf cell culture extract: A split-face, randomized controlled trial. *J. Cosmet. Dermatol.* **2020**, *19*, 671–676. [CrossRef] [PubMed]
10. Zasada, M.; Debowska, R.; Pasikowska, M.; Ostrowska, B.; Budzisz, E. Efficacy of tri-active brightening and anti-aging complex in treatment of facial skin hyperpigmentation. *J. Pharm. Pharmacol.* **2016**, *4*, 564–573.
11. Mustafa, F.H.; Jones, P.W.; McEwan, A.L. Near infrared spectroscopy for body fat sensing in neonates: Quantitative analysis by GAMOS simulations. *Biomed. Eng. Online* **2017**, *16*, 14. [CrossRef] [PubMed]
12. Mizukoshi, K.; Hamanaka, Y.; Niwayama, M. Investigation of oxygen saturation in regions of skin by near infrared spectroscopy. *Ski. Res. Technol.* **2022**, *28*, 695–702. [CrossRef] [PubMed]
13. Miyamae, Y.; Kawabata, M.; Yamakawa, Y.; Ozaki, Y. Review article: Non-invasive assessment of photoaging and physiological aging of human skin and non-invasive estimation of the thickness of mice skin by near infrared diffuse reflectance spectroscopy. *NIR News* **2013**, *24*, 11–14. [CrossRef]
14. Qassem, M.; Kyriacou, P.A. In vivo optical investigation of short term skin water contact and moisturizer application using NIR spectroscopy. In Proceedings of the 35th Annual International Conference of the IEEE Engineering in Medicine and Biology Society, Osaka, Japan, 3–7 July 2013; pp. 2392–2395.
15. Mohammad, M.; Msabbri, A.R.; MatJafri, M.Z. Conceptual design of near infrared spectroscopy instrumentation for skin moisture measurement. In Proceedings of the IEEE Colloquium on Humanities, Science and Engineering, Penang, Malaysia, 5–6 December 2011; pp. 801–804.
16. Egawa, M.; Arimoto, H.; Hirao, T.; Takahashi, M.; Ozako, Y. Regional difference of water content in human skin studied by diffuse-reflectance near-infrared spectroscopy: Consideration of measurement depth. *Appl. Spectrosc.* **2006**, *60*, 24–28. [CrossRef]
17. Swiatczak, B.; Schaeffel, F. Effects of short-term exposure to red or near-infrared light on axial length in young human subjects. *Ophthalmic Physiol. Opt.* **2024**, *44*, 954–962. [CrossRef] [PubMed]
18. Giacci, M.K.; Wheeler, L.; Lovett, S.; Dishington, E.; Majda, B.; Bartlett, C.A.; Thornton, E.; Harford-Wright, E.; Leonard, A.; Vink, R.; et al. Differential effects of 670 nm and 830 nm red near infrared irradiation therapy: A comparative study of optic nerve injury, retinal degeneration, traumatic brain and spinal cord injury. *PLoS ONE* **2014**, *9*, e194565. [CrossRef] [PubMed]
19. Skopin, M.D.; Molitor, S.C. Effects of near-infrared laser exposure in a cellular model of wound healing. *Photodermatol. Photoimmunol. Photomed.* **2009**, *25*, 75–80. [CrossRef]

20. Yadav, A.; Gupta, A. Noninvasive red and near-infrared wavelength-induced photobiomodulation: Promoting impaired cutaneous wound healing. *Photodermatol. Photoimmunol. Photomed.* **2017**, *33*, 4–13. [CrossRef] [PubMed]
21. Tatmatsu-Rocha, J.C.; Ferraresi, C.; Hamblin, M.R.; Maia, F.D.; do Nascimento, N.R.F.; Driusso, P.; Parizotto, N.A. Low-level laser therapy (904 nm) can increase collagen and reduce oxidative and nitrosative stress in diabetic wounded mouse skin. *J. Photochem. Photobiol.* **2016**, *164*, 96–102. [CrossRef]

cosmetics

MDPI

Article

Multifunctional, Novel Formulation for Repairing Photoaged and Sun-Damaged Skin: Insights from In Vitro, Ex Vivo, and In Vivo Studies

María Moneo-Sánchez *, Nagore de Pablo , Leire Arana-Pascual, Itziar Beitia, Sandra Benito-Cid and Raúl Pérez-González

i+Med S. Coop., Alava Technology Park, Hermanos Elhuyar 6, 01510 Vitoria-Gasteiz, Spain; nagore.depablo@imasmed.com (N.d.P.); leire.arana@imasmed.com (L.A.-P.); ibeitia@imasmed.com (I.B.); rperez@imasmed.com (R.P.-G.)
* Correspondence: maria.moneo@imasmed.com

Abstract: Prolonged sun exposure disrupts the skin's structural, mechanical, and functional properties, accelerating aging and contributing to skin disorders. To counteract these effects, we developed a formulation containing potent antioxidant, anti-inflammatory, moisturizing, and reparative ingredients to protect and repair sun-damaged skin. The efficacy of the formulation was evaluated through in vitro, ex vivo, and in vivo studies. Results demonstrated that the formulation reduced oxidative stress and suppressed the production of pro-inflammatory cytokines interleukin-6 (IL-6) and interleukin-8 (IL-8), as well as lactate dehydrogenase (LDH) secretion, in a photodamaged skin model, nearly reaching levels observed in undamaged skin. It also restored collagen levels, improving structural integrity. In vivo, no adverse reactions were observed when used for 28 or 56 days. The formulation improved skin hydration by up to 46%, reduced transepidermal water loss by 20%, increased luminosity by 70%, and reduced hyperpigmented spots by 14%. It also enhanced skin firmness and elasticity by 30% and reduced wrinkle volume and density by up to 53% and 19%, respectively. These findings demonstrate that the formulation's active ingredients effectively target the pathways altered by UV exposure, offering considerable potential for preventing and reversing sun-induced skin damage while improving both the appearance and functionality of the skin.

Keywords: photodamage; UV radiation; sun damage; aging; skin; oxidative stress; inflammation; collagen; pigmentation; hyaluronic acid

1. Introduction

The skin serves as a vital barrier that shields the body from external influences, including physical, chemical, and microbiological challenges, while maintaining hydration by preventing water loss. However, being the body's outermost layer, it is constantly exposed to environmental factors that accelerate its natural aging process. Among these, prolonged exposure to ultraviolet (UV) radiation from sunlight is the most significant contributor to extrinsic aging [1], a process referred to as photoaging [2–4]. Clinically, photoaged skin manifests as dryness, roughness, deep wrinkles, sagging, loss of elasticity, and hyperpigmentation [5]. This appearance results from tissue-level structural changes involving both cellular components and the extracellular matrix (ECM). Key alterations include reduced collagen levels, driven by increased degradation and mediated by matrix metalloproteinases (MMPs), and decreased biosynthesis of new collagen fibers [6]. Additionally, there is an accumulation of abnormal elastin fibers, known as elastosis, and a significant reduction in glycosaminoglycans (GAGs) such as hyaluronic acid, which are critical for maintaining skin hydration and elasticity. At the cellular level, UV radiation impairs the function of keratinocytes, fibroblasts, and melanocytes, leading to compromised barrier integrity, diminished regenerative capacity, hyperpigmentation, and an overall decline in skin resilience. These changes contribute to the structural and functional decline observed in photoaged skin, increasing its susceptibility to skin disorders, including non-melanoma and melanoma skin cancers [3,7,8].

The detrimental effects of photoaging result from complex biological mechanisms triggered by UV radiation, visible light, and infrared radiation, all of which penetrate the skin at varying depths. A primary mechanism is the generation of reactive oxygen species (ROS), which directly damage cellular and extracellular biomolecules, including DNA, proteins, lipids, and polysaccharides [9]. These oxidative events destabilize cellular components, impair repair processes, and activate signaling pathways, including the mitogen-activated protein kinase (MAPK) cascade [10]. This activation leads to the nuclear translocation of transcription factors such as nuclear factor-kappa B (NF-κB) and activator protein-1 (AP-1), which promote the expression of inflammatory mediators, such as interleukin-6 (IL-6) and interleukin-8 (IL-8), and matrix metalloproteinases (MMPs). MMPs are responsible for ECM degradation, particularly of collagen and elastin, which are vital for maintaining the skin's structural integrity and elasticity. In addition to ECM remodeling, UV radiation induces premature cellular senescence, limiting the replicative capacity of fibroblasts, keratinocytes, and melanocytes, and promoting a senescence-associated secretory phenotype (SASP) characterized by the release of inflammatory cytokines and proteases. Furthermore, chronic low-grade inflammation, termed "inflammaging," perpetuates oxidative damage and ECM degradation, creating a self-sustaining cycle of tissue damage [11]. Chronic UV exposure also disrupts the skin microbiota [12]. The microbiota play a vital role in maintaining skin health, including protection against harmful pathogens, modulation of immune responses, and enhancement of skin's barrier function. UV radiation alters this microbial balance, reducing the skin's natural defenses and exacerbating inflammation.

In response to this need, we have designed a reparative product that incorporates our novel OxIR PhotoRescue Technology® to target and counteract the primary mechanisms of photodamage. This advanced formulation combines antioxidants such as encapsulated beta-carotene, a corticosteroid-like anti-inflammatory agent that mimics hydrocortisone's effects without associated adverse reactions, and a unique blend of plant-derived extracts with potent regenerative effects. When synergized with specially selected hyaluronic acid, an advanced combination of natural sugars comprising natural moisturizing factors (NMF) and physiomoisturizers, and a product that enhances skin microbiota protection, this formula is intended to improve signs of aging while providing robust protection against sun exposure-associated skin disorders. In order to validate the combined effects of the formulation in mitigating skin photodamage, a series of in vitro, ex vivo, and in vivo studies have been conducted in which antioxidant and anti-inflammatory capacity, its ability to counteract cellular damage and the effects of sun exposure on the extracellular matrix, as well as its impact on visible signs of skin aging have been evaluated.

2. Materials and Methods

2.1. Investigational Product

The product to be evaluated is a multi-active formula with antioxidant, anti-inflammatory, and regenerative ingredients combined in the novel technology OxIR PhotoRescue Technology®, manufactured by Unikare Bioscience SL (Vitoria-Gasteiz, Spain), part of i+Med Cooperative of Scientists group, for Sibari Republic® brand and commercialized under the trademark name of Cellular Rescue.

The specific ingredients of the product, listed according to the INCI (International Nomenclature of Cosmetic Ingredients), are as follows: AQUA, GLYCERIN, ISOTRIDECYL ISONONANOATE, ETHYL OLEATE, BATYL ALCOHOL, STEARIC ACID, GOSSYPIUM HERBACEUM SEED EXTRACT, PUNICA GRANATUM SEED EXTRACT, HYLOCEREUS UNDATUS FRUIT EXTRACT, KRAMERIA TRIANDRA ROOT EXTRACT, CAPRYLIC/CAPRIC TRIGLYCERIDE, SODIUM ACRYLATES COPOLYMER, PENTYLENE GLYCOL, PHYSALIS ANGULATA EXTRACT, HYDROXYACETOPHENONE, GLYCERYL BEHENATE, HYDROGENATED POLYDECENE, TREHALOSE, 1,2-HEXANEDIOL,CAPRYLYL GLYCOL,

FRUCTOSE, UREA, CITRIC ACID, SODIUM HYDROXIDE, LECITHIN, PHOSPHOLIPIDS, POLYGLYCERYL-10 STEARATE, HELIANTHUS ANNUUS SEED OIL, SODIUM HYALURONATE, HYDROGENATED PHOSPHATIDYLCHOLINE, LYSOLECITHIN, MALTOSE, SODIUM PCA, SODIUM CHLORIDE, SODIUM LACTATE, TREHALOSE, ALLANTOIN, POTASSIUM SORBATE, FRUCTOOLIGOSACCHARIDES, TOCOPHEROL, PHYTIC ACID, GLUCOSE, DAUCUS CAROTA SATIVA ROOT EXTRACT, BETA-CAROTENE.

2.2. Reagents

Ascorbic acid (A4544), hydrogen peroxide solution (H_2O_2, H1009), and Hank's Balanced Salts Solution (H-2387) were all purchased from Sigma-Aldrich (St. Louis, MO, USA). 5 (and-6)-carboxy-2′,7′-difluorodihydrofluorescein diacetate (carboxy-H2DFFDA) (C-13293) was sourced from Invitrogen (Waltham, MA, USA), while PBS (10×) was obtained from Roche (11 666 789001) (Basel, Switzerland). The IL-8 Human ELISA Kit (DY208) and IL-6 Human ELISA Kit (DY206) were provided by R&D Systems (Minneapolis, MN, USA). The Collagen Assay Kit (S1000) was acquired from Biocolor (Belfast, UK), and the CytoTox 96 kit (G1780) was supplied by Promega (Madison, WI, USA). Skin culture medium without animal components with Pen/Strep was supplied by Biopredic International MIL215 (Saint-Grégoire, France).

2.3. Measurement of Antioxidant Capacity

The antioxidant activity of the testing product was determined by using the in-tube DFFDA assay. This assay measures the capacity of the testing product to block the oxidation of DFFDA by H_2O_2. Different concentrations of the product (10%, 1%, and 0.1% v/v, diluted in assay buffer) were incubated in a solution with 5 μM DFFDA and 4 mM H_2O_2. After 1 h, fluorescence was measured at Ex/Em = 485/535 nm. Ascorbic acid (50 mM) was used as reference material (positive control). The percentage of antioxidant activity (AA%) was calculated by comparing the fluorescence values of the control (negative control) and test samples by applying the following formula:

$$AA\% = 100 - \left[\frac{(F^{485/535}\,CX - F^{485/535}\,B - CX) \times 100}{F^{485/535}\,\text{Control}} \right]$$

$F^{485/535}$ CX: fluorescence at Ex/Em = 485/535 nm of the product at a given concentration in DFFDA and H_2O_2 presence.

$F^{485/535}$ B − CX: fluorescence at Ex/Em = 485/535 nm of the product at a given concentration (Blank).

$F^{485/535}$ Control: fluorescence Ex/Em = 485/535 nm of the assay buffer in DFFDA and H_2O_2 presence (C−).

2.4. Ex Vivo Model of Photoaging

Human organotypic skin explant cultures (hOSECs) were obtained from healthy donors undergoing plastic surgery (authorization granted by the French government's ethical committee according to French law L.1245 CSP). Up to 2 h from the surgery, the skin was cut into 0.8 cm^2 pieces and shipped in transport medium. Upon receipt, samples were placed with dermis facing down and epidermis facing up in culture plates containing skin culture medium without animal components supplemented with Pen/Strep (1%). Tissue cultures were incubated for at least 48 h at 37 °C under 5% CO_2 for recovery prior to study initiation.

In order to mimic skin photoaging, UV-Vis (ultraviolet-visible) light irradiation (5 J/cm^2) was applied daily to the hOSEC by using the SOL 500 Solar simulator (Dr. Hönle, Gräfelfing, Germany), based on previously published studies using similar doses and exposure times [13,14]. At the same time, the product was administered topically at 2 mg/cm^2. The product was in contact with the hOSEC throughout this study, for a total of 9 days.

Four replicates of each experimental group (control, photoaged skin, and photoaged skin + product) were carried out.

2.5. *Cytokine Quantification*

The IL-6 and IL-8 cytokine quantification assays were performed in the supernatants of the skin culture. Culture supernatants were collected, clarified by centrifugation, aliquoted, and stored at −70 °C until analysis. Once thawed, cytokine levels were determined by typical sandwich ELISA assay kits according to the manufacturer's recommendations.

2.6. *Lactate Dehydrogenase (LDH) Cytotoxicity Assay*

The LDH Cytotoxicity test is a colorimetric assay that quantitatively measures lactate dehydrogenase (LDH), a stable cytosolic enzyme that is released into the culture medium supernatant upon damage to the cytoplasmic membrane. The released LDH in culture medium supernatants can be measured by a 30 min coupled enzymatic reaction; LDH oxidizes lactate to pyruvate, which then reacts with the tetrazolium salt WST-1 to form formazan. The increase in the amount of formazan measured in the culture supernatant directly correlates to the increase in the number of damage cells in the skin explant. 50 μL of supernatant were removed from each sample and transferred into a 96-well microplate. After that, 50 μL of formazan dye was added to each sample, and, following a 30-min incubation period, absorbance was read using a standard ELISA plate reader at 490 nm.

2.7. *Collagen Content Analysis*

The soluble collagen content was determined using the collagen kit from Biocolor Ltd. (Carrickfergus, UK). The collagen assay is a dye-binding method for the analysis of acid and pepsin-soluble collagens. hOSECs were incubated with a solution of pepsin concentration at 0.1 mg/mL and acetic acid at 0.5 mM at 4 °C overnight. Collagen Dye Reagent (Sircol Dye Reagent) (1 mL) was added to each supernatant and shaken for 30 min. Subsequently, 750 μL ice-cold Acid-Salt Wash Reagent was added to the collagen-dye pellet to remove unbound dye from the surface of the pellet and centrifuged at 13,200× *g* for 10 min. Finally, 250 μL of alkali reagent was added. The dissolved dye (200 μL of each sample in 96 microwell plates) was measured using a standard ELISA plate reader at 556 nm. The reading for the samples was compared against a standard curve to obtain semi-quantitative information and made relative to mg of fresh dermal tissue.

2.8. *Anti-Aging Efficacy Test Under Dermatological Control*

The anti-aging efficacy of the product was evaluated in 20 volunteers with sensitive and photoaged skin for 56 days. During the study days, facial photographs and biometrical measurements were taken before the application of the product and after its continued use (images were processed using Adobe Lightroom software version 8). In addition, after 56 days of product use, the volunteers filled out the subjective evaluation questionnaire, with questions regarding the efficacy of the product. At the end of this study, the dermatologist also evaluated the tolerance of the product.

2.9. *Volunteer Recruitment*

Each volunteer participating in this study was previously informed about the type and procedures of the study, signing an informed consent before the start of the study. The volunteers participating in this study met the inclusion and exclusion criteria, verified through a recruitment questionnaire (Appendix A).

2.10. *Product Application Criteria*

The application area was the face, and the product was applied to clean, dry skin morning and evening or after prolonged exposure to the sun by gently massaging until complete absorption. Volunteers were instructed not to use any other cosmetic products of the same type in the experimental areas during this study and not to change their usual hygiene habits.

One week before starting to use the product, the volunteers were instructed to carry out a correct pre-wash phase. This phase was performed to standardize the skin conditions of the volunteers. The conditions in the week before the examination were to avoid applying any skin care products to the experimental areas and to avoid the application of topical drugs in experimental areas.

2.11. Biometric Evaluation of Efficacy

On the measurement days, the volunteers participating in this study remained for 10 min in an acclimatized room at temperature 20 °C ± 2 °C and a relative humidity of 40–60%. The experimental area needed to be clean, without any product applied.

Moisturizing efficacy was determined by measuring the distribution of near-surface hydration and the microtopography of the skin with MoistureMap MM 100 (Courage & Khazaka electronic, Köln, Germany) the initial day (D0) and after 30 min (T30MIN), 2 h (T2H), 24 h (T24H), 28 days (T28D), and 56 days (T56D) of product use.

Efficacy in improving the barrier function was determined by measuring the water-evaporation density gradient of the skin with the Tewameter® TM 300 probe (Courage & Khazaka electronic, Köln, Germany) on the initial day (D0) and after 2 h (T2H), 24 h (T24H), 28 days (T28D), and 56 days (T56D) of product use.

Luminosity was determined by colorimetric image analysis, measuring the difference between lightness and darkness with the VisioFace® 1000D (Courage & Khazaka electronic, Köln, Germany) on the initial day (D0), at 28 days (T28D), and after 56 days (T56D) of product use.

Depigmenting efficacy was evaluated by measuring the light reflected by the skin, using the Mexameter® MX 18 probe (Courage & Khazaka electronic, Köln, Germany) on the initial day (D0), at 28 days (T28D), and after 56 days (T56D) of product use.

Firming and elasticity efficacy were determined by measuring the mechanical deformation of the skin with the Cutometer® dual MPA 580 probe (Courage & Khazaka electronic, Köln, Germany) on the initial day (D0) and at 30 min (T30MIN), 2 h (T2H), 24 h (T24H), and 28 days (T28D) of product use.

Anti-wrinkle efficacy was determined by measuring volume and density of wrinkles using AEVA^{3D}-HE^{2} VisioHop (Eotech, MI, USA) on the initial day (D0), at 28 days (T28D), and after 56 days (T56D) of product use.

The time points selected were chosen based on their common use in evaluating the parameters described in our study and are typically employed in efficacy studies of similar products [13–15].

Extended information on biometric measurements can be found in Appendix B.

2.12. Dermatological Efficacy Evaluation

At the initial visit and after 28 days and 56 days of product application, the dermatologist evaluated the severity of volunteers' wrinkles from two facial regions: forehead and frown lines. Wrinkles were classified using a scale from 0 to 5, being: 0 none, 1 noticeable, 2 shallow, 3 moderately deep, 4 deep with well-defined borders, and 5 very deep, redundant fold.

2.13. Subjective Evaluation

The volunteers, on their visit to the center at the end, filled out a questionnaire answering questions related to the efficacy (Appendix C).

2.14. Tolerance Evaluation

At the end of this study, the dermatologist assessed the skin tolerance of the product. The alterations to be evaluated were erythema, xerosis/desquamation, edema, exudation, comedogenicity, and pigmentation alterations on a four-point scale (absence, slight/moderate/intense). It was also indicated whether the changes observed could be related to the use of the product, according to the scale: not related, improbable, possible, probable, certain, or not valuable.

2.15. Data Presentation

Data from in vivo experiments are presented in tables as the average percentage of variation from initial day (T0) across all volunteers (% variation from T0) and the percentage of volunteers showing improvement (% responders) at the indicated times: 30 min (T30MIN), 2 h (T2H), 24 h (T24H), 28 days (T28D), and/or 56 days (T56D) of product use.

2.16. Statistical Analysis

In the in vitro and ex vivo assays, values are given as mean ± SEM. The homogeneity of variance was confirmed by the Bartlett's test, and the normality was confirmed by the Anderson–Darling test. One-factor analysis of variance (ANOVA) with Fisher's LSD post-hoc tests were performed to assess differences between groups' means. When homogeneity of variances could not be assumed, Welch's correction was used, and when parametric assumptions were not met, Kruskal–Wallis with Dunn's post-hoc test was used.

For in vivo study, a normality analysis (Shapiro–Wilk) and equal variance test (Brown–Forsythe) were conducted to determine the data distribution and choose the type of analysis: non-parametric tests (Kruskal–Wallis) or parametric tests (one-way repeated measures ANOVA + Holm–Sidak or Student–Newman–Keuls post hoc test). Asterisks indicate a statistically significant difference (* $p < 0.05$; ** $p < 0.01$; *** $p < 0.001$) compared to T0.

3. Results

Firstly, given that oxidative stress is considered one of the central mechanisms of photodamage due to its direct and indirect effects on cells and tissues, in vitro studies were conducted to evaluate the antioxidant capacity of the formula. The DFFDA assay, which measures the capacity of the product to block the oxidation of DFFDA by H_2O_2, revealed that even at a 0.1% concentration, the product inhibited 60% of DFFDA oxidation (Figure 1). This antioxidant capacity increased to over 90% at a 1% concentration, and a complete inhibition (100%) was observed at a 10% concentration, comparable to the antioxidant potential of the positive control. The results were statistically significant for all the analyzed doses. These results confirm the strong antioxidant potential of the formula.

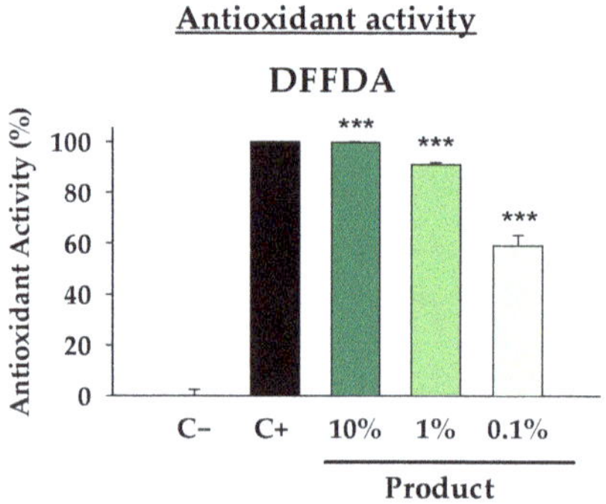

Figure 1. Antioxidant activity of the product measured by DFFDA assay. Negative control (C−) represents a condition in which no product is added (0% antioxidant activity). Positive control (C+) represents a condition in which ascorbic acid 50 mM is added. Concentration values are expressed as % volume/volume. Asterisks (***) indicate a statistically significant difference ($p < 0.001$) compared to the negative control (C−) group.

In addition, solar radiation can both directly and through generating oxidative stress create a pro-inflammatory environment that exacerbates tissue damage. Therefore, it is crucial for a reparative product to effectively reduce the levels of inflammatory mediators. For this reason, we evaluated the anti-inflammatory capacity of the product, using in this case an ex vivo model of photodamage. In this model, human skin explants (hOSECs) were exposed to UV-Vis (ultraviolet-visible) irradiation mimicking the harmful effects of sun exposure, and levels of two key pro-inflammatory cytokines, IL-6 and IL-8, commonly elevated in inflammatory conditions, were measured. As shown in Figure 2, UV-Vis irradiation led to a statistically significant increase in the secretion of both IL-6 and IL-8 compared to the control, non-irradiated group. However, the topical application of the product on hOSECs resulted in an also statistically significant reduction in the levels of these cytokines compared to the photoaged group, reducing them to levels comparable to those of non-irradiated tissue. This result demonstrates the potent anti-inflammatory effect of the product, highlighting its potential to alleviate the inflammatory responses associated with photodamage and aging.

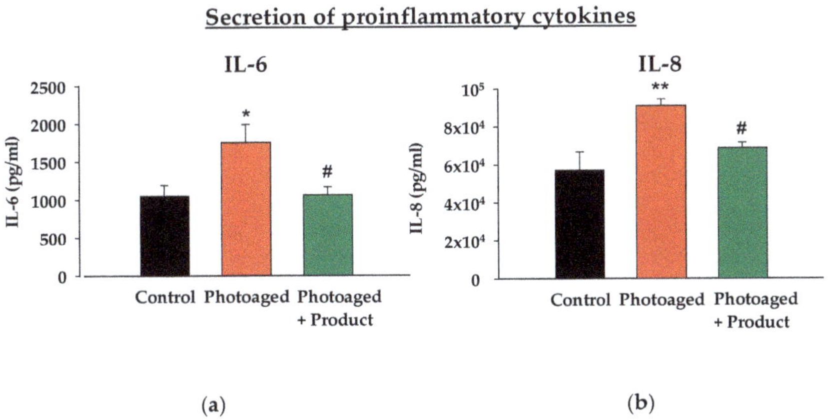

Figure 2. Inflammation in photoaged human skin explants. IL-6 (**a**) and IL-8 (**b**) secretion levels in non-irradiated control human skin explants (Control), explants exposed to UV-Vis light (Photoaged), or skin explants exposed to UV-Vis light and treated with the product (Photoaged + Product) after 9 days. Asterisks indicate a statistically significant difference (* $p < 0.05$; ** $p < 0.01$) compared to the Control group. Pounds indicate a statistically significant difference (# $p < 0.05$) compared to the Photoaged group.

Following the previous observation of the product's anti-inflammatory effects, we decided to study whether the product could also protect cells from damage caused by sun irradiation. To achieve this, we analyzed lactate dehydrogenase (LDH) leakage in the three experimental groups within the photoaged model. The LDH cytotoxicity test is a colorimetric assay that quantitatively measures lactate dehydrogenase, a stable cytosolic enzyme released into the culture medium upon cytoplasmic membrane damage. This assay is widely used as an indicator of cell viability in cytotoxicity studies. As expected, UV-Vis irradiation led to a 50% increase in LDH leakage compared to the non-irradiated group (statistically significant), indicating higher plasma membrane damage (Figure 3). Remarkably, the topical application of the product reduced LDH leakage to levels comparable to those of the non-irradiated group, demonstrating its protective effect against UV-induced damage. This reduction was statistically significant compared to the photoaged condition.

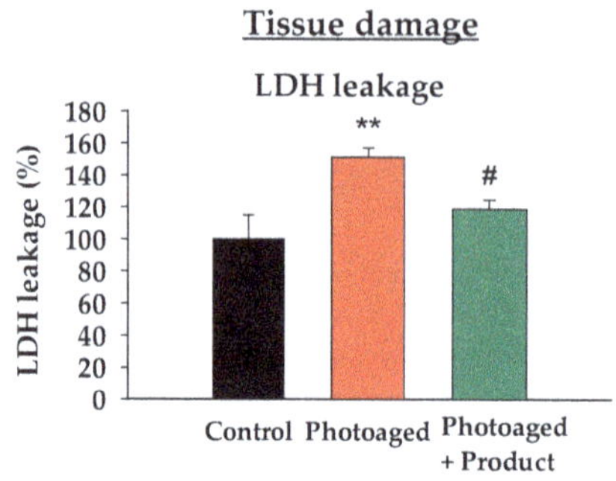

Figure 3. Tissue damage in photoaged human skin explants. Tissue damage, measured by LDH leakage, of human skin explants (hOSECs) that were either non-irradiated (Control), exposed to UV-Vis light (Photoaged), or exposed to UV-Vis light and treated with the product (Photoaged + Product) after 9 days. Results were normalized to 100% LDH concentration based on non-irradiated control skin explants. Asterisks indicate a statistically significant difference (** $p < 0.01$) compared to the Control group. Pounds indicate a statistically significant difference (# $p < 0.05$) compared to the Photoaged group.

To assess if the observed improvements in cellular damage and inflammation led to better tissue function and extracellular matrix recovery, we measured tissue collagen levels. Using a dye-binding method that analyzes acid and pepsin-soluble collagens, which include the predominant types I and III that are crucial for maintaining dermal tissue structure and integrity, we found that, as expected, UV-Vis irradiated (Photoaged) skin explants exhibited a statistically significant reduction in soluble collagen content compared to the non-irradiated controls (Figure 4). Remarkably, topical administration of the product was able to restore collagen levels to 80% of the control, indicating a recovery of 50% of the UV-induced loss. In fact, under the product treatment condition, collagen levels were no longer significantly different from the control group, suggesting that the product effectively restores the collagen loss observed in the absence of its application. These results demonstrate the product's effectiveness in promoting collagen preservation and maintaining matrix integrity.

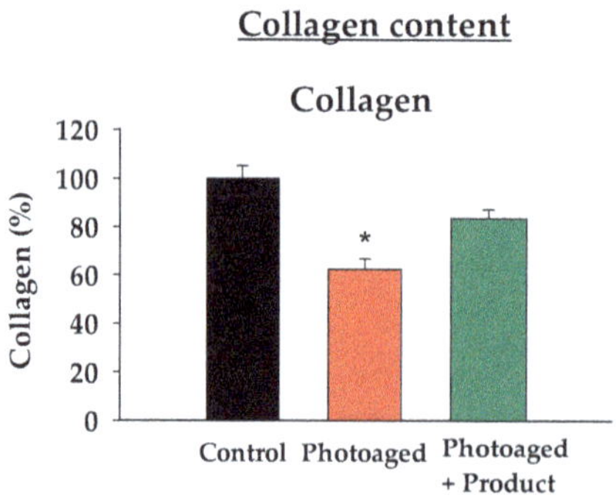

Figure 4. Collagen content of photoaged human skin explants. Collagen levels in non-irradiated control hOSECs (Control), hOSECs exposed to UV-Vis light (Photoaged), and hOSECs exposed to UV-Vis light and treated with the product (Photoaged + Product) after 9 days. Results were normalized to 100% collagen content based on non-irradiated control skin explants. Asterisks indicate a statistically significant difference (* $p < 0.05$) compared to the Control group.

The model was found to be valid, as it effectively reproduced UV-Vis radiation-induced damage such as cellular damage, secretion of inflammatory molecules, and reduction in collagen levels. However, to further test the global anti-photoaging effect, we proceeded to perform a clinical evaluation of the efficacy in a group of volunteers.

A total of 20 volunteers with sensitive and photoaged skin participated in this study, and the product was used for 56 days. Anti-aging efficacy was assessed through biometric evaluations conducted by a dermatology specialist. Parameters such as hydration, barrier function, pigmentation, luminosity, firmness, elasticity, and wrinkles were measured at baseline, after 28 days, and at the end of the 56-day treatment period. At the conclusion of the trial, product tolerance and subjective efficacy through a participant questionnaire were also evaluated.

Regarding the tolerance evaluation, a dermatologist examined the possible alterations present in the volunteers in the experimental area: erythema, xerosis/desquamation, exudation, edema, comedogenicity, pigmentary alterations, and others. No volunteers presented adverse effects after continued use of the product.

The instrumental assessment of efficacy focused on several key parameters: moisturization, barrier function, luminosity, pigmentation, firmness, elasticity, and wrinkle severity.

Moisturizing efficacy was assessed by measuring near-surface hydration distribution and skin microtopography at 30 min, 2 h, and 24 h to evaluate immediate effects, as well as at 28 and 56 days to determine long-term benefits. Results, shown in Table 1, indicated a 47% increase in hydration within the first 30 min, which remained high at 2 h (42%) and decreased only slightly at 24 h (26%). Improvements were sustained over time, with a 20% increase in hydration at 28 days, with 75% of volunteers showing improvement, and a 31% increase at 56 days, also with 75% of volunteers experiencing better hydration. The differences were statistically significant in all cases. This demonstrates a strong short-term and long-lasting moisturizing effect of the product. Barrier function was evaluated by measuring the water-evaporation density gradient of the skin. At 2 h post-application, a slight reduction (5%) in transepidermal water loss (TEWL) was observed, becoming more pronounced at 24 h (9%). Long-term assessments showed an 11% reduction at 28 days and a 20% reduction at 56 days, with 74% of volunteers demonstrating improvement at both time points, indicating a notable enhancement of the epidermal barrier function, especially after prolonged use (Table 1).

Table 1. Moisturizing and barrier function improvement efficacy results. Asterisks indicate a statistically significant difference (* $p < 0.05$; ** $p < 0.01$; *** $p < 0.001$) compared to T0.

	Moisturizing Efficacy				
	T30MIN	**T2H**	**T24H**	**T28D**	**T56D**
% variation from T0	46.62 ***	42.31 ***	25.75 **	19.81 *	30.96 **
% responders	93.75	81.25	68.75	75.00	75.00
	Barrier function improvement efficacy (TEWL)				
		T2H	**T24H**	**T28D**	**T56D**
% variation from T0		−4.77	−9.27	−11.24	−19.56
% responders		63.16	63.16	73.68	73.68

Luminosity was measured using colorimetric image analysis, assessing the difference between lightness and darkness. The results showed that the average improvement among the volunteers that exhibited an enhanced luminosity was 59% and 69% for 28 and 56 days, respectively (Table 2). On the other hand, the ability of the product to reduce existing pigmentation was evaluated by measuring the light reflected by the skin both in spot and control areas. At 28 and 56 days, 67% and 61% of volunteers showed a reduction in melanin content in spot areas, with statistically significant average reductions of 9% and 14% (negative results in the table indicate a reduction in melanin content) and maximum reductions of up to 19% and 53%, respectively (Table 2). Figure 5 presents a representative image of one of the volunteers, highlighting a visible reduction in pigmentation of the larger spots, along with a diminished appearance of the smaller ones. Importantly, melanin content in control areas was not decreased (Supplementary Information, Table S1), indicating that the product does not reduce basal skin tone. Thus, these results evidence the product's ability to even out the skin tone, improving the uniformity and appearance of the skin.

Table 2. Luminosity and depigmenting efficacy results. Asterisks indicate a statistically significant difference (** $p < 0.01$; *** $p < 0.001$) compared to T0.

	Luminosity Efficacy		Depigmenting Efficacy (Spot Areas)	
	T28D	**T56D**	**T28D**	**T56D**
% responders	33.33	33.33	66.67	61.11
% variation from T0	59.17	69.05	−9.03 ***	−14.08 **
Maximum improvement (%)	-	-	−18.95	−52.72

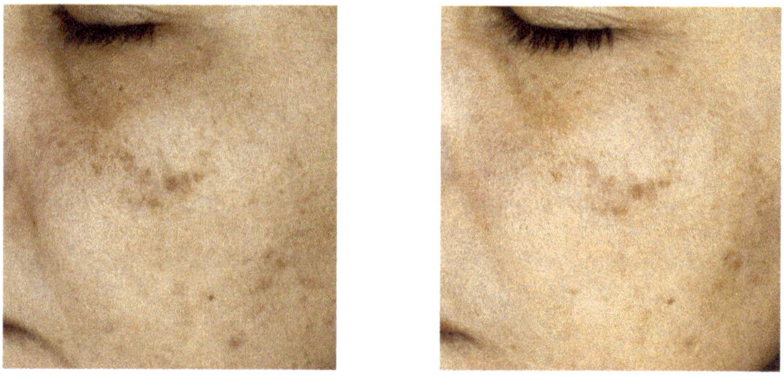

Figure 5. Representative image of depigmenting efficacy in a volunteer. The image depicts a pigmented area at initial day (T0) on the left, and at final day (T56D) on the right.

Firming and elasticity efficacy were also analyzed, in this case by measuring the mechanical deformation of the skin under suction. Skin firmness was evaluated based on resistance to deformation, while elasticity was assessed by the skin's ability to return to its original position. Firmness improved significantly within 30 min of product application (24%), with a further increase observed at 2 h (30%). Although there was a slight reduction at 24 h (14%), long-term treatment demonstrated renewed improvement, with a 21% increase at 28 days (Table 3). Elasticity followed a similar pattern, showing a 24% improvement at 30 min, rising to 31% at 2 h, and slightly decreasing to 16% at 24 h before reaching 28% at 28 days. Notably, up to 88% of volunteers reported enhanced firmness and elasticity. All time points assessed showed statistically significant differences in both parameters.

Wrinkle-improvement efficacy was assessed using both instrumental and dermatological evaluations. The instrumental assessment was conducted by a three-dimensional scanning sensor that measures skin topography based on fringe-projection technology combined with active stereometry, and both wrinkle volume and density were evaluated (Table 4). Wrinkle volume decreased in 44% of volunteers after 28 days, with an average reduction of 27% among those showing improvement and maximum reductions reaching 40%. By day 56, 56% of volunteers exhibited positive changes, with an average reduction in volume of 29% and maximum reductions of 53%. Regarding wrinkle density, 47% and 53% of volunteers showed improvements at 28 and 56 days, respectively, with an average reduction of 8% but reaching maximum reductions of up to 19% in both cases (Table 4). Variation percentages were statistically significant for all cases. Representative images of the three-dimensional scanning sensor AEVA-HE showing wrinkle density can be found in Figure 6.

Table 3. Firming and elasticity efficacy results. Asterisks indicate a statistically significant difference (*** $p < 0.001$) compared to T0.

		Firming Efficacy		
	T30MIN	**T2H**	**T24H**	**T28D**
% variation from T0	−24.20 ***	−30.50 ***	−14.41 ***	−20.63 ***
% responders	88.89	88.89	72.22	77.78
		Elasticity efficacy		
	T30MIN	**T2H**	**T24H**	**T28D**
% variation from T0	−24.42 ***	−31.31 ***	−16.25 ***	−27.66 ***
% responders	88.24	88.24	70.59	88.24

Table 4. Anti-wrinkle efficacy (Biometric evaluation). Asterisks indicate a statistically significant difference (** $p < 0.01$; *** $p < 0.001$) compared to T0.

	Wrinkle Volume		**Wrinkle Density**	
	T28D	**T56D**	**T28D**	**T56D**
% responders	44.44	55.56	47.06	52.94
% variation from T0	−27.04 ***	−29.72 ***	−8.04 **	−7.61 **
Maximum improvement (%)	−39.55	−53.22	−19.23	−18.64

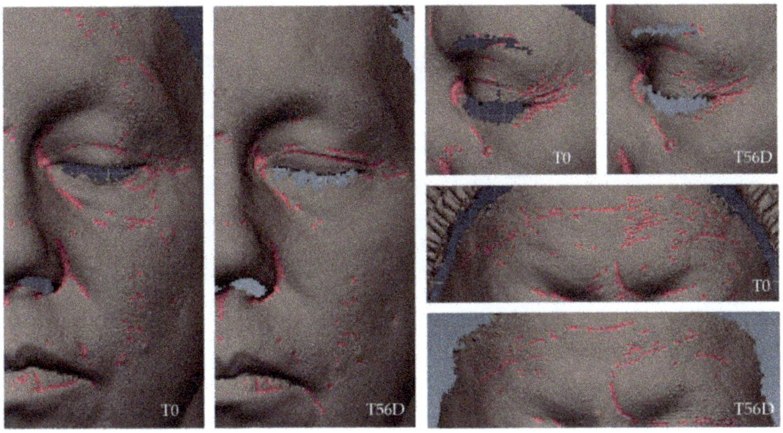

Figure 6. Anti-wrinkle efficacy (AEVA-HE images). Three-dimensional scanning sensor images, where wrinkles are shown in pink for three different volunteers before (T0) and after 56 days of product application (T56D).

In parallel, a dermatological assessment was conducted by a dermatologist who evaluated wrinkles on the forehead and frown lines, using a scoring scale from 0 (no wrinkles) to 5 (very deep, redundant folds). The evaluation revealed that wrinkles were generally classified as less severe after product use (Figure 7). In particular, forehead wrinkles rated at the highest levels showed a 10% reduction at both 28 and 56 days, and frown lines exhibited a 10% reduction at 28 days and a 5% reduction at 56 days.

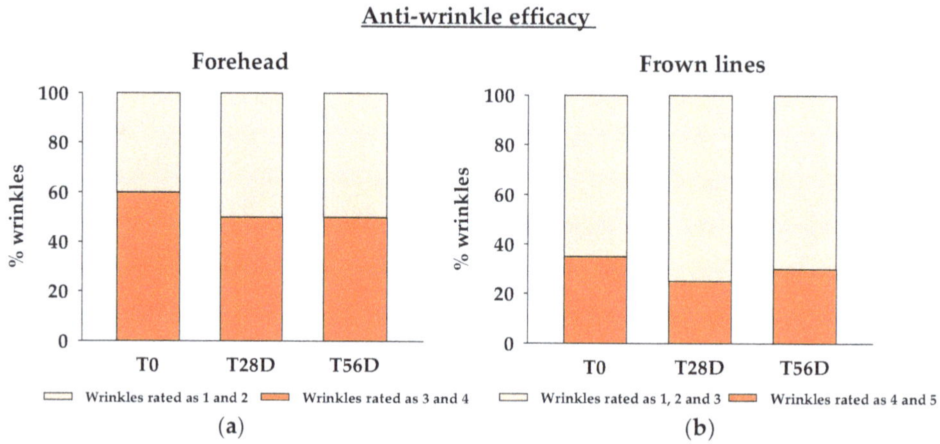

Figure 7. Anti-wrinkle efficacy (Dermatological evaluation). Wrinkle severity was evaluated by a dermatologist. Percentage of forehead wrinkles (**a**) and frown lines (**b**) rated at higher (dark orange) and lower (light orange) severity levels are shown for initial day (T0), day 28 (T28D), and day 56 (T56D).

Lastly, the subjective evaluation revealed that at least 80% of the volunteers felt their skin was significantly more hydrated, nourished, soft, flexible, refreshed, and comfortable after using the product (Figure 8).

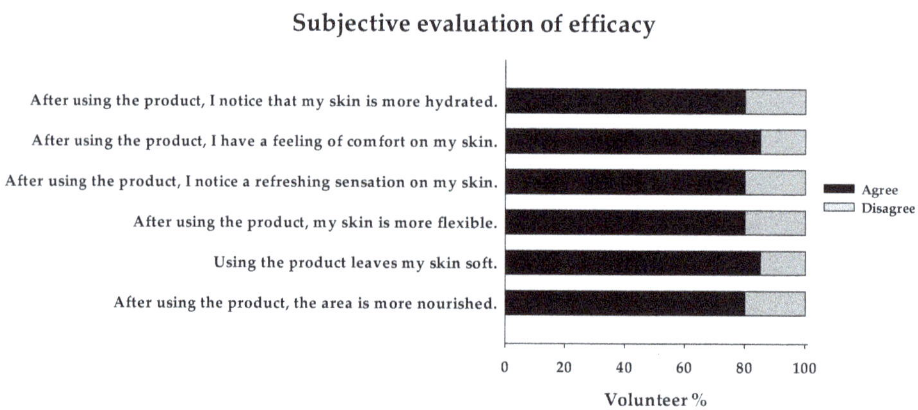

Figure 8. Subjective evaluation of efficacy. The percentage of volunteers who agree or disagree with each statement is shown.

In conclusion, both instrumental indicators and volunteer perceptions confirm that this product, due to its potent antioxidant, anti-inflammatory, protective, and moisturizing properties, effectively repairs sun-damaged skin, improving its appearance and functionality.

4. Discussion

The present work demonstrates, through a series of in vitro, ex vivo, and in vivo studies, that we have achieved a groundbreaking advancement in skincare with the development of a single formulation capable of addressing photodamage at multiple levels. By effectively counteracting key mechanisms such as oxidative stress, inflammation, and extracellular matrix degradation, our product provides a comprehensive solution to the complex challenges of photoaging. This includes restoring collagen levels, improving hydration, repairing the skin barrier, enhancing elasticity and firmness, reducing wrinkles, and correcting uneven skin tone. The results highlight the formulation's ability to not only enhance the skin's appearance but also restore its functionality, offering an advanced approach to the repair of sun-induced damage.

This remarkable achievement is the result of our specific approach that combines a deep understanding of the mechanisms underlying photodamage with the precise selection of scientifically validated ingredients and their integration using our advanced OxIR PhotoRescue Technology®.

Among these mechanisms, oxidative stress plays a critical role in photoaging. Since the skin's natural antioxidants are insufficient to fully counteract the effects of sun exposure [16], the addition of antioxidants to skincare formulations has proven to be an effective strategy in mitigating photoinduced damage [17]. However, the challenge with using antioxidants in topical formulations lies in their low stability and the need for effective delivery into the deeper skin layers to be beneficial. Our product addresses these challenges by incorporating a delivery system that stabilizes and delivers beta-carotene into deep epidermal layers. Beta-carotene is the most abundant and efficient precursor of Vitamin A, also called Pro-Retinol, a natural sun protector and powerful antioxidant [18,19]. Experimental studies have consistently demonstrated its protective effects against UV-induced photodamage [20], with topical applications of beta-carotene providing protective against acute and chronic manifestations of photodamage, including sunburn, premature aging, actinic keratosis, and other skin conditions [21].

Another key active ingredient in our formulation is *Hylocereus undatus* fruit extract, which offers not only hydrating, mineralizing, and vitaminizing properties but also has potent antioxidant activity due to its polyphenolic compounds, pigments, flavonoids, and vitamins B, C, and E [22,23]. These components have demonstrated strong free-radical scavenging activity in various studies related to photoaging [24–26]. The inclusion of all these ingredients has yielded a product with powerful antioxidant activity, as evidenced by our data.

Our results also indicate that the product exerts a potent anti-inflammatory effect on radiation-induced damage due to the inclusion of strong anti-inflammatory agents such as *Physalis angulata* extract, which has demonstrated efficacy as a topical anti-inflammatory and wound-healing agent [27,28]. Actually, *Physalis angulata* extract exhibited corticosteroid-like effects comparable to hydrocortisone in a randomized double-blind placebo-controlled clinical trial [29].

Additionally, our formulation includes a lipid complex composed primarily of batyl alcohol and soy lecithin, which possess anti-inflammatory and moisturizing properties. Batyl alcohol is known for its anti-inflammatory and DNA-protecting properties, while it has also demonstrated significant reparative effects on the epidermal barrier [30,31]. Soy lecithin, a phospholipid, is highly valued in cosmetic formulations due to its role as a key component of cell membranes, which gives it a natural affinity with the skin. This strong biocompatibility and biomimetic properties enhance the bioavailability of active ingredients, leading to more effective results [32]. Additionally, soy lecithin has been found to improve skin hydration, reduce irritation, and promote wound healing [33,34]. The inclusion of *Krameria triandra* root extract, rich in catechin tannins and other components, further contributes to the anti-inflammatory properties of the formulation [35].

As previously mentioned, sun radiation-induced oxidative stress and inflammation lead to cellular damage, induction of senescence, and the degradation of the extracellular matrix. Given the potent ability of our product to mitigate these harmful processes, we could expect a protective effect on cellular and tissue levels in a photodamage model. Indeed, our studies demonstrate that the application of the product significantly reduced cellular damage, while it preserved collagen levels within the tissue.

This protective effect was not only observed in ex vivo models but was also confirmed through instrumental measurements in an in vivo study involving volunteers who applied the product for 56 days. The results demonstrated significant improvements in skin firmness as well as in elasticity, suggesting that the product also protects against degradation and modification of other extracellular matrix components, such as elastin. This beneficial impact on the extracellular matrix explains the reductions in wrinkle density and volume observed in the volunteers. This reduction levels, although they may seem subtle in terms of average numerical improvement, are of considerable significance, since the results are comparable to those obtained with collagen-based products. However, while these products improve signs of aging by providing an external source of collagen, our product is able to act on pre-existing wrinkles by stimulating endogenous extracellular matrix production, thereby facilitating the recovery of collagen levels within the skin.

Another hallmark of photoaging is the loss of hydration due to impaired barrier function and the degradation of natural moisturizing factors caused by sun damage. To restore hydration and maintain the skin's structural integrity and protective function, potent hydrating and nourishing agents were included in our formulation.

Hyaluronic acid (HA), a naturally occurring polysaccharide in the skin, plays a crucial role in maintaining dermal and epidermal hydration due to its hygroscopic nature. Beyond providing hydration, HA is crucial for skin integrity as it influences cell-to-cell and cell-to-matrix interactions [36] and regulates keratinocyte differentiation and the formation of extracellular lipids in the stratum corneum, which are essential for the integrity and functionality of the epidermal barrier [37], thereby contributing to skin resilience and promoting regeneration [38]. Due to its unique viscoelasticity, biocompatibility, biodegradability, and non-immunogenicity properties, HA has been extensively used in dermal applications for skin rejuvenation [39], among multiple other applications [40]. To date, numerous clinical studies have demonstrated the effectiveness and tolerability of HA [39,41]. In cosmetics, the trend has increasingly favored the use of sodium hyaluronate, the salt form of hyaluronic acid, due to its enhanced water solubility and stability, which allows for deeper skin penetration. In addition, it is especially suitable for sensitive and, hence, photodamaged skin.

Another potent hydrating component of the formulation is a complex blend of natural sugars, natural moisturizing factors (NMFs), and physiomoisturizers. This mixture has been specifically designed to deliver both immediate and long-lasting hydration to the skin, acting through a dual mechanism; hygroscopic molecules restore moisture to the skin layers, while filmogenic polymers form a protective barrier to prevent water evaporation [42]. These sugars have low molecular weights, allowing them to penetrate the skin effectively and restore hydration. Together, these components help to maintain the skin's moisture balance, improving its overall resilience and comfort.

The formula also contains a component with phospholipids, which support the regeneration of the skin barrier, thereby reducing transepidermal water loss. Additionally, it enhances the anti-inflammatory and antioxidant effects of the formula due to its content of *Helianthus annuus* (sunflower) seed oil, rich in vitamin E [43,44]. This combination of agents is intended to reinforce skin's protective barrier while improving hydration and provide added protection against oxidative stress and inflammation. Indeed, our TEWL evaluation results reveal that the product significantly repairs the skin's barrier function, restoring its ability to protect against external aggressors and maintaining adequate hydration levels. Coupled with the potent moisturizing agents included in the formulation, the product demonstrates both immediate and long-lasting hydration benefits, which not only improves the skin's appearance but also facilitates better absorption of active ingredients, thereby increasing the overall efficacy of the product.

The product also enhanced facial luminosity and reduced dark spot pigmentation without altering the natural skin tone, thereby promoting a more even skin appearance. This effect can be attributed to an improved cellular function of melanocytes, as well as to the properties of additional active ingredients in the formulation, which reduce excessive melanogenesis and promote the production of protective molecules by the skin microbiota [45–49]. Due to recent evidence highlighting the importance of the sun–microbiota–skin axis, incorporating a product that supports and enhances the skin microbiota is crucial for enhancing the effectiveness of a reparative cream. The skin microbiota plays a key role in protecting against solar radiation by producing beneficial compounds, named as solar postbiotics, which contribute to UV defense, reduce inflammation, and help maintain overall skin homeostasis. However, prolonged sun exposure can disrupt this protective function, leading to increased oxidative stress and inflammation, which accelerates photoaging. By incorporating ingredients that protect and maintain healthy skin microbiota, the product not only reinforces the skin's natural defenses but also mitigates the harmful effects of solar exposure, thus offering a complete protection and repair. In particular, this formulation includes *Gossypium herbaceum* seed extract, a cellular lysate rich in protective factors consisting on a mix of plant chromophores, polyphenolic antioxidants, anti-inflammatory agents, photoprotective compounds [50], *Punica granatum* seed extract, rich in selective membrane lipids and prebiotic polyphenols with anti-inflammatory and antioxidant properties [51–53], and other plant-derived components such as fructooligosaccharides and trehalose. These ingredients enhance the resistance of microbial cell membranes to solar radiation, thereby preventing the formation of harmful molecules produced by prolonged sun exposure while they promote the secretion of postbiotic solar molecules, which act as natural photoprotective agents. As a result, the product restores microbiota balance and, consequently, supports overall skin homeostasis.

While studies on the individual active ingredients explain why the product is effective in addressing multiple aspects of photoaging, it is also important to highlight that our product has delivered better results than similar products. We have found several studies assessing the efficacy of different products. However, these are often less comprehensive, typically targeting specific aspects of photoaging separately. Furthermore, many of these studies demonstrate effects using cellular systems, with far fewer utilizing tissue-based explants, and even fewer conducting clinical trials. Among the studies we reviewed, one investigating the effect of natural retinol analogs [13] reported a 5.6% increase in firmness, a 14% improvement in elasticity, and a 3.8% reduction in wrinkle volume after 28 days, results that fall significantly short of those achieved with our product. Another study on a biorevitalizing facial serum [15] demonstrated improvements in elasticity (10%) and hydration (13%) after 28 days, but, once again, these results are notably less substantial compared to ours. Even a product claiming to repair sun damage [14] showed only a 14.1% improvement in firmness, which is still much lower than the results achieved with our product. Based on these findings, we can confidently assert that our carefully designed strategy, involving the meticulous selection and combination of active ingredients, has proven to be highly effective, resulting in a product that delivers remarkable improvements across multiple parameters of photoaging.

We believe it is crucial to emphasize that, while this product also offers protective benefits, its primary strength lies in its ability to repair the damage caused by cumulative sun exposure. For many years, the focus on UV protection, while essential, has overlooked the harm already caused on the skin, which retains a 'memory' of past exposure with long-lasting effects on its structure and function. This damage is not merely an aesthetic concern but also a matter of long-term skin health. As such, the development of products capable of repairing existing damage is essential, and our product stands out for addressing both the functional and aesthetic dimensions of photoaging, providing protection while effectively restoring the skin's integrity.

While our findings are robust, certain limitations should be considered. First, the in vivo study included a limited number of volunteers, which may affect the statistical robustness of the results. Expanding the sample size in future studies will improve the generalizability of these findings. Second, while significant improvements were observed after 56 days, long-term studies would provide additional insights into the product's sustained effects on deeper dermal remodeling and pigmentation normalization. Furthermore, while our studies focused on key processes relevant to photodamage, additional mechanisms or alternative techniques could have been explored to provide a more comprehensive understanding of the formulation's effects.

5. Conclusions

In conclusion, this work demonstrates through a series of in vitro, ex vivo, and in vivo studies that our product, by addressing multiple pathways involved in the pathogenesis of photodamage, offers comprehensive protection against sun-induced damage. Thanks to a sophisticated blend of potent antioxidant, anti-inflammatory, protective, and hydrating compounds combined in our OxIR PhotoRescue Technology®, this formulation repairs damaged skin, effectively mitigating signs of photoaging while significantly promoting overall skin health.

Supplementary Materials: The following supporting information can be downloaded at: https://www.mdpi.com/article/10.3390/cosmetics11060224/s1, Table S1: Depigmenting efficacy results in control areas.

Author Contributions: Conceptualization, M.M.-S., N.d.P. and I.B.; methodology, N.d.P. and M.M.-S.; validation, M.M.-S.; formal analysis, N.d.P.; investigation, M.M.-S., N.d.P. and L.A.-P.; resources, S.B.-C. and R.P.-G.; data curation, N.d.P.; writing—original draft preparation, N.d.P.; writing—review and editing, M.M.-S. and R.P.-G.; supervision, M.M.-S., S.B.-C. and R.P.-G.; project administration, S.B.-C., M.M.-S. and R.P.-G.; funding acquisition, R.P.-G. and S.B.-C. All authors have read and agreed to the published version of the manuscript.

Funding: This work has been funded by the Provincial Council of Alava (Alava-Innova-Digitaliza program; ICODERMA project, file number INNOEM-2023/00075).

Institutional Review Board Statement: Due to the cosmetic rather than medicinal nature of the formulation described in the manuscript, approval from an ethics committee or institutional review board was not required for the studies conducted. In Spain, cosmetics regulations (Regulation (EC) No. 1223/2009 and Royal Decree 1599/1997) establish that cosmetic products must be evaluated for safety and efficacy, but efficacy trials do not require the intervention of an ethics committee or institutional review board if they do not involve medical interventions or health risks. Our study was performed in line with the Declaration of Helsinki (1964) and its subsequent amendments and following COLIPA guidelines for the Evaluation of the Efficacy of Cosmetic Products, and good clinical practice was maintained throughout this study.

Informed Consent Statement: Informed consent was obtained from all subjects involved in this study. Written informed consent has been obtained from the volunteers to publish this paper.

Data Availability Statement: The data presented in this study are available on request from the corresponding author due to privacy reasons.

Conflicts of Interest: The authors were employed by the company I+Med S. Coop. The authors declare that the research was conducted in the absence of any commercial or financial relationships that could be construed as potential conflicts of interest.

Appendix A

Appendix A.1. Volunteer Recruitment

The specific inclusion criteria, defined in the protocol, were the following:

- Sex: female.
- Age: from 35 to 60 years.
- Sensitive and photoaged skin.
- Agree to voluntarily participate in the study and give their written informed consent.
- Adequate level of understanding of the clinical study.

- Good state of health (physical and mental).
- Having stopped using anti-aging and anti-spots products 7 days before the study in the experimental area.
- Not having applied any type of product in the experimental area on the first day of the test.
- Availability to guarantee visits to research centers.

The specific exclusion criteria, defined in the protocol, were the following:

- Being pregnant or breastfeeding.
- Injuries or infection in the test area.
- Patients with the presence of relevant skin pathologies: eczema, psoriasis, dermatitis, etc.
- History of allergies to cosmetic products.
- Skin hyperreactivity.
- Patients who have had recent surgery or treatments in the study area.
- Patients who have had treatment with botulinum toxin or hyaluronic acid in less than 6 months from the start of the trial.
- Oncology patients.
- Volunteers who are undergoing treatment with antibiotics, antihistamines, corticosteroids, beta-blockers, retinoids, azelaic acid, anti-acne therapies, or whose treatment has ended in the 15 days prior to the start of the study.
- Use a cosmetic product other than the one provided for the study.
- Being a pacemaker holder.
- Sunbathing/UVA rays during the study.
- Be participating in another clinical study.
- Present health problems that may compromise adherence to the study protocol.
- Refusal to sign the informed consent form.

Volunteers were warned of possible unwanted or unpleasant reactions to the product and their reversibility. In case of any unwanted reaction or doubt, the volunteers were advised to contact the center that would inform the specialist responsible for the clinical evaluation.

Appendix A.2. Product Application Criteria

Application area: face.

Study duration: 56 days.

Mode of use: Apply to clean, dry skin morning and evening or after prolonged exposure to the sun. Massage gently until completely absorbed. Volunteers were instructed not to use any other cosmetic products of the same type in the experimental areas during this study. They did not change their usual hygiene habits.

One week before starting to use the product, the volunteers were instructed to carry out a correct pre-wash phase. This phase was necessary to standardize the skin conditions of the volunteers. The conditions in the week before the examination were to avoid applying any skin care products to the experimental areas and to avoid the application of topical drugs in experimental areas.

Appendix B

Biometric Measurements (Effectiveness Evaluation)

Environmental conditions: On the measurement days, the volunteers participating in this study remained for 10 min in an acclimatized room: temperature 20 °C ± 2 °C and relative humidity 40–60%. The experimental area had to be clean, without any product applied.

Moisturizing efficacy

Equipment: MoistureMap® MM 100. The measurement is based on capacitance images. The sensor measures the penetration of the electromagnetic field. Conductive material (water) reflects the signal, making the resulting pixel darker, while non-conductive material causes the signal to go further in, and the resulting pixel will be lighter on a 255-level grey scale.

The experimental times were the initial day (D0), after 30 min (T30MIN), 2 h (T2H), 24 h (T24H), 28 days (T28D), and after 56 days (T56D).

The measurement area was the forehead. The parameter to be statistically analyzed was:

- GrayIdx: the higher the value, the more hydration.

Efficacy in barrier function

Equipment: Tewameter TM 300. The principle is the measurement of transepidermal water loss (TEWL). The density gradient of water evaporation from the skin was measured indirectly by the two pairs of sensors (temperature and relative humidity) located inside the hollow cylinder. This is an "open chamber" measurement. The measured values express the evaporation rate in $g/h/m^2$.

The study area was the cheek.

The experimental times for the evaluation were the initial day (D0), after 2 h (T2H), 24 h (T24H), at 28 days (T28D), and the final day (T56D) of product use.

The parameter analyzed was:

- TEWL: The lower the value, the lower the water loss.

Luminosity efficacy by colorimetric image analysis

Equipment: VisioFace® 1000D. The principle for color measurement is the photographic analysis of the skin. The basis for the calculation is colorimetric analysis, which measures the difference between lightness/darkness.

The experimental times for evaluation are the initial day (D0), at 28 days (T28D), and the final day (T56D) of product use.

The measurement area chosen was forehead.

To evaluate the effectiveness, the following parameters were evaluated:

- ΔL: indicates the increase in skin luminosity. The closer the value is to 0 (the more the value increases), the closer it is to white, and the more luminous the skin is.

Depigmenting efficacy

Equipment: Mexameter® MX 18. The measurement is based on absorption/reflection. The probe emits three specific wavelengths of light. Melanin is measured by specific wavelengths chosen to correspond to different absorption rates of the pigments.

The experimental times for evaluation were the initial day (D0), at 28 days (T28D), and the final day (T56D) of product use.

The measurement areas chosen were a spot area and a control area.

To evaluate the effectiveness, the following parameter was evaluated:

- Mean melanin value. The lower it is, the less pigmented the skin is.

Anti-wrinkle efficacy

Equipment: AEVA^{3D} -HE^2. The AEVA- HE system is a three-dimensional scanning sensor for evaluating skin topography. It is based on fringe-projection technology combined with active stereometry. AEVA- HE is a very sensitive system due to its pixel resolution.

The experimental times for the evaluation were the initial day (D0), at 28 days (T28D), and the final day (T56D) of product use.

The measurement area chosen was forehead for wrinkle volume and the entire face for wrinkle density.

To evaluate the effectiveness, the following parameters were evaluated:

- Wrinkle volume. The lower the value, the lower the wrinkle volume.
- Wrinkle density (%). The lower the value, the lower the wrinkle density.

Firming/elasticity efficacy

Equipment: Cutometer® dual MPA 580. The elasticity of the skin surface was measured following suction of the study area (negative pressure 450 mbar), applying a pressure in the plane perpendicular to the surface that produces a mechanical deformation. The resistance of the skin to suction is considered the firmness and its ability to return to the original position the elasticity. These are represented as curves (mm depth of penetration per unit of time). In each of the visits to the center, a measurement was made, which is made up of 10 cycles, with two seconds of suction and two seconds of relaxation per cycle to tire the skin.

The experimental times for the evaluation were the initial day (D0), after 30 min (T30min), 2 h (T2H), 24 h (T24H), and after 28 days (T28D).

The study area was cheekbone.

The parameters analyzed were:

- R0: Firmness. When the value is lower, the firmness increases.
- F1: Elasticity. When the value is lower, the elasticity increases.

Appendix C

Subjective Evaluation

The volunteers, on their visit to the center at the end, filled out a questionnaire answering questions related to the efficacy of the product.

- After using the product, I notice that my skin is more hydrated.
- After using the product, the area is more nourished.
- Using the product leaves my skin soft.
- After using the product, my skin is more flexible.
- After using the product, I notice a refreshing sensation on my skin.
- After using the product, I have a feeling of comfort on my skin.

A five-point scale is used to rate efficacy:

5: Totally agree. 4: Agree. 3: Neither agrees nor disagrees. 2: Disagree. 1: Totally disagree.

Volunteers with opinions between 5 and 4 are considered in agreement with the aspects of subjective efficacy.

References

1. Friedman, O. Changes Associated with the Aging Face. *Facial Plast. Surg. Clin. N. Am.* **2005**, *13*, 371–380. [CrossRef] [PubMed]
2. Mora Huertas, A.C.; Schmelzer, C.E.H.; Hoehenwarter, W.; Heyroth, F.; Heinz, A. Molecular-Level Insights into Aging Processes of Skin Elastin. *Biochimie* **2016**, *128–129*, 163–173. [CrossRef] [PubMed]
3. Bernerd, F.; Marionnet, C.; Duval, C. Solar Ultraviolet Radiation Induces Biological Alterations in Human Skin in Vitro: Relevance of a Well-Balanced UVA/UVB Protection. *Indian J. Dermatol. Venereol. Leprol.* **2012**, *78*, 15. [CrossRef] [PubMed]
4. Lorencini, M.; Brohem, C.A.; Dieamant, G.C.; Zanchin, N.I.T.; Maibach, H.I. Active Ingredients against Human Epidermal Aging. *Ageing Res. Rev.* **2014**, *15*, 100–115. [CrossRef]
5. Gilchrest, B.A. Photoaging. *J. Investig. Dermatol.* **2013**, *133*, E2–E6. [CrossRef]
6. Scharffetter–Kochanek, K.; Brenneisen, P.; Wenk, J.; Herrmann, G.; Ma, W.; Kuhr, L.; Meewes, C.; Wlaschek, M. Photoaging of the Skin from Phenotype to Mechanisms. *Exp. Gerontol.* **2000**, *35*, 307–316. [CrossRef]
7. Bosch, R.; Philips, N.; Suárez-Pérez, J.; Juarranz, A.; Devmurari, A.; Chalensouk-Khaosaat, J.; González, S. Mechanisms of Photoaging and Cutaneous Photocarcinogenesis, and Photoprotective Strategies with Phytochemicals. *Antioxidants* **2015**, *4*, 248–268. [CrossRef]
8. González, S.; Fernández-Lorente, M.; Gilaberte-Calzada, Y. The Latest on Skin Photoprotection. *Clin. Dermatol.* **2008**, *26*, 614–626. [CrossRef]
9. Naylor, E.C.; Watson, R.E.B.; Sherratt, M.J. Molecular Aspects of Skin Ageing. *Maturitas* **2011**, *69*, 249–256. [CrossRef]
10. Fisher, G.J.; Kang, S.; Varani, J.; Bata-Csorgo, Z.; Wan, Y.; Datta, S.; Voorhees, J.J. Mechanisms of Photoaging and Chronological Skin Aging. *Arch. Dermatol.* **2002**, *138*, 1462–1470. [CrossRef]
11. Franceschi, C.; Bonafè, M.; Valensin, S.; Olivieri, F.; De Luca, M.; Ottaviani, E.; De Benedictis, G. Inflamm-Aging. An Evolutionary Perspective on Immunosenescence. *Ann. N. Y. Acad. Sci.* **2000**, *908*, 244–254. [CrossRef] [PubMed]
12. Nelson, K.L.; Boehm, A.B.; Davies-Colley, R.J.; Dodd, M.C.; Kohn, T.; Linden, K.G.; Liu, Y.; Maraccini, P.A.; McNeill, K.; Mitch, W.A.; et al. Sunlight-Mediated Inactivation of Health-Relevant Microorganisms in Water: A Review of Mechanisms and Modeling Approaches. *Environ. Sci. Process Impacts* **2018**, *20*, 1089–1122. [CrossRef] [PubMed]
13. Brown, A.; Furmanczyk, M.; Ramos, D.; Ribes, A.; Pons, L.; Bustos, J.; De Henestrosa, A.R.F.; Granger, C.; Jourdan, E. Natural Retinol Analogs Potentiate the Effects of Retinal on Aged and Photodamaged Skin: Results from In Vitro to Clinical Studies. *Dermatol. Ther.* **2023**, *13*, 2299–2317. [CrossRef] [PubMed]

14. Narda, M.; Ramos-Lopez, D.; Bustos, J.; Trullàs, C.; Granger, C. A Novel Water-Based Anti-Aging Suncare Formulation Provides Multifaceted Protection and Repair against Environmental Aggressors: Evidence from in Vitro, Ex Vivo, and Clinical Studies. *Clin. Cosmet. Investig. Dermatol.* **2019**, *12*, 533–544. [CrossRef]
15. Sanz, M.T.; Campos, C.; Milani, M.; Foyaca, M.; Lamy, A.; Kurdian, K.; Trullas, C. Biorevitalizing Effect of a Novel Facial Serum Containing Apple Stem Cell Extract, Pro-collagen Lipopeptide, Creatine, and Urea on Skin Aging Signs. *J. Cosmet. Dermatol.* **2016**, *15*, 24–30. [CrossRef]
16. Chung, J.; Cho, S.; Kang, S. Why Does the Skin Age. In *Photoaging;* Marcel Dekker, Inc.: New York, NY, USA, 2004; pp. 1–13.
17. Masaki, H. Role of Antioxidants in the Skin: Anti-Aging Effects. *J. Dermatol. Sci.* **2010**, *58*, 85–90. [CrossRef]
18. Stahl, W.; Sies, H. Bioactivity and Protective Effects of Natural Carotenoids. *Biochim. Biophys. Acta* **2005**, *1740*, 101–107. [CrossRef]
19. Antille, C.; Tran, C.; Sorg, O.; Saurat, J.-H. Topical Beta-Carotene Is Converted to Retinyl Esters in Human Skin Ex Vivo and Mouse Skin in Vivo. *Exp. Dermatol.* **2004**, *13*, 558–561. [CrossRef]
20. Böhm, F.; Edge, R.; Lange, L.; Truscott, T.G. Enhanced Protection of Human Cells against Ultraviolet Light by Antioxidant Combinations Involving Dietary Carotenoids. *J. Photochem. Photobiol. B* **1998**, *44*, 211–215. [CrossRef]
21. Biesalski, H.K.; Obermueller-Jevic, U.C. UV Light, Beta-Carotene and Human Skin--Beneficial and Potentially Harmful Effects. *Arch. Biochem. Biophys.* **2001**, *389*, 1–6. [CrossRef]
22. Le Bellec, F.; Vaillant, F.; Imbert, E. Pitahaya (*Hylocereus* spp.): A New Fruit Crop, a Market with a Future. *Fruits* **2006**, *61*, 237–250. [CrossRef]
23. Chatterjee, D.; Mansuri, S.; Poonia, N.; Kesharwani, P.; Lather, V.; Pandita, D. Therapeutic Potential of Various Functional Components Presents within Dragon Fruit: A Review. *Hybrid. Adv.* **2024**, *6*, 100185. [CrossRef]
24. Farriol, M.; Mourelle, M.; Schwartz, S. Effect of Vitamin C and Vitamin E Analog on Aged Fibroblasts. *Rev. Esp. Fisiol.* **1994**, *50*, 253–257. [PubMed]
25. Chiu, A.; Kimball, A.B. Topical Vitamins, Minerals and Botanical Ingredients as Modulators of Environmental and Chronological Skin Damage. *Br. J. Dermatol.* **2003**, *149*, 681–691. [CrossRef]
26. Tenore, G.C.; Novellino, E.; Basile, A. Nutraceutical Potential and Antioxidant Benefits of Red Pitaya (Hylocereus Polyrhizus) Extracts. *J. Funct. Foods* **2012**, *4*, 129–136. [CrossRef]
27. Wu, J.; Zhao, J.; Zhang, T.; Gu, Y.; Khan, I.A.; Zou, Z.; Xu, Q. Naturally Occurring Physalins from the Genus Physalis: A Review. *Phytochemistry* **2021**, *191*, 112925. [CrossRef]
28. Pinto, N.B.; Morais, T.C.; Carvalho, K.M.B.; Silva, C.R.; Andrade, G.M.; Brito, G.A.C.; Veras, M.L.; Pessoa, O.D.L.; Rao, V.S.; Santos, F.A. Topical Anti-Inflammatory Potential of Physalin E from Physalis Angulata on Experimental Dermatitis in Mice. *Phytomedicine* **2010**, *17*, 740–743. [CrossRef]
29. Pereda, M.D.C.V.; Dieamant, G.; Nogueira, C.; Eberlin, S.; Facchini, G.; Mussi, L.; Polezel, M.A.; Martins-Oliveira, D.; Rosa, P.T.V.; Di Stasi, L.C. Sterol-standardized Phytopharmaceutical from Ground Cherry: Corticoid-like Properties on Human Keratinocytes and Fibroblasts and Its Effects in a Randomized Double-blind Placebo-controlled Clinical Trial. *J. Cosmet. Dermatol.* **2019**, *18*, 1516–1528. [CrossRef]
30. Bernal-Chávez, S.; Pérez-Carreto, L.; Nava-Arzaluz, M.; Ganem-Rondero, A. Alkylglycerol Derivatives, a New Class of Skin Penetration Modulators. *Molecules* **2017**, *22*, 185. [CrossRef]
31. Iannitti, T.; Palmieri, B. An Update on the Therapeutic Role of Alkylglycerols. *Mar. Drugs* **2010**, *8*, 2267–2300. [CrossRef]
32. Puglia, C.; Rizza, L.; Offerta, A.; Gasparri, F.; Giannini, V.; Bonina, F. Formulation Strategies to Modulate the Topical Delivery of Anti-Inflammatory Compounds. *J. Cosmet. Sci.* **2013**, *64*, 341–353. [PubMed]
33. Magalhães, M.S.F.; Fechine, F.V.; Macedo, R.N.D.; Monteiro, D.L.S.; Oliveira, C.C.; Brito, G.A.D.C.; Moraes, M.E.A.D.; Moraes, M.O.D. Effect of a Combination of Medium Chain Triglycerides, Linoleic Acid, Soy Lecithin and Vitamins A and E on Wound Healing in Rats. *Acta Cir. Bras.* **2008**, *23*, 262–269. [CrossRef] [PubMed]
34. De Nardi, A.B.; Rodaski, S.; Sousa, R.S.; Baudi, D.L.K.; Castro, J.H.T. Cicatrização Secundária Em Feridas Dermoepidérmicas Tratadas Com Ácidos Graxos Essenciais, Vitaminas A E E, Lecitina De Soja E Iodo Polivinilpirrolidona Em Cães. *Arch. Vet. Sci.* **2004**, *9*, 1–16. [CrossRef]
35. Baumgartner, L.; Sosa, S.; Atanasov, A.G.; Bodensieck, A.; Fakhrudin, N.; Bauer, J.; Favero, G.D.; Ponti, C.; Heiss, E.H.; Schwaiger, S.; et al. Lignan Derivatives from *Krameria Lappacea* Roots Inhibit Acute Inflammation in Vivo and Pro-Inflammatory Mediators in Vitro. *J. Nat. Prod.* **2011**, *74*, 1779–1786. [CrossRef]
36. Brecht, M.; Mayer, U.; Schlosser, E.; Prehm, P. Increased Hyaluronate Synthesis Is Required for Fibroblast Detachment and Mitosis. *Biochem. J.* **1986**, *239*, 445–450. [CrossRef]
37. Bourguignon, L.Y.W.; Ramez, M.; Gilad, E.; Singleton, P.A.; Man, M.-Q.; Crumrine, D.A.; Elias, P.M.; Feingold, K.R. Hyaluronan-CD44 Interaction Stimulates Keratinocyte Differentiation, Lamellar Body Formation/Secretion, and Permeability Barrier Homeostasis. *J. Investig. Dermatol.* **2006**, *126*, 1356–1365. [CrossRef]
38. Hussain, Z.; Thu, H.E.; Katas, H.; Bukhari, S.N.A. Hyaluronic Acid-Based Biomaterials: A Versatile and Smart Approach to Tissue Regeneration and Treating Traumatic, Surgical, and Chronic Wounds. *Polym. Rev.* **2017**, *57*, 594–630. [CrossRef]
39. Bukhari, S.N.A.; Roswandi, N.L.; Waqas, M.; Habib, H.; Hussain, F.; Khan, S.; Sohail, M.; Ramli, N.A.; Thu, H.E.; Hussain, Z. Hyaluronic Acid, a Promising Skin Rejuvenating Biomedicine: A Review of Recent Updates and Pre-Clinical and Clinical Investigations on Cosmetic and Nutricosmetic Effects. *Int. J. Biol. Macromol.* **2018**, *120*, 1682–1695. [CrossRef]

40. Iaconisi, G.N.; Lunetti, P.; Gallo, N.; Cappello, A.R.; Fiermonte, G.; Dolce, V.; Capobianco, L. Hyaluronic Acid: A Powerful Biomolecule with Wide-Ranging Applications—A Comprehensive Review. *Int. J. Mol. Sci.* **2023**, *24*, 10296. [CrossRef]
41. Bravo, B.; Correia, P.; Gonçalves Junior, J.E.; Sant'Anna, B.; Kerob, D. Benefits of Topical Hyaluronic Acid for Skin Quality and Signs of Skin Aging: From Literature Review to Clinical Evidence. *Dermatol. Ther.* **2022**, *35*. [CrossRef]
42. Rawlings, A.V.; Harding, C.R. Moisturization and Skin Barrier Function. *Dermatol. Ther.* **2004**, *17* (Suppl. S1), 43–48. [CrossRef] [PubMed]
43. Mo, X.; Chen, X.; Pan, X.; Lu, Y.; Pan, G.; Xie, J.; Pan, Z.; Li, L.; Tian, H.; Li, Y. Protective Effect of Helianthus Annuus Seed Byproduct Extract on Ultraviolet Radiation-Induced Injury in Skin Cells. *Photochem. Photobiol.* **2024**, *100*, 756–771. [CrossRef] [PubMed]
44. Michalak, M. Plant Extracts as Skin Care and Therapeutic Agents. *Int. J. Mol. Sci.* **2023**, *24*, 15444. [CrossRef] [PubMed]
45. Mayser, P.; Pape, B. Decreased Susceptibility of Malassezia Furfur to UV Light by Synthesis of Tryptophane Derivatives. *Antonie Van Leeuwenhoek* **1998**, *73*, 315–319. [CrossRef]
46. Youngchim, S.; Nosanchuk, J.D.; Pornsuwan, S.; Kajiwara, S.; Vanittanakom, N. The Role of L-DOPA on Melanization and Mycelial Production in Malassezia Furfur. *PLoS ONE* **2013**, *8*, e63764. [CrossRef]
47. Gaya, P.; Peirotén, Á.; Medina, M.; Álvarez, I.; Landete, J.M. Bifidobacterium Pseudocatenulatum INIA P815: The First Bacterium Able to Produce Urolithins A and B from Ellagic Acid. *J. Funct. Foods* **2018**, *45*, 95–99. [CrossRef]
48. Hug, D.H.; Dunkerson, D.D.; Hunter, J.K. The Degradation of L-Histidine and Trans- and Cis-Urocanic Acid by Bacteria from Skin and the Role of Bacterial Cis-Urocanic Acid Isomerase. *J. Photochem. Photobiol. B* **1999**, *50*, 66–73. [CrossRef]
49. Negari, I.P.; Keshari, S.; Huang, C.-M. Probiotic Activity of Staphylococcus Epidermidis Induces Collagen Type I Production through FFaR2/p-ERK Signaling. *Int. J. Mol. Sci.* **2021**, *22*, 1414. [CrossRef]
50. Larayetan, R.A.; Ayeni, G.; Yahaya, A.; Ajayi, A.; Omale, S.; Ishaq, U.; Abiodun, D.J.; Olisah, C.; Aigbogun, J.; Enyioma-Alozie, S. Chemical Composition of Gossypium Herbaceum Linn and Its Antioxidant, Antibacterial, Cytotoxic and Antimalarial Activities. *Clin. Complement. Med. Pharmacol.* **2021**, *1*, 100008. [CrossRef]
51. Maphetu, N.; Unuofin, J.O.; Masuku, N.P.; Olisah, C.; Lebelo, S.L. Medicinal Uses, Pharmacological Activities, Phytochemistry, and the Molecular Mechanisms of Punica Granatum L. (Pomegranate) Plant Extracts: A Review. *Biomed. Pharmacother.* **2022**, *153*, 113256. [CrossRef]
52. Moradnia, M.; Mohammadkhani, N.; Azizi, B.; Mohammadi, M.; Ebrahimpour, S.; Tabatabaei-Malazy, O.; Mirsadeghi, S.; Ale-Ebrahim, M. The Power of Punica Granatum: A Natural Remedy for Oxidative Stress and Inflammation; a Narrative Review. *J. Ethnopharmacol.* **2024**, *330*, 118243. [CrossRef]
53. Melgarejo-Sánchez, P.; Núñez-Gómez, D.; Martínez-Nicolás, J.J.; Hernández, F.; Legua, P.; Melgarejo, P. Pomegranate Variety and Pomegranate Plant Part, Relevance from Bioactive Point of View: A Review. *Bioresour. Bioprocess.* **2021**, *8*, 2. [CrossRef]

cosmetics

MDPI

Article

Formulation of Biological Sunscreen from *Calendula arvensis* Capitula Extracts: Antioxidant, Anti-Aging, Surface Tension, and UVB Protection Properties Assessed

Najlae El-Otmani [1,*], Ikrame Zeouk [2] and Ahmed Zahidi [1]

1 Therapeutic Chemistry Laboratory, Department of Drug Sciences, Faculty of Medicine, and Pharmacy, Mohammed V University, Rabat 6203, Morocco; a.zahidi@um5r.ac.ma

2 Pharmacology-Toxicology Laboratory, Faculty of Medicine, Pharmacy and Dentistry-Fez, University Sidi Mohamed Ben Abdellah, Fez 1893, Morocco; ikramezeouk20@gmail.com

* Correspondence: najlae.elotmani@um5r.ac.ma or najlae.elotmani@unibas.ch; Tel.: +41-783406185

Abstract: Skin protection against ultraviolet (UV) radiation has long been crucial due to its role in photoaging, sunburn, and wrinkles. This study focuses on developing a bio-based sunscreen from *Calendula arvensis* capitula extract. Various extraction methods (maceration, sonication, and infusion) and solvents (EtOH, EtOH-H_2O, and H_2O) were explored in order to identify the most effective extract for use in a sunscreen formulation. Each extract was analyzed for its phenolic content, as well as antioxidant activities (assessed through DPPH, CAT, and FRAP assays). Enzyme inhibition assays for tyrosinase, elastase, and collagenase highlighted the low IC_{50} values of the hydroethanolic extract. Furthermore, the in vitro sun protection factor (SPF) against UVB radiation was measured using ultraviolet spectrophotometry. A phytochemical analysis showed phenolic levels between 8 and 27 mg GAE/g, flavonoid concentrations of 7–13 mg QE/g, and tannin levels of 1.15–1.68 mg/mL, alongside moderate antioxidant activity. The ethanol maceration extract reduced the interfacial tension to 2.15 mN/m in 600 s, outperforming the conventional emulsifier polysorbate 20. The sonicated hydroethanolic extract demonstrated remarkable SPF efficacy (SPF = 193.65 ± 0.02), far exceeding that of the standard zinc oxide (SPF = 11.88 ± 0.03). The proposed formulations meet the COSMOS standards, suggesting their potential for certification as biological products. Further clinical and in vivo studies are necessary to confirm their safety and commercial viability.

Keywords: *Calendula arvensis*; phytochemical quantification; interfacial tension; UVB protection; antioxidant; anti-aging; bioformulation; enzyme inhibition; COSMOS standard

1. Introduction

Skin disorders, as a major problem exacerbated by UV exposure, are a constant focus of dermatological research [1]. UVB radiation is associated with adverse effects, such as sunburn, skin cancer, and melanoma, while UVA radiation is primarily involved in inducing skin tanning and wrinkles [2]. This radiation stimulates cell surface receptors for cytokines and growth factors such as epidermal growth factor, tumor necrosis factor, interleukin-1, and keratinocyte growth factor. However, biological responses to UV exposure can be immediate and transient (e.g., inflammation, sunburn cell formation, and pruritus) or delayed and chronic (e.g., photoaging, immunosuppression, and carcinogenesis) [3].

In the field of cosmetics, progress is being made to develop products that not only improve appearance, but also protect against and mitigate the effects of aging, merging scientific innovation with skincare solutions. Modern sunscreen products were first developed in the 1930s and, since then, numerous UV filters with specific application characteristics have been introduced. These filters are designed to protect against UV radiation from sunlight, mainly at short UVB wavelengths (290–320 nm). Considerable effort has been made in recent years to develop UV filters that are effective at absorbing the longer wavelengths of UVB or UVA (320–400 nm) [4]. In fact, the development of modern sunscreens has greatly improved the degree of protection against sunlight-related radiation and its associated damaging effects on the skin [5].

For organic and natural cosmetics, the ECOCERT certification standards require formulations not only to provide effective protection but also to adhere to the principles of sustainability and environmental responsibility in production, which can be achieved by exploring and utilizing botanical sources [6]. Furthermore, growing public awareness has led several cosmetic companies to start effectively promoting green marketing strategies [7]. In an effort to meet market demands for fully green-labeled products, there has recently been growing interest in natural emulsifiers. Therefore, as surface-active agents, emulsifiers play a central role in stabilizing emulsions by adsorbing to the interface between oil and water. Emulsifiers reduce interfacial tension, facilitating the formation of fine, uniformly dispersed droplets. Furthermore, the stability of emulsions is enhanced by steric and/or electrostatic repulsion mechanisms between droplets, thus limiting their aggregation and coalescence, which contributes to the persistence and efficacy of emulsified formulations [8].

Calendula arvensis (*C. arvensis*), also known as "field marigold" [9], can be observed in various regions of Morocco, particularly in northern areas around Fes-Meknes, from early March onwards. Previous research has indicated that only a limited number (1%) of herbal cosmetics cooperatives in this region are engaged in the extraction and valorization of this plant [10]. *C. arvensis* extracts have been reported to have antioxidant, antibacterial, enzyme-inhibiting, and immunomodulating activities [11]. These pharmacological properties are probably due to the high content of phenolic compounds and flavonoids in the plant. Polyphenols, in particular, have great potential as photoprotective agents [12], being capable of preventing or attenuating UV-induced photodamage [13]. However, to the best of our knowledge, no research has yet explored the photoprotective activities of *C. arvensis* capitula. Therefore, this study aims to investigate the biological and phytochemical potential of *C. arvensis* capitula.

2. Materials and Methods

2.1. Plant Material

Capitula of the plant material were collected in Taounate, Morocco (N 34°29′54″ W 4°48′17″), in April 2023, during the flowering season. The identification of the species was confirmed as *Calendula arvensis* L., cataloged under reference code RAB 114766.

2.2. Extractions

The dried capitula were ground to a particle size of 300 µm using a mechanical grinder (Industriestrasse 855743 Idar-Oberstein, Germany). The resulting powder was subjected to extraction through three distinct methods, utilizing solvents of varying polarities. Maceration was performed over 24 h with a solid-to-solvent ratio of 1:10 (w/v). Sonication was conducted for 45 min with the same ratio, using either absolute ethanol or a 70–30% ethanol–water mixture. Additionally, an infusion method was applied, where 2.5 g of the powder was infused in 75 mL of water for 1 h. The resulting organic entities were filtered using Whatman No. 1 filter paper, then centrifuged at 3000 rpm over a 10 min period. The crude extracts obtained were then stored in the refrigerator until further use. The percentage extraction yield (w/w) is shown in Table 1 (n = 3).

Table 1. Yield percentage of extraction methods.

Extracts	Yield %
S-EtOH-H_2O	34.78 ± 2.01
S-EtOH	24.76 ± 2.23
M-EtOH-H_2O	22.44 ± 0.57

Table 1. *Cont.*

Extracts	Yield %
M-EtOH	21.36 ± 1.34
IF-H_2O	31.96 ± 1.74

S-EtOH-H_2O: sonication ethanol 70%, S-EtOH: sonication ethanol 100%, M-EtOH-H_2O: maceration ethanol 70%, M-EtOH: maceration ethanol 100%, IF-H_2O: infusion in water.

2.3. *Phenolic Compounds Quantification*

2.3.1. Total Phenolic Contents (TPCs)

The total polyphenol content (TPC) of all extracts was investigated as described by Lister and Wilson [14]. Briefly, 450 μL of Folin–Ciocalteu reagent (10-fold diluted) and 450 μL of sodium carbonate (Na2CO3, 7.5%) were added to 50 μL of the studied extracts. After 2 h of incubation in the dark, absorbance was measured at 765 nm using a UV–vis spectrometer. The results are expressed in terms of gallic acid equivalents per gram of plant material, and the test was performed in triplicate (n = 3).

2.3.2. Total Flavonoid Contents (TFCs)

The total flavonoid content (TFC) of crude extracts was assessed according to the method described by Djeridane et al. [15], with some modifications. In brief, 500 μL of aluminum chloride solution (Alcl3, 10%) was added to 500 μL of diluted extract. After one hour of incubation in the dark, absorbances were measured at 420 nm using a UV–vis spectrometer. A standard range based on quercetin was prepared according to the same conditions, and the test was performed in triplicate.

2.3.3. Tannin Content (TC)

Condensed tannin was analyzed using the Bate–Smith reaction (Stonestreet, 1966) [16]. Technically, 1 mL of each extract (at a concentration of 1 mg/mL) was combined with 0.5 mL of distilled water and 1.5 mL of 37% hydrochloric acid. The mixture was then heated for 30 min in a water bath at 100 °C. Absorbance was measured at 550 nm regarding the control test at room temperature. Thus, the difference in the absorbance between the hydrolyzed tube and the control is the content of tannins contained in the extract, obtained using the following equation:

$$\text{Tannins (mg/mL)} = \left(A_{\text{Hydrolysed}} - A_{\text{Control}}\right) \times 19.33.$$

2.4. *Antioxidant Capacity*

2.4.1. Anti-Free Radical Activity by 1,1-Diphenyl-2-picrylhydrazil (DPPH)

The free radical scavenging effects of the different extracts towards the DPPH° radical were assessed based on the method described by Topçu et al. [17]. Briefly, 50 μL of the diluted extract was added to 1.95 mL of a freshly prepared methanolic solution of DPPH (2.5 mg dissolved in 100 mL methanol under stirring for 3 h). Then, the mixtures were incubated for 30 min in the dark. The absorbance was measured at 517 nm. Ascorbic acid was used as the standard antioxidant and was subject to the same operating conditions.

2.4.2. Ferric Reducing Antioxidant Power (FRAP)

A FRAP test was performed according to the method described by Djeridane et al. [15]. In particular, 0.25 mL of each diluted extract (6 sets of dilutions) was added to 1.25 mL of phosphate-buffer solution and 1.25 mL of an aqueous solution of potassium ferricyanide (K3FeCN6, 1%). The solutions were then incubated for 20 min at 50 °C in a water bath. After cooling to ambient temperature, 1.25 mL of 10% trichloroacetic acid was added to the reaction mixture and centrifuged at 3000 rpm for 10 min. Then, 1.25 mL of supernatant was added to 1.25 mL of distilled water and 0.25 mL of ferric chloride solution (FeCl3, 0.01%).

After 10 min of incubation, absorbance was measured at a wavelength of 700 nm. The results are expressed in terms of equivalents of ascorbic acid per gram of plant material.

2.4.3. Total Antioxidant Capacity (TAC)

The total antioxidant capacity was estimated according to the Prieto method [18], with slight modifications. This method consists of adding 200 µL of each extract at different concentrations combined with 3000 µL of a reagent composed of H_2SO_4 (60 M), Na_2PO_4 (280 mM), and ammonium molybdate (40 mM), then incubating at 95 °C for 90 min. The mixture was then further incubated at room temperature for 20–30 min, and the absorbance was measured at 695 nm. A control was made up using 200 µL of methanol and 3000 µL of the above reagent. The results are expressed as mgGAE/g of dry extract.

2.5. Anti-Aging Effect

In vitro anti-aging assays were performed to assess anti-elastase, anti-tyrosinase, and anti-collagenase activities, following published protocols [19,20]. Kojic acid and epigallocatechin gallate (EGCG) were used as reference standards, while controls without active samples were included for baseline comparison. Each assay was repeated in triplicate and enzyme inhibition percentages were calculated using the following formula:

$$I\,(\%) = \frac{A_C - A_S}{A_C} \times 100$$

A_S: The measured absorbance of enzyme activity in the presence of samples;
A_C: The measured absorbance of enzyme activity in the absence of samples.

2.6. Surface and Interfacial Activities

Interfacial and surface tensions of *C. arvensis* extracts were assessed using the hanging drop method with a fully automated tensiometer (DataPhysics Instruments GmbH, Filderstadt, Germany). Briefly, a 1% (*w*/*w*) extract solution was loaded into a glass syringe and dispensed via a 22-gauge stainless steel needle. The drop was introduced either into air to measure surface tension or into soybean oil inside a glass cell to determine interfacial tension. The drop was allowed to equilibrate for approximately 10 min in the continuous phase, and a high-resolution camera recorded its shape over this time. The interfacial or surface tension was then calculated using the Young–Laplace equation, with automatic processing via the dpiMAX software, version 20.05.2022.

2.7. Photoprotective Activity

2.7.1. Sun Protection Factor (SPF)

All extracts were investigated for their protective activity against UVB rays according to the Mansur method [21]. In particular, the absorbance of the extracts at a concentration of 2 mg/mL was measured at a range of wavelengths between 290 and 320 nm with 5 nm intervals using a UV–vis spectrometer. Each measurement was performed in triplicate and the SPF was calculated using the formula below (i.e., applying the mathematical equation of Mansur):

$$SPF = CF \times \sum_{290}^{320} EE(\lambda) \times (\lambda) \times Abs(\lambda),$$

EE × I values are constants, which were set by Sayre et al. [22];
EE: erythemogenic effect;
I: the radiation intensity;
Abs: absorbance of extract;
CF: correction factor.

2.7.2. Sunscreen O/W Emulsion Formulas Based on *C. arvensis* Capitula Extracts

O/W emulsions were developed to optimize the photoprotective activity of the *C. arvensis* extracts. The IF-H_2O, M-EtOH-H_2O, S-EtOH, M-EtOH, and S-EtOH-H_2O extracts were used at SPF ratios of 1:200 (0.05% *w*/*w*), 1:100 (0.1% *w*/*w*), and 1:50 (0.2% *w*/*w*), testing different concentrations to identify the concentration offering the best sun protection (Table 2). A low ratio of the synthetic emulsifier to the lipid phase was chosen in order to increase the biological percentage of the formulation and make the most of the potential emulsifying activity of *C. arvensis* extracts. Therefore, a hydrolat (73.95% *w*/*w*, 73.9% *w*/*w*, or 73.8% *w*/*w*) was used as the main aqueous phase to increase the biological percentage.

Table 2. Sunscreen O/W emulsion formula.

Phase	INCI	Function	Weight (%, *w*/*w*)
Phase A	GLYCERIN	Humectant	3
	C. arvensis FW *	Dispersion agent	Qsp 100
Phase B	CETEARYL ALCOHOL	Emulsifier, thickening agent, stabilizer	5
	STEARIC ACID	Emulsifier, thickener, consistency enhancer	2
	CETEARETH-20	surfactant	5.5
	C. arvensis OIL MACERATE	Dispersed agent	10
Phase C	*C. arvensis* ORGANIC EXTRACTS	SPF, antioxidant, anti-aging, stabilizer	0.2, 0.1, 0.05
	P. graveolens EO **	Antimicrobial, fragrance	0.5

Phase A: aqueous phase, Phase B: oil phase, Phase C: cooling phase, INCI: International Nomenclature Cosmetic Ingredient, Qsp 100: quantum satis pro 100%. * *C. arvensis* floral water, ** *Pelargonium graveolens* essential oil.

The aqueous phase, in which glycerin (3% *w*/*w*) and *C. arvensis* floral water were mixed, was placed in a pot and heated to a temperature of 75 °C, with continuous stirring until completely homogenized. At the same time, the oil phase was prepared by combining cetearyl alcohol, stearic acid, ceteareth-20, and *C. arvensis* oil macerate in another pot. This mixture was also heated to 75 °C, allowing the components to melt and mix completely. Emulsification was then carried out by gradually adding the oily phase to the aqueous phase under intense agitation for 10 min at 12,000 rpm using a rotor–stator homogenizer (T 25 digital ULTRA-TURRAX®, IKA, Staufen, Germany). This process continued until a stable emulsion was obtained, characterized by a homogenous consistency. Once the emulsion stabilized, the mixture was cooled to below 40 °C before the cooling phase (Phase C) was incorporated. The organic extract of *C. arvensis* was added at concentrations of 0.2%, 0.1%, or 0.05%. *P. graveolens* essential oil (0.5% *w*/*w*) was then introduced due to its antimicrobial and olfactory properties. As an alternative, 0.5% phenoxyethanol can be used. The final formulations were subjected to efficacy tests in order to assess their sun protection factor (SPF) using the Mansur method in a similar way to the crude extracts. Triplicate measurements were carried out to ensure the reproducibility of the results.

The biological percentage of the sunscreen formulation was determined in accordance with the COSMOS standards version 4.0 (1 January 2023) [23]. According to the principles of natural and organic cosmetics certification, this method calculates the percentage of organic ingredients in the final product. Each ingredient was examined to identify those certified as organic, and their organic content percentages were obtained from supplier certifications. Water and minerals were excluded when developing the formulas, as they are not considered organic according to COSMOS standards. The organic content of the dried extracts was calculated by measuring the quantity of each extract used, excluding the weight of the solvent, and applying the COSMOS Equations (1) and (2). The final organic percentage was then calculated using the COSMOS Equation (3), ensuring compliance with COSMOS requirements; that is, at least 20% organic content for leave-on products.

$$\text{Ratio} = \frac{\text{Organic freach plant}}{\text{final extract} - \text{solvents}}, \quad (1)$$

$$\%\text{Bio of extract} = \left[\text{Ratio} \times \left(\frac{\text{Final extract} - \text{solvants}}{\text{Final extract}} + \frac{\text{Organic solvants}}{\text{Final extract}}\right)\right] \times 100, \quad (2)$$

$$\text{Bio \% of product} = \frac{\sum \text{\% of Bio ngredients}}{\text{Total ingredients}}. \quad (3)$$

2.8. Stability Tests and Characterization of the Formulas

Following the formulation of the sunscreens, their pH, conductivity, peroxide value, and viscosity were assessed first at the production stage and then after 30, 60, and 90 days of storage at room temperature, in order to confirm their stability over time. pH and conductivity were measured and recorded directly from each sample using a BOECO Bench-CT-676 pH meter and conductivity meter (BOECO, Germany). The peroxide value was quantified using a CDR FoodLab® Junior (Perugia, Italy), and the results are expressed in meqO2/kg.

The dynamic viscosity was measured using a Brookfield DV2T viscometer (Brookfield Ametek, Middleboro, MA, USA). All samples (80 g) underwent equilibration at 25 °C prior to testing. A Helipath T-Bar spindle was immersed in a glass vessel containing sunscreen products at a speed of 20 rpm. The viscosity was recorded in centipoise (cP).

2.9. Statistical Analysis

Data were calculated using Microsoft Excel, and each experiment was performed in triplicate. Data are presented as mean values with corresponding standard deviations (SDs). Statistical analysis involved A2.4NOVA analysis using the Origin Pro 2024 software, with significance determined at $p < 0.05$.

3. Results

3.1. Phenolic Compounds

Table 3 summarizes the contents of the *C. arvensis* extracts in terms of TPC, TFC, and TC. Statistically, there was no significant difference in total phenolic content (TPC) between the S-EtOH-H_2O, S-EtOH, M-EtOH-H_2O, and IF-H_2O extracts ($p > 0.05$). Meanwhile, the M-EtOH extract showed the lowest TPC (8.24 ± 1.17 mg GAE/g) and TFC (6.47 ± 0.49 mg QE/g), implying that maceration in ethanol alone is less efficient for the extraction of these compounds. However, the results for the IF-H_2O extract, with a TPC of 16.91 ± 1.11 mg GAE/g and a TFC of 10.69 ± 0.68 mg QE/g, suggest that water-based infusion allows for the extraction of a substantial content of phenolic compounds, albeit slightly less efficiently than mixed solvent systems. Regarding TC, there was no significant difference between those extracts obtained by maceration with either 100% ethanol or 70% ethanol, yielding values of TC = 1.13 ± 0.01 mg/mL and TC = 1.10 ± 0.06 mg/mL, respectively, for M-EtOH-H_2O and M-EtOH. However, the highest concentration of tannins was detected in the S-EtOH-H_2O extract, with a TC of 1.66 ± 0.02 mg/mL.

Table 3. Contents of total phenolics, flavonoids, and condensed tannins.

Extracts	Phenolic Compounds		
	TPC (mgGAE/g)	TFC (mgQE/g)	TC (mg/mL)
S-EtOH-H_2O	27.20 ± 2.76 ^{a}	13.37 ± 1.04 ^{a}	1.66 ± 0.02 ^{a}
S-EtOH	14.96 ± 1.65 ^{b}	11.40 ± 0.32 ^{ab}	1.32 ± 0.04 ^{bc}
M-EtOH-H_2O	16.38 ± 1.15 ^{b}	7.72 ± 1.40 ^{c}	1.13 ± 0.01 ^{cd}
M-EtOH	8.24 ± 1.17 ^{c}	6.47 ± 0.49 ^{c}	1.10 ± 0.06 ^{d}
IF-H_2O	16.91 ± 1.11 ^{b}	10.69 ± 0.68 ^{b}	1.46 ± 0.12 ^{b}

TPC—total polyphenol content; TFC—total flavonoids content; TC—tannin content; GAE—gallic acid equivalent; QE—quercetin equivalent. The letters a–d are used to indicate a statistically significant difference ($p < 0.05$) between the samples that were tested.

3.2. Antioxidant Activity

The antioxidant activity of *C. arvensis* capitula extracts was assessed in vitro using four methods: the DPPH, ABTS, FRAP, and TAC assays (Table 4). The findings revealed that the antioxidant capacity varied significantly, depending on the extraction method and solvent used.

Table 4. In vitro antioxidant activity of *C. arvensis* capitula extracts using DPPH, ABTS, TAC, and FRAP methods.

Extracts	Antioxidant Activity			
	DPPH IC_{50} mg/mL	ABTS IC_{50} mg/mL	FRAP mg AAE/g pm	TAC mg AAE/g pm
S-EtOH-H_2O	3.77 ± 0.44 [a]	3.21 ± 0.08 [a]	18.69 ± 0.46 [ab]	12.46 ± 0.61 [a]
S-EtOH	13.57 ± 0.41 [b]	6.84 ± 0.03 [b]	14.67 ± 0.55 [b]	12.34 ± 0.82 [a]
M-EtOH-H_2O	7.33 ± 0.33 [c]	2.71 ± 0.05 [c]	21.90 ± 1.63 [a]	8.99 ± 0.85 [b]
M-EtOH	15.27 ± 0.90 [d]	15.22 ± 0.01 [d]	16.95 ± 0.84 [b]	10.66 ± 0.84 [ab]
IF-H_2O	26.98 ± 0.93 [e]	8.55 ± 0.02 [e]	15.58 ± 3.15 [b]	9.72 ± 0.45 [b]
Vit-C	0.14 ± 0.03 [f]	0.004 ± 0.001 [f]	-	-

pm: plant material; the letters a–f are used to indicate a statistically significant difference ($p < 0.05$) between the samples that were tested. Values are presented as means ± standard error (n = 3). Different letters in the same row indicate statistical significance at $p < 0.05$. DPPH refers to DPPH free radical scavenging activity; FRAP denotes the ferric reducing antioxidant power assay; ABTS stands for ABTS radical scavenging assay; TAC represents total antioxidant capacity. mg GAE/g pm: mg gallic acid equivalent per gram of plant material. Vit-C: vitamin C.

The S-EtOH-H_2O extract demonstrated the highest overall antioxidant potential. In the DPPH assay, it exhibited a remarkably low IC_{50} value of 3.77 ± 0.44 mg/mL, indicating strong free radical scavenging activity. Similarly, in the ABTS assay, it maintained its superiority with an IC_{50} of 3.21 ± 0.08 mg/mL. These results suggest that the hydroethanolic solvent combination is particularly effective at extracting bioactive compounds—likely phenolics and flavonoids—which are responsible for these activities. The M-EtOH-H_2O extract also showed promising antioxidant capacity, especially in the ABTS assay, where it recorded an IC_{50} of 2.71 ± 0.05 mg/mL. Moreover, this extract achieved the highest reducing power, as determined by the FRAP assay, with a value of 21.90 ± 1.63 mg AAE/g pm. These observations highlight the beneficial role of water in the solvent mixture, which enhances the extraction of polar compounds contributing to antioxidant activity. On the other hand, the IF-H_2O and M-EtOH extracts displayed relatively weaker antioxidant effects. The IF-H_2O extract had the poorest performance in both DPPH and ABTS assays (IC_{50} values of 26.98 ± 0.93 mg/mL and 8.55 ± 0.02 mg/mL, respectively), reflecting the limited efficacy of aqueous infusion in extracting potent antioxidants. The M-EtOH extract also showed less significant activity in the DPPH (15.27 ± 0.09 mg/mL) and ABTS (15.22 ± 0.01 mg/mL) assays, indicating that pure ethanol might not be optimal for extracting the most effective phenolic compounds from *C. arvensis*. Regarding total antioxidant capacity (TAC), both the S-EtOH-H_2O (12.46 ± 0.61 mg AAE/g pm) and S-EtOH (12.34 ± 0.82 mg AAE/g pm) extracts displayed high and comparable capacities.

3.3. *Effect of* C. arvensis *on Cell Aging-Inducing Enzymes*

The inhibitory effects of the *C. arvensis* extracts against enzymes involved in cell aging—namely, elastase, tyrosinase, and collagenase—were evaluated in vitro at various concentrations, with positive standards used as comparisons. The inhibition capacity was observed to be dose-dependent (Table 5). The S-EtOH-H_2O and M-EtOH-H_2O extracts exhibited maximum inhibition rates of 82.91% and 77.72%, respectively, with relatively low IC_{50} values of 3.58 mg/mL and 3.89 mg/mL, indicating potent anti-tyrosinase activity. These inhibition percentages were comparable to those achieved by the standard kojic acid at a concentration of 1 mg/mL ($p > 0.05$), which outperformed all *C. arvensis* extracts obtained through different extraction methods (Table 5).

Table 5. Inhibitory activity of *C. arvensis* capitula against tyrosinase, elastase, and collagenase.

Extracts	Concentration mg/mL	Tyrosinase		Elastase		Collagenase	
		I% ± SD	CI_{50} ± SD mg/mL	I% ± SD	CI_{50} ± SD mg/mL	I% ± SD	CI_{50} ± SD mg/mL
M-EtOH	10	68.04 ± 3.09 ^{bc}		57.95 ± 1.75 ^{cd}		36.84 ± 5.54 ^{bc}	
	5	46.79 ± 8.64 ^{bc}		48.95 ± 0.44 ^{b}		30.00 ± 2.78 ^{bc}	
	2.5	21.76 ± 1.88 ^{c}	6.71 ± 0.06 ^{a}	29.15 ± 0.94 ^{c}	7.56 ± 0.19 ^{b}	25.26 ± 2.41 ^{b}	-
	1.25	9.37 ± 2.06 ^{c}		14.97 ± 0.69 ^{d}		-	
	0.625	-		3.02 ± 0.64 ^{e}		-	
M-EtOH-H_2O	10	77.72 ± 0.65 ^{ab}		64.47 ± 4.95 ^{bc}		35.51 ± 3.66 ^{bc}	
	5	58.12 ± 4.20 ^{b}		43.94 ± 0.97 ^{cd}		30.71 ± 2.80 ^{b}	
	2.5	45.47 ± 2.88 ^{b}	3.89 ± 0.76 ^{bc}	30.07 ± 1.16 ^{c}	6.87 ± 0.38 ^{b}	11.49 ± 2.48 ^{c}	-
	1.25	32.32 ± 4.30 ^{b}		21.37 ± 2.33 ^{c}		8.00 ± 5.96 ^{b}	
	0.625	20.75 ± 4.96 ^{a}		8.20 ± 2.25 ^{cd}		-	
S-EtOH	10	57.31 ± 3.64 ^{cd}		57.54 ± 1.03 ^{cd}		36.90 ± 1.57 ^{bc}	
	5	40.63 ± 3.28 ^{c}		42.84 ± 1.95 ^{d}		24.80 ± 1.91 ^{c}	
	2.5	29.14 ± 2.16 ^{c}	7.91 ± 0.50 ^{a}	36.07 ± 1.67 ^{b}	7.45 ± 0.06 ^{b}	14.68 ± 3.27 ^{c}	-
	1.25	14.87 ± 1.82 ^{c}		28.62 ± 0.44 ^{b}		-	
	0.625	-		12.15 ± 1.22 ^{bc}		-	
S-EtOH-H_2O	10	82.91 ± 4.74 ^{a}		70.40 ± 1.06 ^{b}		41.13 ± 2.02 ^{b}	
	5	59.12 ± 7.71 ^{b}		47.63 ± 1.69 ^{bc}		28.08 ± 1.94 ^{bc}	
	2.5	51.80 ± 5.57 ^{b}	3.58 ± 0.56 ^{c}	38.28 ± 2.81 ^{b}	5.93 ± 0.13 ^{c}	14.79 ± 2.59 ^{c}	-
	1.25	38.42 ± 2.56 ^{b}		21.78 ± 2.61 ^{c}		-	
	0.625	27.40 ± 1.18 ^{a}		12.87 ± 0.99 ^{b}		-	
IF-H_2O	10	56.78 ± 7.06 ^{d}		55.90 ± 3.49 ^{d}		31.16 ± 6.27 ^{c}	
	5	47.67 ± 3.18 ^{bc}		36.19 ± 2.56 ^{e}		25.13 ± 1.11 ^{c}	
	2.5	43.60 ± 3.88 ^{b}	6.62 ± 2.24 ^{ab}	28.38 ± 0.68 ^{c}	8.30 ± 0.34 ^{a}	14.76 ± 2.24 ^{c}	-
	1.25	12.24 ± 7.95 ^{c}		21.70 ± 1.17 ^{c}		-	
	0.625	-		5.03 ± 1.91 ^{de}		-	
Standards *	1	86.94 ± 1.50 ^{a}		78.39 ± 1.39 ^{a}		81.82 ± 1.87 ^{a}	
	0.25	81.90 ± 1.57 ^{a}		66.75 ± 0.63 ^{a}		75.22 ± 0.26 ^{a}	
	0.125	64.00 ± 3.14 ^{a}	0.071 ± 0.01 ^{d}	52.94 ± 1.39 ^{a}	0.11 ± 0.00 ^{d}	60.37 ± 0.88 ^{a}	0.08 ± 0.00
	0.06	51.53 ± 3.07 ^{a}		46.12 ± 0.65 ^{a}		51.86 ± 1.02 ^{a}	
	0.03	25.05 ± 1.91 ^{a}		32.82 ± 1.54 ^{a}		31.47 ± 0.83 ^{a}	

Similar letters in the same column and row indicate that there is no significant difference, with a $p < 0.05$ threshold (n = 3). * Kojic acid is used as a standard for the tyrosinase enzyme, while epigallocatechin gallate is used as a standard for elastase and collagenase.

Conversely, the extracts demonstrated relatively low inhibition against both elastase and collagenase. The S-EtOH-H_2O extracts showed significant elastase inhibition at 70.40% with an IC_{50} of 5.93 mg/mL, while the M-EtOH-H_2O extracts achieved 64.47% inhibition with an IC_{50} of 6.87 mg/mL. At a concentration of 10 mg/mL, the S-EtOH-H_2O extract exhibited the highest collagenase inhibition among the tested extracts (41.13 ± 2.02%; $p < 0.05$), followed by the S-EtOH (41.13 ± 2.02%) and M-EtOH extracts (36.84 ± 5.54%). However, collagenase inhibition by the extracts was generally moderate, with the S-EtOH-H2O extract showing the highest activity at 41.13% inhibition.

The S-EtOH-H_2O extract showed the most potent inhibition against elastase (70.40% inhibition, IC_{50}: 5.93 mg/mL), followed by the M-EtOH-H_2O extract (64.47% inhibition, IC_{50}: 6.87 mg/mL; $p < 0.05$). Meanwhile, the extracts demonstrated promising inhibition levels of the enzyme tyrosinase, which is crucial for regulating melanin synthesis and treating hyperpigmentation.

3.4. Interfacial and Surface Tension Characteristics

C. arvensis capitula extracts (1% *w*/*w*) were evaluated for their ability to reduce surface tension (ST) and interfacial tension (IT) over a time frame of 600 s (Figure 1). The results demonstrate a varied activity when compared to that of Polysorbate 20—a commonly used synthetic emulsifier. Initially (0 s), the surface tensions of the extracts varied considerably, with values ranging from 38.63 mN/m for the methanol–water extract (M-EtOH-H_2O) to 55.79 mN/m for the aqueous infusion (IF-H_2O). These initial differences suggest that the methanol and ethanol extracts have a higher affinity for the air–water interface than the aqueous infusion.

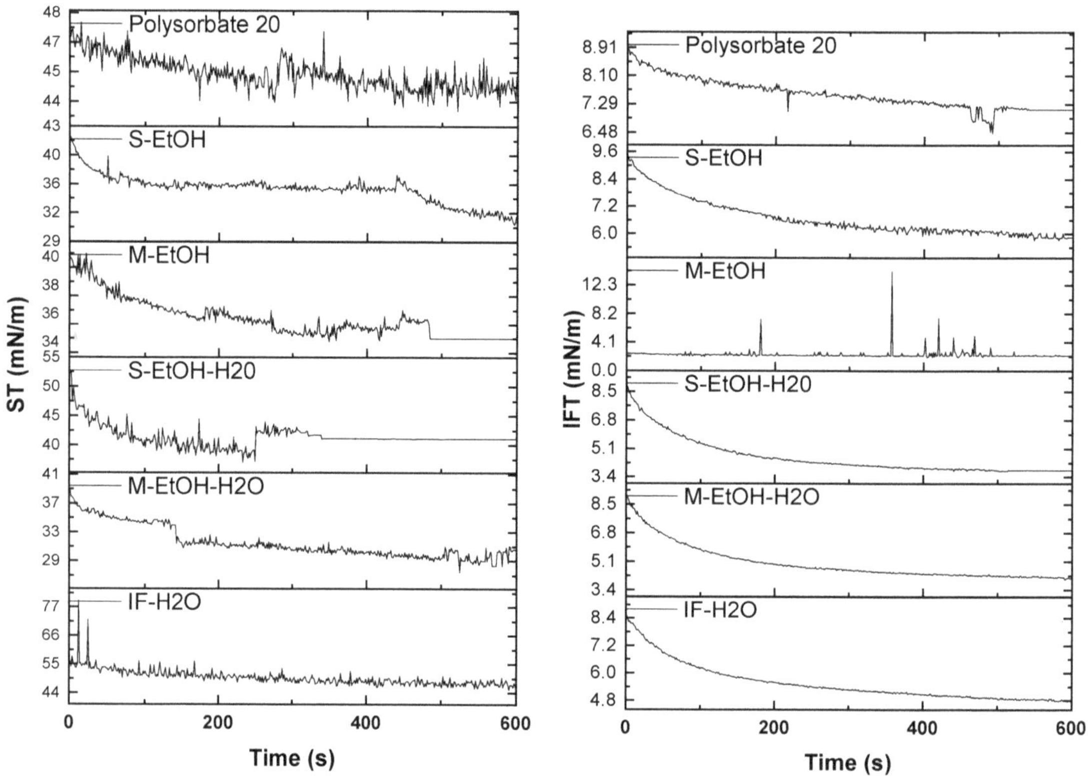

Figure 1. Surface tension (ST) and interfacial tension (IFT) profiles of *C. arvensis* extracts as a function of time and solvent conditions.

Following 600 s, a significant decrease in TS was observed for all extracts. M-EtOH-H_2O showed a notable drop, from 38.63 mN/m to 31.14 mN/m, reflecting rapid and efficient adsorption of active compounds to the interface. The S-EtOH-H_2O extract followed a similar trend, dropping from 52.38 mN/m to 41.08 mN/m. The pure methanolic (M-EtOH) and pure ethanolic (S-EtOH) extracts also showed significant reductions in ST, reaching 34.647 mN/m and 31.64 mN/m, respectively. Comparatively, Polysorbate 20, although having an initial TS of 47.65 mN/m, showed only a slight reduction to 43.96 mN/m after 600 s. This indicates that, although Polysorbate 20 is effective in reducing ST, the *C. arvensis* extracts—particularly those containing ethanol or methanol—are more effective in this regard.

Interfacial tension measurements also revealed the considerable activity of the extracts. At 0 s, M-EtOH (2.46 mN/m) already exhibited a remarkably low IFT, indicating a high affinity of the compounds for the water–oil interface. After 600 s, the IFT for M-EtOH decreased again slightly to 2.16 mN/m, confirming its exceptional efficiency. The other extracts also showed a substantial reduction in IFT, notably M-EtOH-H_2O (from 9.22 mN/m to 4.11 mN/m) and S-EtOH-H_2O (from 8.99 mN/m to 3.82 mN/m), suggesting that these extracts have a high capacity to interact with the water–oil interface. Polysorbate 20, in contrast, showed an initial IFT of 8.93 mN/m, which decreased to 7.14 mN/m after 600 s. Although Polysorbate 20 is effective, the methanolic and ethanolic extracts of *C. arvensis* outperformed this synthetic emulsifier in terms of IFT reduction. These results indicate that *C. arvensis* extracts, particularly those obtained with ethanol, have a superior ability to reduce surface and interfacial tension, when compared with Polysorbate 20. These extracts could, therefore, represent an interesting natural alternative to synthetic emulsifiers in formulations where an effective reduction in surface tension is required.

3.5. Sun Protection Factor (SPF-UVB)

The *C. arvensis* extracts were then evaluated for their protective properties (Table 6). The results suggest that, at a concentration of 2 mg/mL, all of the tested extracts displayed extremely high SPF-UVB values, ranging from 193.67 ± 0.04 to 192.19 ± 0.08 for S-EtOH-H_2O and IF-H_2O extracts, respectively, as no significant difference was observed between the different extraction methods ($p > 0.05$). Significantly, ZnO—a commonly used ingredient in sunscreen formulations—has an SPF of 11.88 ± 0.03 ($p < 0.05$) at the same concentration of 2 mg/mL. However, the UV-B absorbance measurements (290–320 nm; Figure 2b) revealed distinct absorbance profiles for each extract. The aqueous infusion (IF-H_2O) showed a relatively stable absorbance across the UVB range, with values fluctuating around 20.09, suggesting a low capacity for UVB absorption by the hydrophilic compounds present in this extract. This trend is similar to that observed for the ethanol–water extract (M-EtOH-H_2O), although the latter showed a slight increase in absorbance around 300 nm, reaching a maximum value of 20.22, which could indicate the presence of ethanol-soluble compounds with a better interaction with UVB. The S-EtOH-H_2O extract showed an overall higher absorbance profile than the other water-soluble extracts, with values close to 20.20 at 320 nm, indicating greater UVB absorption efficiency. The M-EtOH and S-EtOH extracts showed more complex absorbance profiles, with significant fluctuations in the UVB range (Figure 2b). M-EtOH presented a peak absorbance at around 300 nm, reaching 20.15, followed by a slight decrease, which could be attributed to a higher concentration.

Table 6. SPF values of *C. arvensis* extracts against UVB.

Extracts	SPF-UVB
S-EtOH-H_2O	193.67 ± 0.04 ^{a}
S-EtOH	193.46 ± 0.19 ^{a}
M-EtOH-H_2O	193.49 ± 0.24 ^{a}
M-EtOH	193.60 ± 0.11 ^{a}
IF-H_2O	192.19 ± 0.08 ^{a}
ZnO	11.88 ± 0.03 ^{b}

Values are presented as means ± standard error (n = 3). Different letters in the same row indicate statistical significance at $p < 0.05$.

Figure 2a highlights the resulting SPF values for various cream formulation-based *C. arvensis* extracts at different concentrations (0.05%, 0.10%, and 0.20%). The data suggest a dose-dependent pattern for all the formulas, with the sunscreen-based S-EtOH-H_2O extract standing out as the most promising candidate for further development in photoprotective skincare products. Notably, while no statistically significant difference was found between the formulations at a concentration of 0.1%, at 0.2%, the formula based on the S-EtOH-H_2O extract had a significantly higher SPF value (SPF = 194.49 ± 0.18; $p < 0.05$). At a lower concentration of 0.05%, there was no significant difference between those with IF-H_2O extract (SPF = 33.40 ± 4.25) and S-EtOH extract (SPF = 32.93 ± 2.65) ($p > 0.05$). Both of these formulations achieved significantly higher SPF values than those based on M-EtOH (SPF = 26.33 ± 5.35) and S-EtOH-H_2O (SPF = 20.38 ± 0.66).

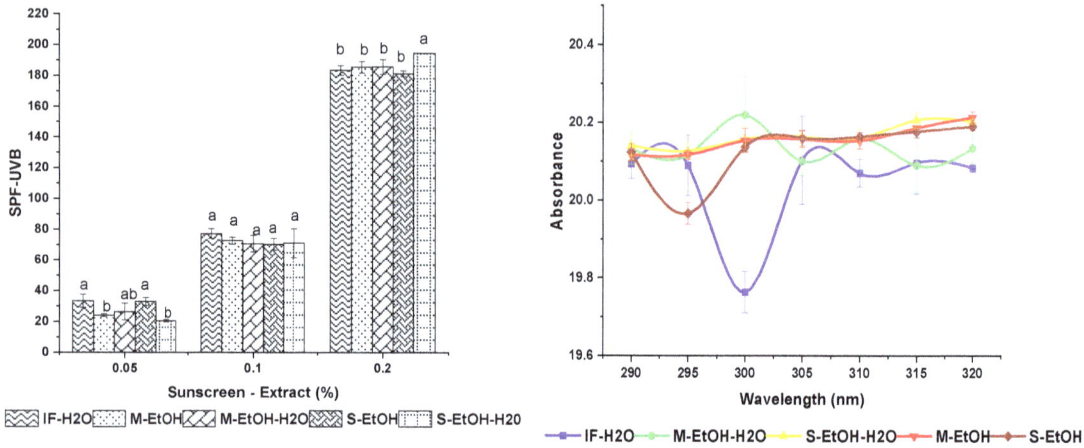

(a) (b)

Figure 2. Sun Protection Factor (SPF) screening of sunscreens based on *C. arvensis* extracts, with similar letters in the same concentration (%) indicate that there is no significant difference, $p < 0.05$ (n = 3) (**a**), and analysis of the UVB absorbance spectra of the various *C. arvensis* extracts (**b**).

COSMOS offers several important criteria for the certification of organic cosmetics; for example, organic ingredients must be grown without the use of synthetic pesticides or fertilizers. Additionally, the formulation must strictly exclude certain contentious substances, such as parabens, silicones, GMOs, synthetic colorants, nanoparticles, and other ingredients which are potentially harmful to both the environment and human health. The production methods should prioritize ecological and sustainable practices, in order to minimize the overall impact on the environment. Moreover, processes that operate at lower temperatures and provide optimal yields to minimize waste and rejection are recommended by COSMOS. Based on this, sonication was identified as a suitable method, as it requires a minimal amount of time, no increase in temperature, and offers the highest % yield of extraction. Additionally, the fact that the plant is wild and has not been subjected to pesticides further supports the possibility of certifying this ingredient as organic by COSMOS.

Following the method of calculation mentioned in point (6.2) "to calculate the percentage of organic ingredients" in Cosmos Standards Version 3.0 (1st September 2018), we were able to determine the percentage of the organic ingredient (5) of the dried extract (without solvent) used in the formulation. As the ratio (4) is greater than 1, it is considered equal to 1, according to the standards.

$$\text{Ratio} = \frac{\text{Organic freach plant}}{\text{final extract} - \text{solvents}} = \frac{300}{2-0} > 1, \tag{4}$$

$$\%\text{Bio} = \left[\text{Ratio} \times \left(\frac{\text{Final extract} - \text{solvants}}{\text{Final extract}} + \frac{\text{Organic solvants}}{\text{Final extract}}\right)\right] \times 100, \tag{5}$$

$$\%\text{Bio} = \left[1 \times \left(\frac{2-0}{2} + \frac{0}{2}\right) \times 100\right] = 100\%.$$

The results indicate an organic percentage of 100% for the *C. arvensis* active ingredient. The steps used to calculate the organic active ingredient percentage of the extract are depicted in Figure 3.

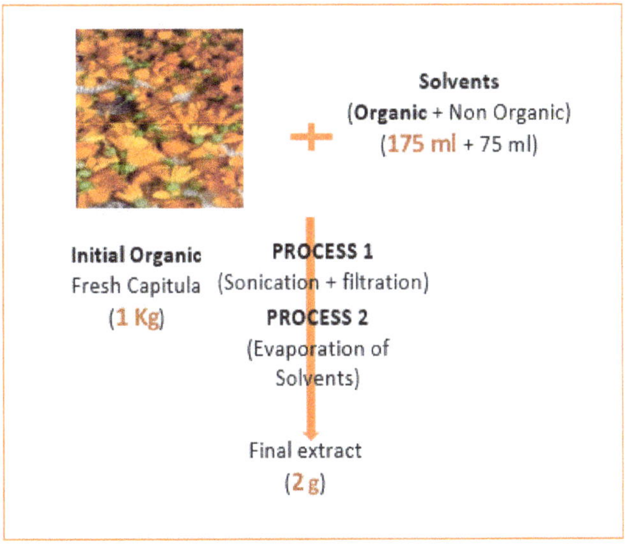

Figure 3. A guideline illustration, inspired by the COSMOS technical reference, for calculating the biological percentage of plant extracts through an extraction process.

In the present formulation, the ingredients were selected to conform the strict standards defined for COSMOS certification [24]. By applying Equation (2) defined in the COSMOS standard, we determined that the formula is based on 87.5% biological ingredients. This percentage exceeds the minimum threshold of 20% required for organic certification.

$$\text{Bio\%} = \frac{\sum \text{\% of Bio ngredients}}{\text{Total ingredients}} = \frac{87.5}{100} \times 100 = 87.5\% > 20\%.$$

3.6. Sunscreen Characteristics and Stability

Conductivity

Among the produced W/O emulsions, the 0.05% S-EtOH-H_2O sunscreen recorded an initial conductivity of 251.67$\pm$ 1.53 µS/cm, which decreased slightly to 235 $\pm$ 2 µS/cm by day 90 (Figure 4c). Conversely, the 0.05% S-EtOH formulation showed a progressive increase from 117 $\pm$ 3.60 µS/cm to 139.67 $\pm$ 2.08 µS/cm. This slight increase could be attributed to an increase in free ions, possibly due to interactions between the components. Furthermore, the ethanolic extract-based formulations (0.05% M-EtOH-H_2O and 0.05% M-EtOH) also showed slight fluctuations, suggesting reasonable stability over time. However, at concentrations of 0.1% and 0.2%, formulations based on aqueous and ethanolic extracts of *C. arvensis* showed good stability over the storage period. Among the microemulsion types, W/O structures exhibit the lowest conductivity, which gradually increases with increasing water content.

pH stability

All sunscreens based on *C. arvensis* extracts exhibited minimal pH changes over time (Figure 4a). For instance, the pH of the 0.05% S-EtOH-H_2O formula decreased slightly between day 1 and day 90, from 5.15 $\pm$ 0.03 to 5.03 $\pm$ 0.01, respectively. In contrast, at a concentration of 0.2%, the formulations were shown to be particularly stable, especially in the ethanolic extract, which achieved pH values of 6.00 $\pm$ 0.13 on the first day of production, 6.53 $\pm$ 0.16 after 30 days, 6.24 $\pm$ 0.01 after 60 days, and 6.24 $\pm$ 0.03 after a total of 90 days.

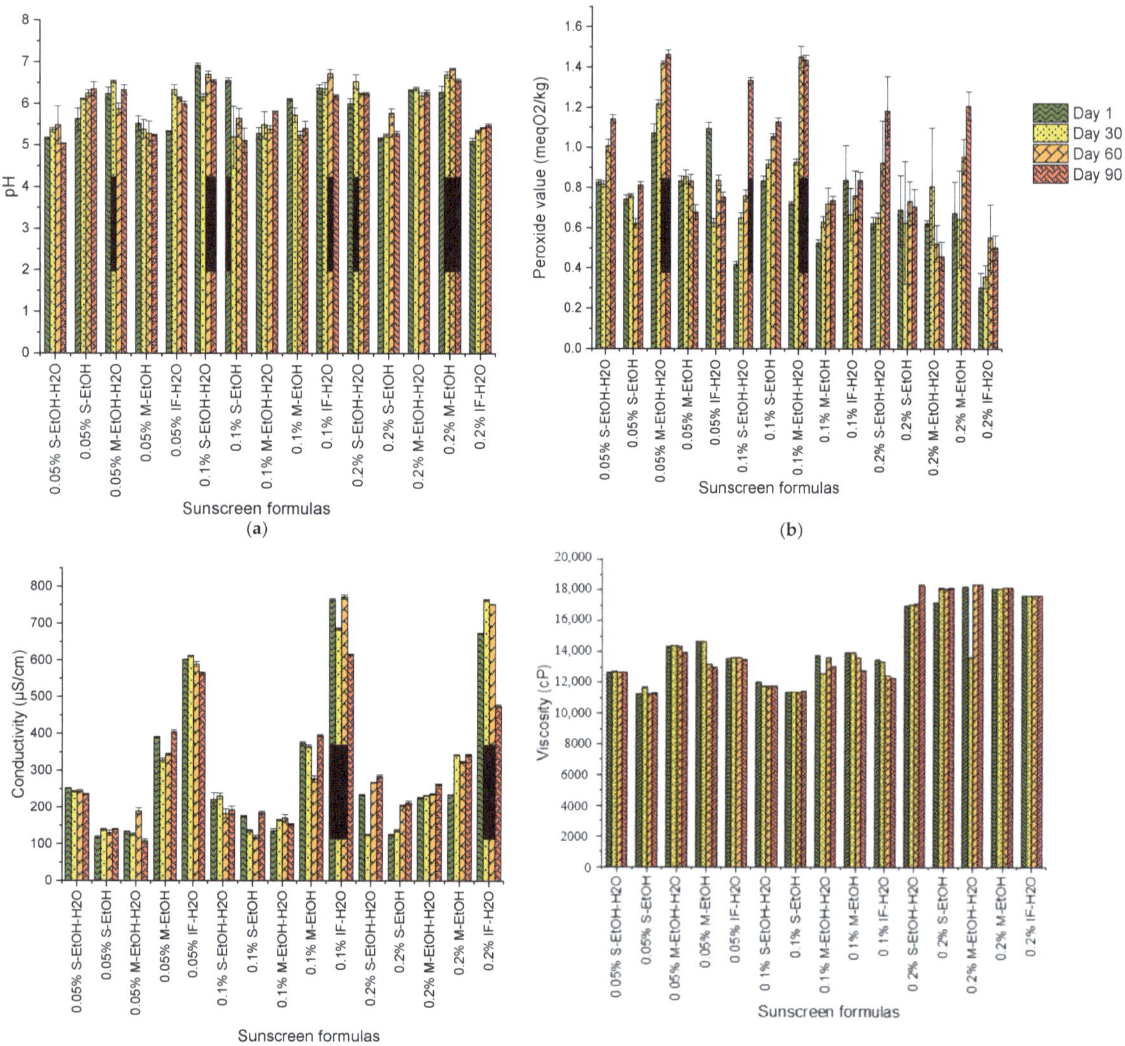

Figure 4. Characterization of sunscreen formulation parameters at different time intervals (day 1, day 30, day 60, and day 90) for assessment of stability over time (n = 3): (**a**) pH values, (**b**) peroxide value, (**c**) conductivity, and (**d**) viscosity analysis.

Dynamic viscosity

Remarkably, as can be seen from Figure 4d, formulations with 0.05% *C. arvensis* recorded lower initial viscosities. For instance, the formula based on 0.05% S-EtOH recorded a value of 11,254 ± 4.36 cP initially, with minimal variations over the 90 days. In contrast, formulations containing 0.2% *C. arvensis* extract exhibited higher initial viscosities, measured as 17,183.67 ± 2.08 cP for the 0.2% S-EtOH formulation, and outstanding stability over 90 days. These findings confirm that *C. arvensis* is a highly effective stabilizer and reducer of surface and interfacial tension.

Peroxide Value (PV)

The formulated sunscreen based on 0.05% M-EtOH exhibited a slight increase in PV (Figure 4b), from 0.83 ± 0.02 meq O_2/kg on day 1 to 0.85 ± 0.03 meq O_2/kg on day 60, followed by a slight increase to 0.68 ± 0.04 meq O_2/kg on day 90. These fluctuations suggest minor oxidation without major instability.

However, at the 0.1% and 0.2% concentrations, a stable range of values were found among the formulations. For example, the S-EtOH-H_2O formula at 0.1% showed consistent values from day 1 (0.41 ± 0.01 meq O_2/kg) to 0.65 ± 0.03 meq O_2/kg on day 60 and 0.76 ± 1.33 meq O_2/kg on day 90. The 0.2% formulations showed better PV stability, particularly for the ethanolic extracts.

4. Discussion

To date, numerous cosmetic and pharmaceutical products based on *Calendula* species—particularly *C. officinalis*—have been introduced to the market in a wide range of formulations and applications. In this context, our study focuses on the development of an innovative biological anti-aging sunscreen based on *C. arvensis* capitula extracts. This approach targets the bioactive properties of *C. arvensis* in order to provide effective sun protection while targeting skin aging concerns. Among all extracts, S-EtOH-H_2O exhibited the highest TC (1.66 ± 0.02 mg/mL), while M-EtOH displayed the lowest TPC (8.24 ± 1.17 mg GAE/g) and TFC (6.47 ± 0.49 mg QE/g). Previous studies, such as that published by Abudunia et al. [25], revealed significantly higher flavonoid contents in their methanolic and aqueous extracts (174.93 ± 5.21 mg RE/g dry extract and 74.93 ± 1.50 mg RE/g dry extract, respectively). The disparity between these values and our results could be attributed to differences in extraction conditions; for instance, a plant material–solvent ratio of 0.4 g dw/mL was used in their study, whereas a much lower ratio of 0.003 g dw/mL was used for our infusion. Furthermore, Ercetin et al. [26] reported even higher phenolic (118.18 ± 10.29 mg/g extract) and flavonoid (74.14 ± 3.09 mg/g extract) concentrations in methanolic extracts of *C. arvensis* flowers. These variations highlight the significant impacts of extraction methods, raw material sources, and other biotic and abiotic factors on the phytochemical compositions observed in the various studies. In this study, S-EtOH-H_2O provided the highest antioxidant potential, as well as a low IC_{50} value in the DPPH (3.77 ± 0.44 mg/mL) and ABTS (3.21 ± 0.08 mg/mL) assays, thus underscoring the suitability of hydroethanol solvents for the extraction of phenolics and flavonoids. Conversely, extracts such as IF-H_2O and M-EtOH showed more limited antioxidant effects. These findings revealed considerably higher values than those reported by Abudunia et al. [25], with IC_{50} values of 20.9 mg/mL and 33.2 mg/mL for the methanolic and aqueous extracts, respectively, demonstrating a radical scavenging effect. The variations observed between the different extracts can be attributed to the phenolic content, extraction efficiency, and the polarities of the used solvents, which influence the bioavailability and stability of the antioxidant compounds. The structural characteristics of phenolic compounds are characterized by the presence of one or more hydroxyl groups (-OH) bonded directly to an aromatic benzene ring [27], and the arrangement and number of these hydroxyl groups play an essential role in determining the reactivity, stability, and overall biological activity of the compound. Hydroxyl groups enable hydrogen bonding and electron donating interactions, which further enhance the antioxidant capacity of the compound by enabling the scavenging of free radicals [27]. In fact, -OH groups contribute to the ability of phenol compounds to chelate metal ions and modulate enzymatic activities, which are crucial mechanisms affecting their antioxidant, anti-aging, and photoprotective effects. The diversity of substitution patterns on the aromatic ring also characterizes various subclasses of phenolic compounds, influencing their polarity, solubility, and interactions with biological targets.

Furthermore, the *C. arvensis* capitula extracts demonstrated notable inhibitory effects against tyrosinase and elastase, with moderate activity against collagenase. Elastase is a target for anti-aging formulations, a serine protease responsible for the breakdown of elastin in skin and connective tissue [28]. The S-EtOH-H_2O extract showed the most potent inhibition against elastase (70.40% inhibition, IC_{50}: 5.93 mg/mL), followed by the M-EtOH-H_2O extract (64.47% inhibition, IC_{50}: 6.87 mg/mL; $p < 0.05$). Meanwhile, the extracts demonstrated promising inhibition levels for the enzyme tyrosinase, which is crucial for regulating melanin synthesis and treating hyperpigmentation. These findings highlight the potential of *C. arvensis* extracts for cosmetics and pharmaceutical applications targeting skin aging and pigmentation issues. Previous studies have reported that an *C. arvensis* aerial part extract with 80% ethanol presented a 51.66 ± 0.57% inhibition rate at 0.666 mg/mL, with an IC_{50} of 0.621 ± 3.47 mg/mL [29]. The present research showed that the S-EtOH-H_2O capitula extract exhibited a lower inhibition rate of 27.40 ± 1.18% at a similar concentration (0.625 mg/mL). This discrepancy may be linked to the synergistic effects of bioactive compounds in the aerial parts, which may provide stronger tyrosinase inhibition than those present in the capitula. However, the findings on collagenase inhibition are in line with the previous work of Deniz et al. [29], who reported no inhibitory activity in the hydroethanol extract of the aerial parts of *C. arvensis* at 0.666 mg/mL. Previous evidence has demonstrated that phenolic compounds, such as flavonoids and tannins, are key factors in modulating the enzymes responsible for the degradation of skin structures [30].

The interfacial tension of surfactants plays a crucial role in determining their effectiveness in stabilizing emulsions. The surface and interfacial tension reduction abilities of 1% (*w/w*) *C. arvensis* capitula extracts were evaluated over 600 s and their superior performance was observed, when compared to Polysorbate 20. Generally, surfactants exhibit stronger emulsifying properties when they achieve lower interfacial tension at the boundary between two immiscible liquids, such as oil and water [31]. This reduction in interfacial tension facilitates the formation and stabilization of smaller droplets, leading to more stable and homogeneous emulsions [32]. Consequently, the ability of a surfactant to lower interfacial tension directly influences its capacity to enhance emulsion stability. Interestingly, M-EtOH (2.46 mN/m) already exhibited a remarkably low IFT, indicating a high affinity of the compounds for the water–oil interface. After 600 s, the TI for M-EtOH decreased again slightly to 2.16 mN/m, confirming its efficiency. This relatively high interfacial tension increase could be assigned to their higher flavonoid content, which could also increase the interfacial tension. Nonetheless, flavonoids could simultaneously improve the physical stability of O/W emulsions. In fact, Jafari et al. [33] have reported that some flavonoids could act as stabilizers of oil-in-water emulsions through their adsorption as water-insoluble particles to the surface of oil droplets, thus forming Pickering emulsions.

Plant extracts rich in tannins have been demonstrated to provide UV protection by scavenging free radicals and preventing oxidative damage [34]. UVB rays (320–290) activate melanocytes and increase melanin production. Receptors on the surface of melanocytes are activated by UVB rays. An increase in melanin in the skin is the body's protective response against the damaging effects of UVB, such as sunburn. However, excessive exposure to UVB can also lead to damage such as accelerated photo-aging and skin cancer [35]. The SPF ingredient may promote direct blocking of UV photons, or neutralization of the direct or indirect effects of UV radiation through DNA repair systems and antioxidants [36]. The protective properties of *C. arvensis* extracts were evaluated (Table 6). At 2 mg/mL, all extracts exhibited high SPF-UVB values (193.67 ± 0.04 to 192.19 ± 0.08), significantly surpassing ZnO (SPF = 11.88 ± 0.03, $p < 0.05$). Furthermore, the formulations demonstrated dose-dependent SPF increases; in particular, S-EtOH-H_2O at 0.2% achieving the highest SPF (194.49 ± 0.18, $p < 0.05$). Previously, Mishra et al. conducted an assessment of the sun protection factor (SPF) value for a cream formulated with *C. officinalis* oil, resulting in a value of 14.84 ± 0.16 [37]. In contrast, Lohani et al. [38] reported a significantly lower SPF value of 8.36 for *C. officinalis* essential oil.

In view of the growing consumer awareness of the risks associated with the synthetic products [39–41], demand for biological sunscreens has increased [40], due to their ability to offer effective UV protection while protecting the skin from potential damage [42]. Organic labels awarded by Ecocert and COSMOS ensure the natural originality of products respecting their standards. Ecocert is widely recognized for its rigorous criteria and commitment to promoting ecologically sound and sustainable practices in the production of cosmetics; however, even if natural ingredients are considered to be safe, they may cause skin reactions. For example, skin toxicity tests of the product are essential. In the present formulations, all ingredients were selected to conform the strict standards defined for COSMOS certification [24]. Calculations based on COSMOS standards confirmed a 100% organic content for *C. arvensis* extracts and 87.5% biological ingredients in the final formulation, exceeding the percentage threshold required for organic certification.

In order to assess the stability of the sunscreen formulations based on *C. arvensis* extracts, the pH, conductivity, viscosity, and peroxide values were analyzed each month over a period of 90 days.

Conductivity measurements allow for the characterization of microemulsion systems. In systems containing ionic surfactants or salts, variations in conductivity serve as indicators of structural differences [43]. The W/O emulsions demonstrated stability over 90 days. The 0.05% S-EtOH-H_2O sunscreen showed a slight conductivity decrease (251.67 to 235 μS/cm), while 0.05% S-EtOH increased slightly (117 to 139.67 μS/cm). As the water content reaches a critical point, interactions between water droplets via charge hopping or transient droplet aggregation can result in a continuous water network within the oil phase, forming a bicontinuous phase that enhances conductivity [44]. However, it has been demonstrated previously that higher pH promotes smaller droplet formation and enhances stability. The studied sunscreens showed minimal pH changes. For instance, the pH value of S-EtOH-H_2O formula at 0.05% decreased slightly (from 5.15 to 5.03) over 90 days. At 0.2%, the ethanolic extract remained stable, with pH values ranging from 6.00 to 6.24. This can be attributed to increased surfactant aggregation at high pH. Carboxylic acids in crude oil, which ionize at higher pH, contribute to the stability of emulsions by forming micelles that reduce droplet size; however, effects vary with the system's composition [45].

Variations in viscosity over time may indicate structural degradation or aggregation of the ingredients. Formulations with 0.05% extract, such as 0.05% S-EtOH, had lower initial viscosities (11,254 $\pm$ 4.36 cP) with minimal changes over 90 days. Higher concentrations (0.2% S-EtOH) showed higher viscosities (17,183.67 $\pm$ 2.08 cP) and stability. Nevertheless, a low emulsifier content leads to larger droplets due to insufficient surface coverage, thus promoting droplet coalescence. Increased surfactant concentration reduces droplet size by enhancing interfacial coverage, thus improving stability. Non-ionic surfactants are required at higher concentrations for stability when compared to anionic surfactants, which can maintain stability even at lower concentrations [45]. In contrast, the sunscreen with 0.05% M-EtOH showed minor fluctuations in PV, while the 0.1% and 0.2% formulations—particularly with ethanolic extracts—showed consistent PV stability over 90 days. The peroxide value (PV) in an emulsion reflects the oxidation of lipids, where high levels of PV indicate increased oxidative degradation, adversely affecting the stability of the emulsion [27].

5. Limitations

The findings of this study are primarily based on in vitro analyses which, while informative, do not fully replicate the complex dynamics of skin interactions. In vivo studies are necessary to validate the efficacy and long-term effects of the formulations, particularly in diverse populations with varying skin types. Furthermore, the SPF results—although highly promising—require further evaluation using standardized in vivo protocols in order to confirm their reproducibility and reliability.

Addressing these limitations will be critical for the commercial translation and certification of these formulations as effective and sustainable sunscreens.

6. Conclusions

Based on the results of the present study, *C. arvensis* flower extracts demonstrated significant potential as biological sunscreen, targeting both UVB protection and skin aging. Furthermore, the M-EtOH extract showed significant surface and interfacial tension-lowering activities. This property improves the stability of oil–water emulsions, enhancing their application in biological formulations. The combined effects of UV protection, enzyme inhibition, and emulsion stability underline the multi-functional potential of *C. arvensis* extracts in the development of advanced, bio-based products that meet the growing consumer demand for natural, eco-certified ingredients.

However, future research should focus on identifying and characterizing the specific active compounds that are responsible for these activities.

Author Contributions: Conceptualization: N.E.-O. and A.Z.; Methodology: N.E.-O.; Software: N.E.-O.; Validation: N.E.-O.; Formal Analysis: N.E.-O.; Investigation: N.E.-O.; Resources: N.E.-O.; Data Curation: N.E.-O.; Writing—Original Draft Preparation: N.E.-O.; Writing—Review and Editing: N.E.-O. and I.Z.; Visualization: N.E.-O.; Supervision: A.Z.; Project Administration: A.Z.; Funding Acquisition: A.Z and N.E.-O. All authors have read and agreed to the published version of the manuscript.

Funding: This research received no external funding.

Institutional Review Board Statement: Not applicable.

Informed Consent Statement: Not applicable.

Data Availability Statement: The project provides the following underlying datasets: Raw_ Material_C_Arvensis.xlsx DOI: 1010.6084/m9.figshare.27643428 [46], and Sunscreen-characterisation.xlsx DOI: 1010.6084/m9.figshare.27643425 [47]. Data are available under the terms of the Creative Commons Zero "No rights reserved" data waiver (CC0 1.0 Public domain dedication).

Acknowledgments: The authors extend their heartfelt appreciation to Benkhnigue Ouafae at the Botanics and Plant Ecology Departement of the Scientific Institute of Rabat, Morocco. and also, Zineelabidine Triqui for their crucial support in identifying *Calendula arvensis* L.

Conflicts of Interest: The authors declare that there are no conflicts of interest.

References

1. Passeron, T.; Krutmann, J.; Andersen, M.; Katta, R.; Zouboulis, C. Clinical and biological impact of the exposome on the skin. *J. Eur. Acad. Dermatol. Venereol.* **2020**, *34*, 4–25. [CrossRef] [PubMed]
2. Diffey, B. Solar ultraviolet radiation effects on biological systems. *Phys. Med. Biol.* **1991**, *36*, 299. [CrossRef] [PubMed]
3. Muthusamy, V.; Piva, T.J. The UV response of the skin: A review of the MAPK, NFκB and TNFα signal transduction pathways. *Arch. Dermatol. Res.* **2010**, *302*, 5–17. [CrossRef] [PubMed]
4. Gasparro, F.P.; Mitchnick, M.; Nash, J.F. A review of sunscreen safety and efficacy. *Photochem. Photobiol.* **1998**, *68*, 243–256. [CrossRef]
5. Nash, J. Human safety and efficacy of ultraviolet filters and sunscreen products. *Dermatol. Clin.* **2006**, *24*, 35–51. [CrossRef]
6. Bozza, A.; Campi, C.; Garelli, S.; Ugazio, E.; Battaglia, L. Current regulatory and market frameworks in green cosmetics: The role of certification. *Sustain. Chem. Pharm.* **2022**, *30*, 100851. [CrossRef]
7. Ottman, J. *The New Rules of Green Marketing: Strategies, Tools, and Inspiration for Sustainable Branding*; Routledge: London, UK, 2017.
8. Tran, E.; Richmond, G.L. Interfacial steric and molecular bonding effects contributing to the stability of neutrally charged nanoemulsions. *Langmuir* **2021**, *37*, 12643–12653. [CrossRef]
9. Arora, D.; Rani, A.; Sharma, A. A review on phytochemistry and ethnopharmacological aspects of genus Calendula. *Pharmacogn. Rev.* **2013**, *7*, 179. [CrossRef]
10. El-Otmani, N.; Zeouk, I.; Hammani, O.; Zahidi, A. Analysis and Quality Control of Bio-actives and Herbal Cosmetics: The Case of Traditional Cooperatives from Fes-Meknes Region. *Trop. J. Nat. Prod. Res. (TJNPR)* **2024**, *8*, 7181–7195. [CrossRef]
11. Khouchlaa, A.; El Baaboua, A.; El Moudden, H.; Lakhdar, F.; Bakrim, S.; El Menyiy, N.; Belmehdi, O.; Harhar, H.; El Omari, N.; Balahbib, A. Traditional uses, bioactive compounds, and pharmacological investigations of *Calendula arvensis* L.: A Comprehensive review. *Adv. Pharmacol. Pharm. Sci.* **2023**, *2023*, 2482544. [CrossRef]
12. Nichols, J.A.; Katiyar, S.K. Skin photoprotection by natural polyphenols: Anti-inflammatory, antioxidant and DNA repair mechanisms. *Arch. Dermatol. Res.* **2010**, *302*, 71–83. [CrossRef] [PubMed]
13. de Lima Cherubim, D.J.; Buzanello Martins, C.V.; Oliveira Fariña, L.; da Silva de Lucca, R.A. Polyphenols as natural antioxidants in cosmetics applications. *J. Cosmet. Dermatol.* **2020**, *19*, 33–37. [CrossRef] [PubMed]
14. Lister, E.; Wilson, P. Measurement of total phenolics and ABTS assay for antioxidant activity (personal communication). *Crop Res. Inst. Linc. New Zealand* **2001**, *7*, 235–239.
15. Djeridane, A.; Yousfi, M.; Nadjemi, B.; Boutassouna, D.; Stocker, P.; Vidal, N. Antioxidant activity of some Algerian medicinal plants extracts containing phenolic compounds. *Food Chem.* **2006**, *97*, 654–660. [CrossRef]

16. RibéreauGayon, P.; Stonestreet, E. Dosage des tanins du vin rouge et détermination de leur structure. *Chim. Anal.* **1966**, *48*, 188–196.
17. Topçu, G.; Ay, M.; Bilici, A.; Sarıkürkcü, C.; Öztürk, M.; Ulubelen, A. A new flavone from antioxidant extracts of Pistacia terebinthus. *Food Chem.* **2007**, *103*, 816–822. [CrossRef]
18. Prieto, P.; Pineda, M.; Aguilar, M. Spectrophotometric quantitation of antioxidant capacity through the formation of a phosphomolybdenum complex: Specific application to the determination of vitamin E. *Anal. Biochem.* **1999**, *269*, 337–341. [CrossRef]
19. Kim, Y.-J.; Uyama, H.; Kobayashi, S. Inhibition effects of (+)-catechin–aldehyde polycondensates on proteinases causing proteolytic degradation of extracellular matrix. *Biochem. Biophys. Res. Commun.* **2004**, *320*, 256–261. [CrossRef]
20. Khatib, S.; Mahdi, I.; Drissi, B.; Fahsi, N.; Bouissane, L.; Sobeh, M. Tetraclinis articulata (Vahl) Mast.: Volatile constituents, antioxidant, antidiabetic and wound healing activities of its essential oil. *Heliyon* **2024**, *10*, e24563. [CrossRef]
21. Mansur, J.d.S.; Breder, M.N.R.; Mansur, M.C.d.A.; Azulay, R.D. Determinação do fator de proteção solar por espectrofotometria. *An. Bras. Dermatol.* **1986**, *63*, 121–124.
22. Sayre, R.M.; Agin, P.P.; LeVee, G.J.; Marlowe, E. A comparison of in vivo and in vitro testing of sunscreening formulas. *Photochem. Photobiol.* **1979**, *29*, 559–566. [CrossRef] [PubMed]
23. COSMOS. Standard 2024. Available online: https://www.cosmos-standard.org/en/documents/ (accessed on 5 February 2024).
24. AISBL C-s. COSMOS-Standard Technical Guide. Available online: https://media.cosmos-standard.org/filer_public/06/29/0629 8b4e-83cb-4064-ae46-f4578f9fc9f5/cosmos-standard_technical_guide_v40.pdf (accessed on 20 February 2024).
25. Abudunia, A.-M.; Marmouzi, I.; Faouzi, M.; Ramli, Y.; Taoufik, J.; El Madani, N.; Essassi, E.; Salama, A.; Khedid, K.; Ansar, M. Anticandidal, antibacterial, cytotoxic and antioxidant activities of *Calendula arvensis* flowers. *J. De Mycol. Medicale* **2017**, *27*, 90–97. [CrossRef] [PubMed]
26. Ercetin, T.; Senol, F.S.; Orhan, I.E.; Toker, G. Comparative assessment of antioxidant and cholinesterase inhibitory properties of the marigold extracts from *Calendula arvensis* L. and *Calendula officinalis* L. *Ind. Crops Prod.* **2012**, *36*, 203–208. [CrossRef]
27. Ghelichi, S.; Hajfathalian, M.; Yesiltas, B.; Sørensen, A.D.M.; García-Moreno, P.J.; Jacobsen, C. Oxidation and oxidative stability in emulsions. *Compr. Rev. Food Sci. Food Saf.* **2023**, *22*, 1864–1901. [CrossRef] [PubMed]
28. Porwal, M.; Rastogi, V.; Chandra, P.; Shukla, S. An Updated Review on the Role of Phytoconstituents in Modulating Signalling Pathways to Combat Skin Ageing: Nature's Own Weapons and Approaches. *Nat. Prod. J.* **2024**, *14*, 55–71. [CrossRef]
29. Deniz, F.S.S.; Orhan, I.E.; Duman, H. Profiling cosmeceutical effects of various herbal extracts through elastase, collagenase, tyrosinase inhibitory and antioxidant assays. *Phytochem. Lett.* **2021**, *45*, 171–183. [CrossRef]
30. Działo, M.; Mierziak, J.; Korzun, U.; Preisner, M.; Szopa, J.; Kulma, A. The potential of plant phenolics in prevention and therapy of skin disorders. *Int. J. Mol. Sci.* **2016**, *17*, 160. [CrossRef]
31. Ravera, F.; Dziza, K.; Santini, E.; Cristofolini, L.; Liggieri, L. Emulsification and emulsion stability: The role of the interfacial properties. *Adv. Colloid Interface Sci.* **2021**, *288*, 102344. [CrossRef]
32. Ho, T.M.; Razzaghi, A.; Ramachandran, A.; Mikkonen, K.S. Emulsion characterization via microfluidic devices: A review on interfacial tension and stability to coalescence. *Adv. Colloid Interface Sci.* **2022**, *299*, 102541. [CrossRef]
33. Jafari, S.M.; Doost, A.S.; Nasrabadi, M.N.; Boostani, S.; Van der Meeren, P. Phytoparticles for the stabilization of Pickering emulsions in the formulation of novel food colloidal dispersions. *Trends Food Sci. Technol.* **2020**, *98*, 117–128. [CrossRef]
34. Fraga-Corral, M.; Otero, P.; Echave, J.; Garcia-Oliveira, P.; Carpena, M.; Jarboui, A.; Nuñez-Estevez, B.; Simal-Gandara, J.; Prieto, M.A. By-products of agri-food industry as tannin-rich sources: A review of tannins' biological activities and their potential for valorization. *Foods* **2021**, *10*, 137. [CrossRef] [PubMed]
35. Rouzaud, F.; Kadekaro, A.L.; Abdel-Malek, Z.A.; Hearing, V.J. MC1R and the response of melanocytes to ultraviolet radiation. *Mutat. Res. /Fundam. Mol. Mech. Mutagen.* **2005**, *571*, 133–152. [CrossRef] [PubMed]
36. Wang, T.; Zhao, J.; Yang, Z.; Xiong, L.; Li, L.; Gu, Z.; Li, Y. Polyphenolic sunscreens for photoprotection. *Green Chem.* **2022**, *24*, 3605–3622. [CrossRef]
37. Mishra, A.; Mishra, A.; Chattopadhyay, P. Assessment of in vitro sun protection factor of *Calendula officinalis* L. (asteraceae) essential oil formulation. *J. Young Pharm.* **2012**, *4*, 17–21. [CrossRef]
38. Lohani, A.; Mishra, A.K.; Verma, A. Cosmeceutical potential of geranium and calendula essential oil: Determination of antioxidant activity and in vitro sun protection factor. *J. Cosmet. Dermatol.* **2019**, *18*, 550–557. [CrossRef]
39. Mitterer-Daltoé, M.; Bordim, J.; Lise, C.; Breda, L.; Casagrande, M.; Lima, V. Consumer awareness of food antioxidants. Synthetic vs. Natural. *Food Sci. Technol.* **2020**, *41*, 208–212. [CrossRef]
40. Sajinčič, N.; Gordobil, O.; Simmons, A.; Sandak, A. An Exploratory Study of Consumers' Knowledge and Attitudes about Lignin-Based Sunscreens and Bio-Based Skincare Products. *Cosmetics* **2021**, *8*, 78. [CrossRef]
41. Reis-Mansur, M.C.P.P.; da Luz, B.G.; dos Santos, E.P. Consumer Behavior, Skin Phototype, Sunscreens, and Tools for Photoprotection: A Review. *Cosmetics* **2023**, *10*, 39. [CrossRef]
42. Fonseca-Santos, B.; Corrêa, M.A.; Chorilli, M. Sustainability, natural and organic cosmetics: Consumer, products, efficacy, toxicological and regulatory considerations. *Braz. J. Pharm. Sci.* **2015**, *51*, 17–26. [CrossRef]
43. JA, S.M.; Khalid, R.M.; Othaman, R. Coconut oil based microemulsion formulations for hair care product application. *Sains Malays.* **2019**, *48*, 599–605.
44. Li, L.; Qu, J.; Liu, W.; Peng, B.; Cong, S.; Yu, H.; Zhang, B.; Li, Y. Advancements in Characterization Techniques for Microemulsions: From Molecular Insights to Macroscopic Phenomena. *Molecules* **2024**, *29*, 2901. [CrossRef]

45. Al-Sakkaf, M.K.; Onaizi, S.A. Effects of emulsification factors on the characteristics of crude oil emulsions stabilized by chemical and Biosurfactants: A review. *Fuel* **2024**, *361*, 130604. [CrossRef]
46. EL-OTMANI N. Raw_Material_C_Arvensis.xlsx. Figshare. Dataset. 2024. Available online: https://figshare.com/articles/dataset/Raw_Material_C_Arvensis_xlsx/27643428/1?file=50338740 (accessed on 26 November 2024).
47. EL-OTMANI N. Sunscreen- Characterisation.xlsx. Figshare. Dataset. 2024. Available online: https://figshare.com/articles/dataset/Sunscreen-_characterisation_xlsx/27643425/1?file=50338743 (accessed on 26 November 2024).

cosmetics

MDPI

Article

1,2-Dihydroxy-9H-Xanthen-9-One, a Multifunctional Nature-Inspired Active Ingredient

Ana Jesus [1,2], Sara F. Vieira [3,4], Gonçalo Brites [5,6], Mylène Carrascal [6,7], Helena Ferreira [3,4], Nuno M. Neves [3,4], Honorina Cidade [8,9,10], Madalena Pinto [8,9], Emília Sousa [8,9,*], Isabel F. Almeida [1,2,*] and Maria T. Cruz [5,6]

[1] UCIBIO—Applied Molecular Biosciences Unit, Faculty of Pharmacy, University of Porto, 4050-313 Porto, Portugal; anaaimjesus@gmail.com
[2] Associate Laboratory i4HB—Institute for Health and Bioeconomy, Faculty of Pharmacy, University of Porto, 4050-313 Porto, Portugal
[3] 3B's Research Group, I3BS—Research Institute on Biomaterials, Biodegradables and Biomimetics, University of Minho, Headquarters of the European Institute of Excellence on Tissue Engineering and Regenerative Medicine, AvePark, Parque de Ciência e Tecnologia, Zona Industrial da Gandra, Barco, 4805-017 Guimarães, Portugal; saravieira2409@gmail.com (S.F.V.); helenaferreira@i3bs.uminho.pt (H.F.); nuno@i3bs.uminho.pt (N.M.N.)
[4] ICVS/3B's–PT Government Associate Laboratory, 4710-057 Braga, Portugal
[5] Faculty of Pharmacy, University of Coimbra, 3004-531 Coimbra, Portugal; g.sousabrites3@gmail.com (G.B.); trosete@ff.uc.pt (M.T.C.)
[6] Center for Neuroscience and Cell Biology—CNC, 3004-504 Coimbra, Portugal; mylenecarrascal87@gmail.com
[7] Labor Qualitas, Tecnimede Group, Rua da Tapada Grande, Abrunheira, 2710-089 Sintra, Portugal
[8] CIIMAR—Interdisciplinary Centre of Marine and Environmental Research, Avenida General Norton de Matos, S/N, 4450-208 Matosinhos, Portugal; hcidade@ff.up.pt (H.C.); madalena@ff.up.pt (M.P.)
[9] Laboratory of Organic and Pharmaceutical Chemistry, Department of Chemical Sciences, Faculty of Pharmacy, University of Porto, 4050-313 Porto, Portugal
[10] UNIPRO—Oral Pathology and Rehabilitation Research Unit, University Institute of Health Sciences (CESPU), 4585-116 Gandra, Portugal
* Correspondence: esousa@ff.up.pt (E.S.); ifalmeida@ff.up.pt (I.F.A.)

Abstract: Incorporating antioxidants into cosmetics is the mainstay for developing new products to mitigate skin aging. However, identifying novel multifunctional antioxidant ingredients with additional relevant properties that block the skin hallmarks of aging is a very striking strategy. Many natural compounds, including xanthones, have demonstrated biologically notable properties. In particular, 1,2-dihydroxy-9*H*-xanthen-9-one (1,2-DHX) has inhibitory activity against skin enzymes, and metal-chelating and radical-scavenging activities. Therefore, 1,2-DHX is an attractive molecule for cosmetic purposes. With this goal in mind, the anti-inflammatory, antioxidant, and anti-allergic potentials of 1,2-DHX were investigated. 1,2-DHX demonstrated anti-inflammatory properties by inhibiting the synthesis of specific pro-inflammatory mediators, including interleukin-6 (IL-6) and prostaglandin E2 (PGE2), in human macrophages. This xanthone did not elicit sensitization reactions and did inhibit allergic reactions triggered by a strong skin allergen, suggesting its potential as an anti-allergic compound. 1,2-DHX also revealed mitochondrial antioxidant activity by mitigating rotenone-induced oxidative stress in macrophages by up to 40%. Overall, 1,2-DHX displayed a safety profile and noteworthy biological activities, highlighting its multifunctional profile as an active cosmetic ingredient with anti-inflammatory, antioxidant, and anti-allergic properties.

Keywords: nature-inspired xanthone; anti-inflammatory; anti-allergic; antioxidant; multifunctional ingredient

1. Introduction

The appearance of deep and coarse wrinkles, compromised skin integrity, and age spots are clinical signs of skin aging [1]. These events mainly originate from external factors, including solar radiation, contact with allergens, pollutants, and tobacco smoke [2].

These elements collectively lead to an increase in reactive oxygen species (ROS) and reactive nitrogen species (RNS) [3]. They are both involved in oxidative damage to skin cells and in the pathophysiology of inflammation, resulting in the activation of the immune system [4,5]. Among the events evoked by oxidative stress are the degradation of several skin extracellular matrix proteins; the oxidation of lipids, carbohydrates, and nucleic acids [6]; and the production of pro-inflammatory mediators, triggering inflammatory cascade reactions [7]. Contact with allergens could also accelerate skin aging. The release of danger signals evoked after contact with skin allergens drives intracellular signaling pathways in antigen-presenting cells, enabling them to mature and prime naïve T lymphocytes [4]. Upon activation, these T cells transform into allergen-specific effector T cells, which circulate throughout the body, triggering a severe inflammatory response in following exposure. Incorporating antioxidants into cosmetic formulations is a key strategy for developing anti-aging products, since it helps to delay and prevent skin aging. This makes the discovery of innovative multifunctional antioxidant compounds highly relevant for enhancing skin protection. Natural products, especially xanthones, have been reported for their numerous biological activities [8]. Only two xanthones (α-mangostin and mangiferin) have been identified with potential to be used in cosmetic products, due to their antioxidant, anti-inflammatory, and anti-aging properties (Figure 1). Both xanthones are the main compounds identified in the respective natural extracts (α-mangostin: *Garcinia mangostana* extract; mangiferin: *Mangifera indica* extract) present in cosmetic products [9–11]. The presence of hydroxyl groups in this scaffold is responsible for their powerful antioxidant activity. It has already been reported that 1,2-dihydroxy-*9H*-xanthen-9-one (1,2-DHX) presents inhibitory activity against enzymes involved in the degradation of collagen (collagenase) and elastin (elastase) fibers and hyaluronic acid (hyaluronidase) [12]. Additionally, 1,2-DHX demonstrates metal-chelating effects [Fe(III) and Cu(II)] and DPPH radical-scavenging activity. Furthermore, the stability demonstrated at skin pH, together with the absence of phototoxicity in irradiated keratinocyte cells [12], reinforces its potential as a cosmetic active ingredient. Considering all the studies previously reported highlighting 1,2-DHX as a hit compound for cosmetic applications, the anti-inflammatory, anti-allergic, and mitochondrial antioxidant activities and skin sensitization potential were investigated, aiming to disclose the in vitro effectiveness and safety profile of 1,2-DHX (Figure 1).

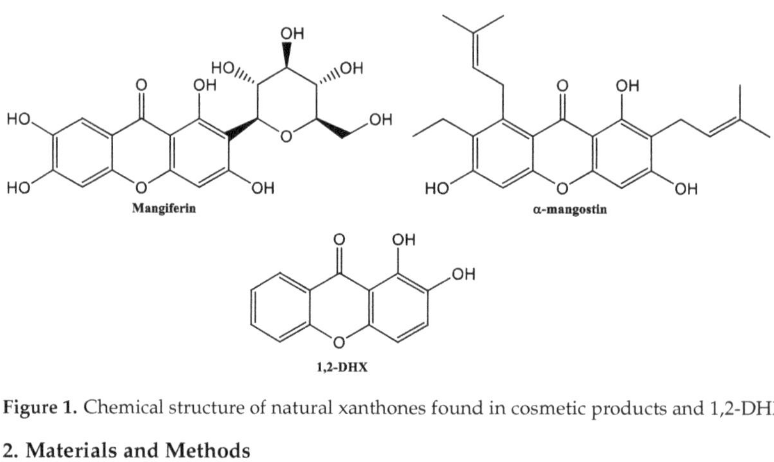

Figure 1. Chemical structure of natural xanthones found in cosmetic products and 1,2-DHX.

2. Materials and Methods

2.1. Materials

The compound 1,2-dihydroxy-*9H*-xanthen-9-one (1,2-DHX) was synthetized according to previously described procedures [13,14]. Briefly, the synthesis of 1,2-DHX started with the reaction between the two starting materials, methyl 2-bromobenzoate and 3,4-dimethoxyphenol, which gave methyl 2-(3,4-dimethoxyphenoxy)benzoate, which was subsequently hydrolyzed to 2-(3,4-dimethoxyphenoxy)benzoic acid. Then, this intermediate was cyclized into 1,2-dimethoxy-*9H*-xanthen-9-one. Finally, the methoxy groups were deprotected, which gave 1,2-dihydroxy-*9H*-xanthen-9-one. Figure 2 illustrates the steps followed to synthesize 1,2-DHX.

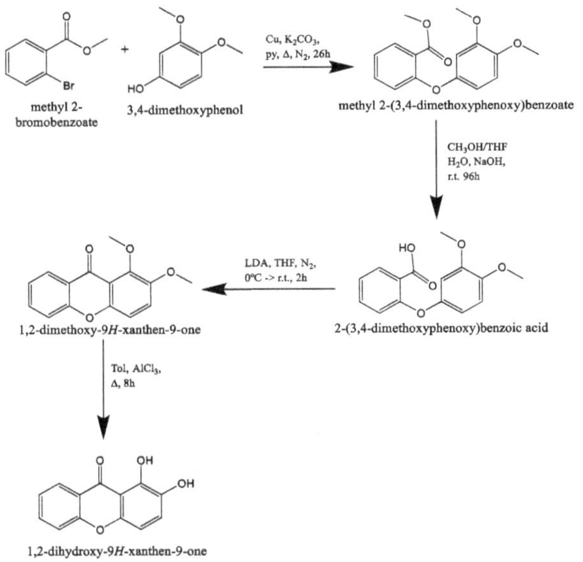

Figure 2. Synthesis of 1,2-DHX. Py: pyridine; Δ: heating; r.t.: room temperature; LDA: lithium diisopropylamide; THF: tetrahydrofuran; Tol: toluene.

For the biological assays, the materials and reagents used were obtained from different companies: RPMI-1640 media and Quant-iT PicoGreen dsDNA Kit from Thermo Fisher Scientific (Waltham, MA, USA); fetal bovine serum (FBS) from Gibco (Waltham, MA, USA); AlamarBlue from Bio-Rad (Hercules, CA, USA); human interleukin-6 (IL-6) DuoSet Enzyme-linked immunosorbent assay (ELISA) from R&D Systems (Minneapolis, MN, USA); prostaglandin E2 (PGE2) ELISA Kit from Abcam (Cambridge, UK); ascorbic acid (AA), dexamethasone (DEX), lipopolysaccharide (LPS; *Escherichia coli* O26:B6), rotenone (Rot), and 1-fluoro-2,4-dinitrobenzene (DNFB) from Sigma-Aldrich (Darmstadt, Germany); celecoxib (CEL) from GmbH (Darmstadt, Germany); anti-CD86 (clone IT2.2) and anti-CD54 (clone HA58) antibodies from Biolegend (San Diego, CA, USA); and MitoSOX™ Mitochondrial Superoxide Indicators (MitoSOX Red Kit) from Invitrogen (Waltham, MA, USA).

2.2. Methods

2.2.1. Cell Culture

Healthy adult volunteers' blood donation units were used to acquire human peripheral blood mononuclear cells (PBMCs) through a collaborative agreement with the Portuguese Institute of Blood and Transplantation (IPST; Instituto Português do Sangue e Transplantação, Portugal). The collection of peripheral blood from healthy volunteers at Hospital of Braga, Portugal, was approved on 14 December 2018 by the Ethics Subcommittee for Life and Health Sciences (SECVS) of University of Minho, Portugal (No. 014/015). The principles expressed in the Declaration of Helsinki were followed, and participants provided their signed informed consent. Monocytes were isolated from human PBMCs according to the previous procedure [15]. Briefly, Histopaque-1077 solution and blood were mixed in a ratio of 1:1 and centrifuged at 400× *g*, for 30 min. Then, the PBMC ring was carefully harvested and washed with PBS. CD14 microbeads were used to isolate the monocytes from PBMCs through positive magnetic separation, in accordance with the manufacturer's instructions. Isolated monocytes were resuspended in RPMI medium (cRPMI; RPMI-1640 medium containing 2 mM glutamine, 10% human serum, 1% penicillin/streptomycin, and 1% HEPES) and seeded (1 × 10^6 cells/mL) in a multi-well plate (MW-24) for 7 days, exposed to 20 ng/mL of recombinant human GM-CSF at 37 °C in a 5% CO_2 saturated atmosphere. The cell culture medium was substituted every 3 days. The morphology of human primary monocyte-derived macrophages (hMDMs) was verified by visualization with an inverted microscope (Axiovert 40; Zeiss, Oberkochen, Germany).

The human monocytic cell line THP-1 (ATCC® TIB-202™; American Type Culture Collection, Manassas, VA, USA) was cultivated in RPMI medium containing 25 mM (*D*)-glucose, 10% inactivated FBS, 10 mM HEPES, 100 U/mL penicillin, 100 µg/mL streptomycin, and 1 mM sodium pyruvate and kept at a cell density of 0.4 × 10^6 cells/mL. According to ATCC® instructions, THP-1 cells were sub-cultured every 2 or 3 days and kept in culture for a maximum of three months.

The mouse macrophage cell line RAW 264.7 (ATCC TIB-71; Manassas, VA, USA) was cultivated in DMEM containing 10% inactivated FBS, 1% antibiotic solution, 1 mM sodium pyruvate, and 1.5 g/L sodium bicarbonate. The cell culture of macrophages was mechanically separated with a cell scraper, sub-cultured in a 1:10 proportion (1 mL of cellular suspension to 10 mL of final volume of DMEM), and used after reaching 80–90% confluence.

2.2.2. Anti-Inflammatory Activity

A 50 mM stock solution of 1,2-DHX was prepared in DMSO and filtered by using a 0.22 µm sterile filter. Serial dilutions were made with cRPMI medium to obtain the final concentrations of 5, 12.5, 25, 50, and 100 µM in the wells. The hMDMs were exposed to 100 ng/mL of LPS in fresh cRPMI medium for 2 h. Afterwards, 1,2-DHX solutions at different concentrations were added to the LPS-stimulated hMDMs and incubated for 24 h. After that, the culture medium was collected, homogenized, and stored at −80 °C until analysis. Then, the metabolic activity of the cells and DNA content were determined. Controls containing 0.2% DMSO were also tested. Non-stimulated and LPS-stimulated hMDMs were used as negative and positive controls of the production of pro-inflammatory mediators, respectively. Dexamethasone (10 µM) and celecoxib (10 µM) were dissolved in ethanol and used as positive controls of the inhibition of the production of the pro-inflammatory mediators.

- Metabolic activity

The metabolic activity of the non-stimulated or LPS-stimulated hMDMs was determined by using the alamarBlue assay, based on the manufacturer's instructions. After collecting the culture medium, the hMDM cells were gently washed with warm PBS. RPMI medium containing 10% AlamarBlue was added to each well and left to incubate at 37 °C for 3 h. At the end, the absorbance for each sample in triplicate was measured at 570 nm and 600 nm with a microplate reader (Synergy HT; BioTek, Winooski, VT, USA). The results were expressed in percentage in relation to the control.

- DNA quantification

DNA quantification was assessed by using the Quant-iT PicoGreen assay, according to the manufacturer's instructions. After the determination of metabolic activity, the hMDM cells were washed with DPBS. Then, 1 mL of ultrapure water was added to each condition, and the samples were frozen at −80 °C. For the analysis, samples were defrosted, collected into an eppendorf, and sonicated for 15 min. The samples or standards at concentrations between 0 and 2 µg/mL, in triplicate, were added to a 96-well plate, followed by PicoGreen solution and Tris-EDTA (TE) buffer. The plate was incubated for 10 min protected from light. Fluorescence was read with a microplate reader (excitation wavelength (EX) of 485 nm and emission wavelength (EM) of 528 nm). The DNA concentration in µg/mL was extrapolated by the standard curve plotted from the fluorescence intensity versus the DNA concentration. The results were expressed in relative DNA concentration of the control.

- Pro-inflammatory mediator quantification

The amounts of PGE2 and IL-6 were determined through ELISA kits, following the manufacturer's instructions. The PGE2 and IL-6 concentrations in pg/mL were obtained from a standard curve of the absorbance intensity versus the specific pro-inflammatory mediator. The values obtained were normalized to the respective DNA concentration. The results were presented in percentages relative to the positive control.

2.2.3. Sensitization Potential and Anti-Allergic Activity

The human monocyte THP-1 cell line was also employed as a dendritic cell (DC) surrogate for the skin sensitizing hazard, following OECD guidelines. By adjusting OECD test Guideline No. 442E [16], the THP-1 cell concentration was set to 0.5×10^6 cells/mL, and they were incubated overnight at 37 °C in a 5% CO_2 saturated atmosphere. On the following day, cells were plated in a 12-well plate (1.5 mL/well), and 1.2-DHX was applied at the final concentration of 100 µM. After pre-incubation (1 h at 37 °C) with the active compound, a solution of DNFB, prepared in DMSO, was added (final concentration of 8 µM), and cells were additionally incubated for another 24 h under the previously mentioned conditions. Afterwards, cells were transferred to eppendorfs and centrifuged at 300× *g* for 5 min at 4 °C. The supernatant was rejected, and the cell pellet was washed with PBS/1% FBS (2 × 1 mL) and centrifuged at 300× *g* for 5 min at 4 °C. Then, cells were further resuspended in 1 mL of PBS/1% FBS, and 200 µL was collected for two 1.5 mL eppendorfs equally divided for unstained (UNST) and stained (CD5486) conditions. Each stained condition contained 3 µL of each antibody, anti-CD86 and anti-CD54, and was further incubated in the dark for 30 min at 4 °C. Afterward, cells were washed with PBS/1% FBS and centrifuged at 300× *g* for 5 min at 4 °C. Then, the supernatant was discarded, and 100 µL of PBS/1% FBS solution was added to the cell pellet and gently homogenized for subsequent analysis through the flow cytometry technique (BD Accuri™ C6 cytometer (San Jose, CA, USA)). Flow cytometry was used to examine the levels of the cellular membrane markers CD86 and CD54 by using acquisition channels FL1 and FL4, respectively. For control cells and chemical-treated cells, the relative fluorescence intensity (RFI) of the membrane markers CD86 and CD54 was calculated by utilizing the geometric mean fluorescence intensity (MFI) and the following equation:

$$RFI = \frac{MFI\ of\ chemical\ treated\ stained\ cells\ -\ MFI\ of\ chemical\ treated\ unstained\ cells}{MFI\ of\ control\ stained\ cells\ -\ MFI\ of\ control\ unstained\ cells} \times 100$$

The findings from a minimum of three independent experiments were presented as percentages of the RFI values observed in relation to the control, which was set to 100%. Samples were categorized as skin sensitizers if the RFI values for CD54 and CD86 exceeded the thresholds outlined in the 442E OECD guideline, precisely 200% and 150%, respectively.

2.2.4. Mitochondrial Antioxidant Activity

Mitochondrial superoxide (O_2^{-}) generation was determined by using MitoSOX fluorescent probe according to the manufacturer's instructions. Briefly, 0.6×10^6 cells/mL were plated in a 12-well plate (1.2 mL/well) and incubated overnight at 37 °C. The next day, cells were exposed to 1,2-DHX (100 µM) and ascorbic acid (antioxidant control)) (500 µM) and pre-incubated for 1 h, using the conditions previously mentioned. Then, rotenone (final concentration of 20 µM) was added and incubated for 5 h at 37 °C. After the incubation period, the cells were carefully detached with a scrapper from each well, transferred to 2 mL eppendorfs, and centrifuged at 500× *g* for 3 min at room temperature. The cell supernatant was collected and rejected, and the cell pellet was washed with HBSS solution (2 × 1 mL) and centrifuged at 500× *g* for 3 min. Afterwards, the cell pellet was resuspended in 400 µL of PBS solution, and 300 µL was collected for two 1.5 mL eppendorfs: 150 µL for unstained and 150 µL for stained conditions. Stained cells were incubated with 100 µL of freshly prepared MitoSOX working solution (final concentration of 2.5 µM) at 37 °C for

15 min protected from light. Then, 400 µL of HBSS solution was added to the dilute stained condition and was further centrifuged. Then, the stained conditions were washed with HBSS solution (2 × 1 mL). Finally, the cell pellet was resuspended in 100 µL of PBS solution and analyzed by flow cytometry (BD Accuri™ C6 cytometer (San Jose, CA, USA)).

2.2.5. Statistical Analysis

GraphPad Prism8 for Windows (GraphPad Software, San Diego, CA, USA; www.graphpad.com; accessed on 5 December 2024) was utilized to conduct the statistical analysis. The results were presented as means ± standard deviations (SDs) or means ± standard errors of the means (SEMs), based on a minimum of three independent experiments, in triplicates. Statistical comparisons were carried out by using one-way ANOVA, followed by Sidak's multiple comparisons test (anti-inflammatory activity and levels of IL-6 and PGE2) and Tukey's multiple comparisons test (sensitization potential, anti-allergic activity, and mitochondrial antioxidant activity). The value of $p < 0.05$ was considered significant, and the statistical levels considered were * $p < 0.05$; ** $p < 0.01$, *** $p < 0.005$, and **** $p < 0.001$ relative to the respective control.

3. Results and Discussion

3.1. Anti-Inflammatory Activity

The metabolic activity and relative DNA content achieved for hMDMs that were stimulated by LPS in the absence or presence of 1,2-DHX are shown in Figure 3. 1,2-DHX did not affect hMDM metabolic activity (Figure 3A) or DNA content (Figure 3B) at the concentrations tested.

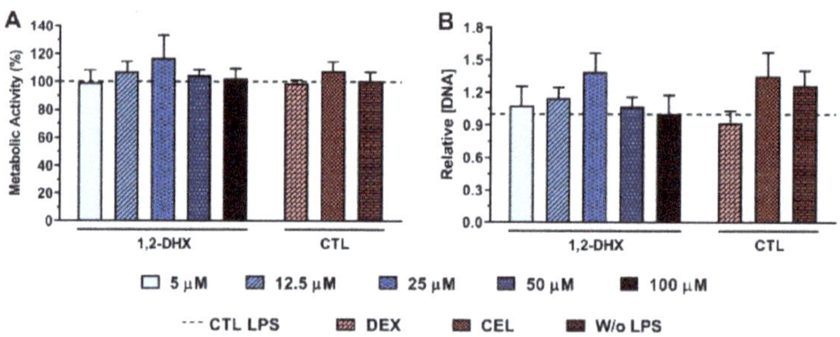

Figure 3. The metabolic activity (**A**) and relative DNA concentration (**B**) of non-stimulated (W/o LPS) and LPS-stimulated hMDMs cultured or not in the presence of different concentrations (5, 12.5, 25, 50, and 100 µM) of 1,2-DHX and clinically used anti-inflammatory compounds (dexamethasone (DEX) and celecoxib (CEL) at 10 µM). The dotted line represents the metabolic activity and the DNA concentration of the positive control (LPS-stimulated hMDMs without treatment (CTL LPS)). The results of three independent assays (n = 3) are expressed as means ± SDs (standard deviations).

The anti-inflammatory activity of 1,2-DHX was assessed by its ability to reduce the pro-inflammatory PGE2 and IL-6 mediators, produced by LPS-stimulated hMDMs (Figure 4). As expected, when the hMDMs were stimulated with LPS, we observed a significant increase in the levels of both PGE2 and IL-6 pro-inflammatory molecules. Celecoxib significantly decreased (29.4 ± 9.1%) PGE2 production, and dexamethasone effectively reduced IL-6 production (58.7 ± 5.1%). When LPS-stimulated hMDMs were treated with the 1,2-DHX, a decrease in the PGE2 level (34.2 ± 9.5% at 100 µM) in the culture medium was observed (Figure 4A). 1,2-DHX showed similar anti-inflammatory activity at all tested concentrations, leading to similar or inferior levels of PGE2 if compared with celecoxib, a clinically employed non-steroidal anti-inflammatory drug (NSAID). 1,2-DHX at 100 µM significantly decreased IL-6 production (34.7 ± 3.7%) by LPS-stimulated hMDMs (Figure 4B).

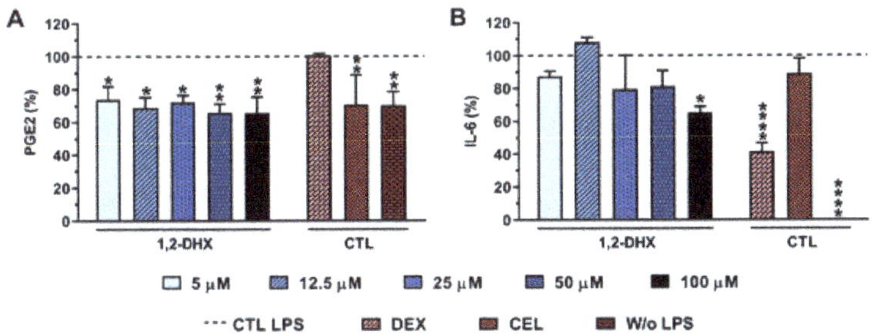

Figure 4. The PGE2 (**A**) and IL-6 (**B**) percentages of non-stimulated (W/o LPS) and LPS-stimulated hMDMs cultured in the presence or in the absence of different concentrations of 1,2-DHX and clinically used anti-inflammatory drugs (dexamethasone (DEX) and celecoxib (CEL) at 10 µM). The dashed line indicates the highest amount of generated pro-inflammatory mediators in the positive control (LPS-stimulated hMDMs without treatment (CTL LPS)). The results of three independent assays (n = 3) are expressed as means ± SDs, and statistically significant differences are indicated by * $p < 0.05$, ** $p < 0.01$, and **** $p < 0.0001$ in comparison with the respective positive control.

Cytocompatible concentrations severely reduced the PGE2 and IL-6 levels in LPS-stimulated hMDMs (Figure 4), being the highest concentration the most effective. This capacity is of utmost relevance, since PGE2 increases vasodilation, vascular leakage, hyperalgesia, and fever [17], and IL-6 promotes T-cell growth and activation, increases B-cell development, and modulates the chemokines that activate neutrophils [18]. Therefore, reducing the synthesis of these pro-inflammatory mediators ameliorates inflammation, specifically skin inflammation, avoiding skin premature aging. Moreover, 1,2-DHX demonstrated anti-inflammatory activity similar to or better than clinically used anti-inflammatory drugs. Considering the results obtained herein at the maximum concentration of 1,2-DHX used, the following biological activities explored for the compound were performed at 100 µM.

3.2. Sensitization Potential and Anti-Allergic Activity

The necessity for non-animal alternative studies has been recognized for the evaluation of the toxicity of chemicals to human health overall. It has become especially demanding for cosmetic active ingredients as a result of the European Union Commission's regulation banning the testing and marketing of products in which animal experiments have been used. Following an alternative validated test, OECD guideline 442E [16], the up-regulation of the co-stimulatory molecules CD54 and CD86 in THP-1 cells was used to evaluate the potential of the molecules to allay skin sensitization and xanthone's ability to mitigate skin allergy by using the strong allergen DNFB (Figure 5).

At the non-cytotoxic concentration of 100 µM, 1,2-DHX revealed a promising safety profile with RFI (%) values of 179.05 ± 29.45% and 90.85 ± 37.86%, with both values being lower than the thresholds for CD54 of 200% and CD86 of 150%, respectively, already established in the guidelines [16]; thus, a non-sensitizer label could be assigned to 1,2-DHX. As expected, the treatment with DNFB increased the levels of CD54 (RFI (%) = 621.34 ± 31.81%) and CD86 (RFI (%) = 367.00 ± 35.92%). 1,2-DHX showed inhibitory activity towards DNFB-induced cell maturation with RFI (%) values of 294.82 ± 34.00% for CD54 and 238.08 ± 35.92% for CD86. For the CD54 membrane cell marker, 1,2-DHX (294.82 ± 34.00%) demonstrated a stronger inhibitory effect (around 50%) towards the DNFB-induced cell maturation (621.34 ± 31.81%) than for the CD86 cell marker (238.08 ± 35.92%). The anti-allergic action of 1,2-DHX allows us to hypothesize that this compound could eliminate and prevent the harmful effects of other allergens that come into contact with the skin.

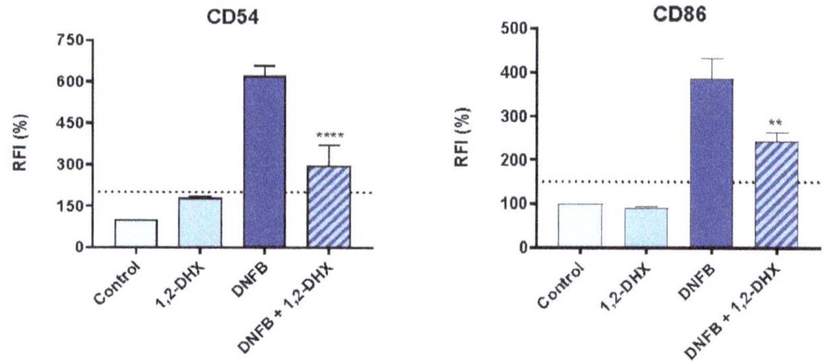

Figure 5. The non-sensitization potential (compared with control) and anti-allergic activity of 1,2-DHX towards the maturation of THP-1 cells induced by the strong allergen DNFB (compared with DNFB). The relative fluorescence intensity (RFI) of CD54 and CD86 expression was determined. The results of seven independent assays ($n = 7$) are expressed as means ± SEMs (standard errors of the means), and statistically significant differences observed are indicated by ** $p < 0.01$ and **** $p < 0.0001$ compared with the positive control (DNFB).

3.3. Mitochondrial Antioxidant Activity

Mitochondria are cellular organelles rich in oxidant species where the cellular oxidative stress process can be initiated more easily, triggered by an external stimulus capable to provoke an increase in oxidant species [19]. Oxidative stress activates other molecular events, including inflammatory and allergic reactions. Considering that mitochondria have been identified as a major source of ROS, mostly of superoxide anion ($O_2^{\bullet -}$), the mitochondrial antioxidant activity of 1,2-DHX was assessed in macrophage cells stimulated by rotenone (Figure 6).

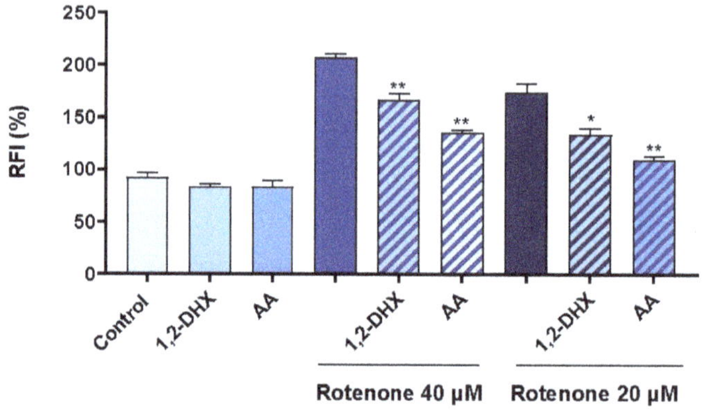

Figure 6. The pro-oxidant potential (compared with control; 100%) and mitochondrial antioxidant activity of 1,2-DHX (100 µM) under rotenone-induced oxidative stress in RAW264.7 cells (compared with rotenone). The antioxidant control ascorbic acid (AA) was also tested. The relative fluorescence intensity (RFI) of the fluorescent probe was measured. The results of eight independent assays (n = 8) are expressed as means ± SEMs, and statistically significant differences observed are indicated by * $p < 0.05$ and ** $p < 0.01$ compared with the positive control (Rot) at the tested concentrations.

Two control molecules, AA and rotenone, with antioxidant activity and potential to induce oxidative stress, respectively, were used. 1,2-DHX (84.14 ± 5.07%) and AA (83.97 ± 4.88%) did not demonstrate the ability to induce oxidative stress in macrophages, with RFI (%) values inferior to the control (100%) (Figure 6). The antioxidant response of 1,2-

DHX was further investigated by using rotenone as a positive control. Rotenone triggered mitochondrial oxidative stress, increasing mitochondrial $O_2^{\bullet -}$ production [20], reflected by an increase in fluorescence. 1,2-DHX showed inhibitory effect on rotenone-induced mitochondrial $O_2^{\bullet -}$ production with RFI (%) values of 167.59 ± 5.41% and 134.57 ± 4.92% compared with rotenone at 40 μM (207.55 ± 4.25%) and rotenone at 20 μM (110.18 ± 7.05%), respectively (Figure 6). Mitochondrial $O^{2\bullet -}$ is one of the most reactive species that contribute to the oxidation of biomolecules, the acceleration of skin aging, the regulation of apoptosis, and senescent events [21,22]. Thus, a reduction in $O^{2\bullet -}$ production can help to avoid cellular oxidative processes, preventing the acceleration of the aging of the skin. Other antioxidant molecules are being used in cosmetic products with the purpose of neutralizing ROS [23]; however, there is not specific information about their antioxidant activity directed to the mitochondria. Ascorbic acid is one of the antioxidants used in cosmetic products that has been already reported for its antioxidant activity, including as an untargeted and mitochondria-targeted antioxidant [24]. With our results, it is also possible to observe the effective inhibitory response of AA towards rotenone-induced mitochondrial $O_2^{\bullet -}$ production at different concentrations of rotenone used as a stimulus (40 μM: 135.42 ± 3.08%; 20 μM: 110.18 ± 6.37%), corroborating the results already reported in the literature [24]. The existence of a few mitochondria-targeted antioxidants in cosmetic products, aligned with the results obtained for 1,2-DHX, highlights it as a potential active cosmetic ingredient.

4. Conclusions

Skin is constantly exposed to external factors that contribute to the aging process, so its protection and equilibrium maintenance are essential. Sources of new active ingredients with multifunctional action are must-haves in the development of new cosmetic products, specifically anti-aging formulations. Nature has been used as a valuable source in the discovery of new active compounds with diverse daily cosmetic applications. 1,2-DHX is a nature-inspired xanthone with promising anti-aging and antioxidant activities and a physicochemical profile suitable for topical application. The evaluation of skin sensitization potential and anti-inflammatory, anti-allergic, and mitochondrial antioxidant activities is unveiled herein for the first time. Non-cytotoxic concentrations of 1,2-DHX demonstrated the ability to reduce PGE2 and IL-6 pro-inflammatory mediator release. 1,2-DHX did not elicit skin sensitization and inhibited DNFB-triggered allergic reactions, highlighting its preventive potential for use after skin contact with allergens. Furthermore, 1,2-DHX significantly reduced mitochondrial superoxide anion production. Based on these results, the application of this ingredient could be diversified in cosmetic products that help to fight skin inflammation, such as those activated by exposure to solar radiation and, in this particular case, aftersun products. 1,2-DHX could also help to minimize the effects of allergic contact dermatitis, due to its ability to mitigate allergen-induced skin sensitization. By acting as an untargeted and mitochondria-targeted antioxidant ingredient, 1,2-DHX could prevent premature skin aging by impeding the deregulation of the biological and metabolic processes of the skin that depend on the equilibrium state of mitochondria. Overall, 1,2-DHX can block relevant hallmarks of skin aging, namely, inflammaging and oxidative stress, and is thus a promising anti-aging ingredient.

Author Contributions: Conceptualization, H.F., N.M.N., M.T.C., E.S. and I.F.A.; methodology, H.F., S.F.V., M.T.C., A.J., G.B. and M.C.; data analysis: A.J., S.F.V. and M.C.; investigation, A.J. and S.F.V.; writing—original draft preparation and final manuscript, A.J. and S.F.V.; writing—review and editing, H.F., N.M.N., H.C., M.P., E.S., I.F.A. and M.T.C.; supervision, H.F., N.M.N., M.T.C., H.C. and I.F.A. All authors have read and agreed to the published version of the manuscript.

Funding: This work was financed by national funds from FCT (Fundação para a Ciência e a Tecnologia), I.P., in the scope of projects UIDP/04378/2020 and UIDB/04378/2020 of the Research Unit on Applied Molecular Biosciences (UCIBIO) and project LA/P/0140/2020 of the Associate Laboratory Institute for Health and Bioeconomy (i4HB). This research was also supported by national funds through FCT (Foundation for Science and Technology) within the scope of UIDP/04539/2020 and UIDP/04423/2020 (Group of Marine Natural Products and Medicinal Chemistry—CIIMAR), as well as a structured program of R&D&I ATLANTIDA (NORTE-01-0145-FEDER-000040), supported by NORTE2020, through the ERDF. This work was also financed by the European Regional Development Fund (ERDF), through Centro 2020 Regional Operational Programme under project CENTRO-01-0145-FEDER-000012 (HealthyAging2020). This research was funded by national funds through FCT in the scope of projects UIDP/04539/2020 and LA/P/0058/2020 of CIBB. The authors of this work would like to thank FCT for the Ph.D. grant awarded to Sara F. Vieira (PD/BD/135246/2017 and COVID/BD/152012/2021), the project PATH (NORTE-08-5369-FSE-000037), and the project "TERM RES Hub—Scientific Infrastructure for Tissue Engineering and Regenerative Medicine" (reference PINFRA/22190/2016 (Norte-01-0145-FEDER-022190)), funded by FCT in cooperation with the Northern Portugal Regional Coordination and Development Commission (CCDR-N). Ana Jesus and Gonçalo Brites acknowledge their Ph.D. grants (UI/BD/151319/2021 with https://doi.org/10.54499/UI/BD/151319/2021, and PD/BDE/142926/2018, respectively).

Institutional Review Board Statement: The study was conducted in accordance with the Declaration of Helsinki and was approved on 14 December 2018 by the Ethics Subcommittee for Life and Health Sciences (SECVS) of University of Minho, Portugal (No. 014/015).

Informed Consent Statement: Informed consent was obtained from all subjects involved in the study.

Data Availability Statement: Data is contained within the article. Part of this work was presented as a poster communication at the 33rd IFSCC Congress 2023 in Barcelona.

Conflicts of Interest: The authors declare no conflicts of interest.

References

1. Zhang, S.; Duan, E. Fighting against Skin Aging: The Way from Bench to Bedside. *Cell Transpl.* **2018**, *27*, 729–738. [CrossRef] [PubMed]
2. Wong, Q.Y.A.; Chew, F.T. Defining skin aging and its risk factors: A systematic review and meta-analysis. *Sci. Rep.* **2021**, *11*, 22075. [CrossRef] [PubMed]
3. Chen, J.; Liu, Y.; Zhao, Z.; Qiu, J. Oxidative stress in the skin: Impact and related protection. *Int. J. Cosmet. Sci.* **2021**, *43*, 495–509. [CrossRef]
4. Brites, G.S.; Ferreira, I.; Sebastiao, A.I.; Silva, A.; Carrascal, M.; Neves, B.M.; Cruz, M.T. Allergic contact dermatitis: From pathophysiology to development of new preventive strategies. *Pharmacol. Res.* **2020**, *162*, 105282. [CrossRef] [PubMed]
5. Lugrin, J.; Rosenblatt-Velin, N.; Parapanov, R.; Liaudet, L. The role of oxidative stress during inflammatory processes. *Biol. Chem.* **2014**, *395*, 203–230. [CrossRef] [PubMed]
6. Chao, M.R.; Evans, M.D.; Hu, C.W.; Ji, Y.; Moller, P.; Rossner, P.; Cooke, M.S. Biomarkers of nucleic acid oxidation—A summary state-of-the-art. *Redox Biol.* **2021**, *42*, 101872. [CrossRef]
7. Pilkington, S.M.; Bulfone-Paus, S.; Griffiths, C.E.M.; Watson, R.E.B. Inflammaging and the Skin. *J. Invest. Dermatol.* **2021**, *141*, 1087–1095. [CrossRef]
8. Pinto, M.M.M.; Palmeira, A.; Fernandes, C.; Resende, D.; Sousa, E.; Cidade, H.; Tiritan, M.E.; Correia-da-Silva, M.; Cravo, S. From Natural Products to New Synthetic Small Molecules: A Journey through the World of Xanthones. *Molecules* **2021**, *26*, 431. [CrossRef]
9. Abate, M.; Pagano, C.; Masullo, M.; Citro, M.; Pisanti, S.; Piacente, S.; Bifulco, M. Mangostanin, a Xanthone Derived from Garcinia mangostana Fruit, Exerts Protective and Reparative Effects on Oxidative Damage in Human Keratinocytes. *Pharmaceuticals* **2022**, *15*, 84. [CrossRef]
10. Telang, M.; Dhulap, S.; Mandhare, A.; Hirwani, R. Therapeutic and cosmetic applications of mangiferin: A patent review. *Expert Opin. Ther. Pat.* **2013**, *23*, 1561–1580. [CrossRef]
11. Pan-In, P.; Wongsomboon, A.; Kokpol, C.; Chaichanawongsaroj, N.; Wanichwecharungruang, S. Depositing α-mangostin nanoparticles to sebaceous gland area for acne treatment. *J. Pharmacol. Sci.* **2015**, *129*, 226–232. [CrossRef] [PubMed]
12. Resende, D.; Almeida, M.C.; Maciel, B.; Carmo, H.; Sousa Lobo, J.; Dal Pozzo, C.; Cravo, S.M.; Rosa, G.P.; Kane-Pages, A.; do Carmo Barreto, M.; et al. Efficacy, Stability, and Safety Evaluation of New Polyphenolic Xanthones Towards Identification of Bioactive Compounds to Fight Skin Photoaging. *Molecules* **2020**, *25*, 2782. [CrossRef] [PubMed]
13. Sousa, E.P.; Silva, A.M.S.; Pinto, M.M.M.; Pedro, M.M.; Cerqueira, F.A.M.; Nascimento, M.S.J. Isomeric Kielcorins and Dihydroxyxanthones: Synthesis, Structure Elucidation, and Inhibitory Activities of Growth of Human Cancer Cell Lines and on the Proliferation of Human Lymphocytes In Vitro. *Helv. Chim. Acta* **2002**, *85*, 2862–2876. [CrossRef]
14. Gale, L.; Sousa, M.E.; Pinto, M.M.M.; Kijoa, A.; Damas, A.M. Naturally occurring 1,2,8-trimethoxyxanthone and biphenyl ether intermediates leading to 1,2-dimethoxyxanthone. *Acta Crystallogr.* **2001**, *C57*, 1319–1323.
15. Goncalves, S.M.; Duarte-Oliveira, C.; Campos, C.F.; Aimanianda, V.; Ter Horst, R.; Leite, L.; Mercier, T.; Pereira, P.; Fernandez-Garcia, M.; Antunes, D.; et al. Phagosomal removal of fungal melanin reprograms macrophage metabolism to promote antifungal immunity. *Nat. Commun.* **2020**, *11*, 2282. [CrossRef]

16. Organisation for Economic Cooperation and Development. Test Guideline No. 442E In Vitro Skin Sensitisation: In Vitro Skin Sensitisation assays addressing the Key Event on activation of dendritic cells on the Adverse Outcome Pathway for Skin Sensitisation. In *OECD Guidelines for the Testing of Chemicals, Section 4*; Organisation for Economic Cooperation and Development: Paris, France, 2022. [CrossRef]
17. Dennis, E.A.; Norris, P.C. Eicosanoid storm in infection and inflammation. *Nat. Rev. Immunol.* **2015**, *15*, 511–523. [CrossRef]
18. Hunter, C.A.; Jones, S.A. IL-6 as a keystone cytokine in health and disease. *Nat. Immunol.* **2015**, *16*, 448–457. [CrossRef]
19. Campos Chisté, R.; Freitas, M.; Zerlotti Mercadante, A.; Fernandes, E. Superoxide Anion Radical: Generation and Detection in Cellular and Non-Cellular Systems. *Curr. Med. Chem.* **2015**, *22*, 4234–4256. [CrossRef]
20. Heinz, S.; Freyberger, A.; Lawrenz, B.; Schladt, L.; Schmuck, G.; Ellinger-Ziegelbauer, H. Mechanistic Investigations of the Mitochondrial Complex I Inhibitor Rotenone in the Context of Pharmacological and Safety Evaluation. *Sci. Rep.* **2017**, *7*, 45465. [CrossRef]
21. Aranda-Rivera, A.K.; Cruz-Gregorio, A.; Arancibia-Hernández, Y.L.; Hernández-Cruz, E.Y.; Pedraza-Chaverri, J. RONS and Oxidative Stress: An Overview of Basic Concepts. *Oxygen* **2022**, *2*, 437–478. [CrossRef]
22. Redza-Dutordoir, M.; Averill-Bates, D.A. Activation of apoptosis signalling pathways by reactive oxygen species. *Biochim. Biophys. Acta* **2016**, *1863*, 2977–2992. [CrossRef] [PubMed]
23. Kusumawati, I.; Indrayanto, G. Chapter 15—Natural Antioxidants in Cosmetics. In *Studies in Natural Products Chemistry*; Atta ur, R., Ed.; Elsevier: Amsterdam, The Netherlands, 2013; Volume 40, pp. 485–505.
24. Fiorani, M.; Guidarelli, A.; Cantoni, O. Mitochondrial reactive oxygen species: The effects of mitochondrial ascorbic acid vs untargeted and mitochondria-targeted antioxidants. *Int. J. Radiat. Biol.* **2021**, *97*, 1055–1062. [CrossRef] [PubMed]

cosmetics

MDPI

Article

Willaertia Lysate: A Hydrobiome-Biosourced Ingredient with Multi-Site Antioxidative and Antiaging Properties

Morgan Dos Santos [1], Julie Rorteau [1], Kilian Laho [1], Hanan Osman-Ponchet [2], Manon Barthe [2], Benjamin Quelard [3], Antoine Carlino [3], Adeline Saha [3] and Sandrine Troussieux [3,*]

1 LabSkin Creations, 69004 Lyon, France; morgan.dossantos@labskincreations.fr (M.D.S.)
2 Laboratoires PKDERM, 06130 Grasse, France; hanan.osman.ponchet@pkderm.com (H.O.-P.)
3 Amoéba, 69680 Chassieu, France
* Correspondence: s.troussieux@amoeba-nature.com

Abstract: Aging is synonymous with the skin becoming increasingly thin and fragile, which is associated with a decrease in epidermal cell layers. Beyond this intrinsic aging process, the skin is continually exposed to environmental stressors such as UV radiation that accelerate aging. To fight the signs of aging, a comprehensive program was implemented in this study to evaluate the efficacy of an innovative ingredient, *Willaertia* lysate, through a multi-scale approach encompassing cellular and advanced 3D skin models. The results show that *Willaertia* lysate, initially sourced from French Alps thermal spring waters, is able to (i) promote cell migration; (ii) improve the quality and abundance of the extracellular matrix in aged skins and in young skins exposed to UV radiation to a similar level to that in unexposed young skins; (iii) decrease tyrosinase activity and melanin content; and (iv) reduce oxidative stress after UV exposure by decreasing exposome markers such as protein carbonylation and lipid peroxidation expression. This complete set of coherent results demonstrates the global protective efficacy of *Willaertia* lysate against the effects of photoaging. This study is the first to report the use of a protozoan lysate as a natural and biosourced postbiotic active ingredient in the fields of cosmetics and dermocosmetics.

Keywords: cosmetics; antiaging; antioxidant; *Willaertia magna* C2c Maky; postbiotic; thermal spring water

1. Introduction

With increasing life span, skincare has become an ever-growing sector. As we age, our skin becomes increasingly thin and fragile. The number of epidermal cell layers decreases, as do the size and quantity of dermal fibroblasts. Elastin and collagen fibers, which are essential for maintaining skin structure, fail to assemble properly and become rigid, leading to a loss of elasticity and firmness in the skin. Cells that once played a vital role in synthesis and renewal are now transitioning toward senescence and degradation [1]. Beyond this intrinsic aging process, the skin is continually exposed to environmental stressors that accelerate aging. Among these extrinsic factors, UV radiation is the primary cause of cutaneous aging, known as photoaging [2].

To help maintain youthful and healthy skin over time, the cosmetic industry offers a broad spectrum of antiaging active ingredients with very different modes of action and sourcing. Public demand for improved skin health and beauty products has led to the emergence of a new type of products called cosmeceuticals, which lie at the interface of cosmetics and pharmaceuticals [3]. Cosmeceuticals encompass, for example, vitamins recommended for preventing photoaging, including retinol, which has been the gold standard in antiaging treatments for many years. However, retinol is not always well tolerated [4], despite formulations designed to reduce its toxicity [5]. Moreover, face creams and serums containing more than 0.3% retinol, and body care products containing more than 0.05% retinol, will be banned from over-the-counter sale in Europe from 2026 onwards, following widespread concern about the potentially dangerous side effects of high retinol concentrations (EU Commission Regulation 2024/996 of 3 April 2024).

More recently, bakuchiol, a meroterpene phenol abundant in the seeds and leaves of the plant *Psoralea corylifolia*, has been used in place of retinol, showing similar effects but with better tolerance even though there is no structural resemblance [4,6]. However, bakuchiol was demonstrated to be allergenic in cosmetics [7]. There is a wide range of other molecules with antiaging and/or antioxidant effects, such as hydroxy acids, growth factors, peptides, microbial compounds, and botanical extracts [3,8–11]. In recent years, biotechnology has emerged as a key driver of innovation, particularly through the development of numerous functional compounds derived from microorganisms such as bacteria, fungi, and algae. These advancements have played a significant role in promoting and facilitating the sustainable growth of the cosmetic industry [12,13]. Microbial fermentation, thus, has great potential in cosmetics, with prebiotics referring to fermentation metabolites, probiotics containing living microorganisms, and postbiotics corresponding to cell lysates or non-viable microorganisms. Pre/pro/postbiotics are usually used to maintain a healthy skin microbiome [14]; however, postbiotics can have antiaging properties due to the selection of skin microbiota linked with the aging phenotype of the skin [15]. Postbiotics are usually members of the *Lactobacillus* genera and/or yeasts such as *Saccharomyces cerevisiae* [16], but to the best of our knowledge, the domain of protozoa has never yet been exploited.

Exploration of the hydrobiome of the thermal waters of Aix-les-Bains (Pr Bodennec, Université Lyon 1, personal communication) has led to the isolation of a microorganism with highly interesting biological properties [17]. Several studies have characterized this organism, deposited with the ATCC under the name *Willaertia magna* C2c Maky (PTA-7824). In particular, this organism demonstrated an absence of pathogenicity in culture experiments that were corroborated by omics analyses [18,19]. This strain is a unicellular microorganism belonging to the Protista kingdom. It is characterized by a deformable cell body emitting shape-changing extensions called pseudopodia, which enable it to capture microscopic prey via phagocytosis. In its natural environment, it is a natural predator and regulator of populations of environmental bacteria and waterborne microorganisms [19]. These kinds of microorganisms, such as *W. magna* C2c Maky, are ubiquitous protozoan that inhabit common aquatic environments [19,20].

A novel natural and biosourced ingredient based on the patented use of a lysate of *W. magna* C2c Maky was developed as a bioactive hydrobiome-derived postbiotic for use in cosmetics and dermocosmetics. The aim of this study was to investigate the intrinsic properties of this *Willaertia* lysate for use in the field of cosmetics, particularly to fight the signs of aging. An initial study identified promising potential properties associated with the stimulation of specific gene expression. These findings were further confirmed through in vitro experiments, including wound healing, skin lightening, and assessments using three-dimensional (3D) reconstructed skin models. These models were used to study both intrinsic and extrinsic aging factors, enabling an evaluation of the effects of this ingredient on a range of established markers within the epidermis, dermal–epidermal junction, and dermis.

2. Materials and Methods

2.1. Expression of Genes of Interest

Target gene induction analysis was performed on human epithelial cells (ATCC CCL-2, BioTK Consulting & Services, Marseille, France) cultured in DMEM (Thermo-Fisher Scientific, Waltham, MA, USA) supplemented with fetal calf serum (10%), with penicillin and streptomycin (5000 U/mL and 5000 μg/mL, respectively) at 37 °C and under 5% CO_2 atmosphere in a 6-well plate with a flat bottom.

Cells at a concentration of 1×10^6 cells/well were either stimulated or not stimulated with *Willaertia* lysate lot 22.LY.040 (Amoéba, Chassieu, France) containing 7.79% dry matter (*w*/*v*) after its dilution at 0.2% and 0.1% for 8 h at 37 °C under 5% CO_2 atmosphere.

Total RNA was isolated using Qiagen's RNAeasy protocol, quantified via NanoDrop (260 nm), and normalized to 100 ng/µL (stored at −20 °C). Only samples with a 260/280 ratio above 2 were used for analysis.

The mRNA contained in the total RNA samples was transformed into cDNA via reverse transcription using oligo-dt 18 and then stored at −20 °C. Quantitative PCR was performed on a CFX96 Bio-Rad using the sybergreen method with primers specific to the targeted genes of interest. The results were expressed as the fold change (FC) according to Formula (1) (n = 3, mean ± standard deviation):

$$FC = 2^{-\Delta\Delta Ct} \tag{1}$$

where $\Delta\Delta Ct = (\text{Ct target gene} - \text{Ct constitutive gene})_{test} - (\text{Ct target gene} - \text{Ct constitutive gene})_{control}$.

The constitutive gene was GAPDH and the target genes were TLR2 (immunity), TGF-beta (cell multiplication, anti-inflammation), RunX1 (cell multiplication, tissue cohesion, cell differentiation), PiwiL1 (cell renewal), SGK1 (UV protection, antioxidative stress, photoaging), FOXO1 (UV protection, antioxidative stress, photoaging), BCL2 (anti-cell death), HMMR (hyaluronic acid receptor, cell mobility), HAS (hyaluronic acid synthesis), COL1A1 (collagen synthesis), and FBL5 (wound healing).

2.2. Antioxidant Activity

The antioxidant activity was measured using the 2,2-diphenyl-1-picrylhydrazil (DPPH) method [21–23].

- Amoeba Extracts

Willaertia lysate lot 22-LY-130 (Amoéba, Chassieu, France) containing 10.64% dry weight of the active substance was mixed with methanol at 4 concentrations (1, 5, 10, and 15 mg dry weight/mL). The mixtures were then shaken for 30 s and extracted with ultrasound (45 kHz) for 30 min (Ultrasonic cleaner, USC600-TH—VWR). The mixtures were centrifuged at 10,000× *g* for 15 min. The supernatants were stored at −20 °C until use.

- DPPH Assay

The DPPH reagent (Sigma–Aldrich, Saint Quentin Fallavier, France) was dissolved in methanol to a concentration of 0.1 mM. A volume of 50 µL of *Willaertia* lysate extracts was placed in a microplate (with 96 flat-bottomed wells), and the procedure was performed in triplicate. Then, 200 µL of DPPH at 0.1 mM was added to the wells. As a positive control, 50 µL of ascorbic acid solution mixed with methanol at 4 concentrations ranging from 10 to 25 µg/mL was placed in the microplate, and 200 µL of DPPH was added. As a blank control, 50 µL of methanol was used, and 200 µL of DPPH was added. Finally, the plate was incubated in a microplate reader (Victor® Nivo, PerkinElmer, Villebon-sur-Yvette, France), and the absorbance was recorded every minute at 521 nm for 30 min at room temperature in the dark.

The inhibition percentage (I%) corresponding to the quantity of DPPH radicals reduced by the extract at a given concentration was calculated according to Formula (2):

$$I\% = \left[\frac{AB - AS}{AB}\right] \times 100 \tag{2}$$

where AB is the absorbance of the blank, and AS is the absorbance of the sample.

The inhibitory concentration IC50, which is the concentration of the sample required to eliminate 50% of DPPH free radicals, was also calculated graphically using a linear regression analysis.

The results were expressed as the mean of three independent replicates.

2.3. Dermal Fibroblast and Keratinocyte Cell Migration Measure

- Cell culture

Normal human dermal fibroblasts (NHDFs, Laboratoires PKDERM, Grasse, France) in primary culture were seeded with fibroblast basal medium (Innoprot, Derio, Spain), and the human epidermal keratinocyte cell line HaCaT (Laboratoires PKDERM, Grasse, France) was seeded with DMEM-F12 (VWR, Radnor, PA, USA), both supplemented with 2% fetal bovine serum, 1% fibroblast growth supplement, and 1% penicillin/streptomycin. Cell incubation was conducted in a humidified cell incubator set at 37 °C and 5% CO_2.

- In vitro scratch test

NHDF or HaCaT cells were seeded in Ibidi® 2-well silicone culture inserts (ibidi GmbH, Gräfelfing, Germany) in a 24-well plate. The growth area of each well insert was 0.22 cm^2, the width of the cell-free gap was 500 µm ± 100 µm, the cell density was 70,000 cells per well insert, and the volume of the culture medium was 70 µL per well insert. The NHDF and HaCaT cells were allowed to proliferate until confluency for 24 h in the humidified cell incubator set at 37 °C and 5% CO_2.

The Ibidi culture inserts were then removed, and the cells were exposed to different treatments (500 µL/well of 24-well plate) for 25 h or 50 h, with three replicates per treatment condition (n = 3). The three treatments—the control (untreated group), and the *Willaertia* lysate lot 22.LY.077 (Amoéba, Chassieu, France) containing 9.24% of dry matter (*w*/*v*) diluted at 1×10^{-5}% and at 1×10^{-4}%—were compared.

- Image acquisition

The culture plate was placed on a CytoSMART OMNI brightfield device (Axion Biosystems, Atlanta, GA, USA) inside the culture incubator. Photographic images were captured every hour over the culture period using the CytoSMART OMNI brightfield live-cell imager equipped with a high-resolution digital camera

The results were expressed as the closure area and the area under the curve (AUC).

- Statistical analysis

Statistical significance was determined through analysis of variance (ANOVA single factor). Statistically significant differences were indicated by asterisks as follows: * $p < 0.05$, ** $p < 0.01$, and *** $p < 0.001$).

2.4. Murine Melanoma Cell Pigmentation Assays

- Cell culture

Cells from the murine melanoma cell line B16-F10 (Laboratoires PKDERM, Grasse, France) were cultured in DMEM supplemented with fetal calf serum (10%) and with penicillin and streptomycin (5000 U/mL and 5000 µg/mL, respectively) at 37 °C and under 5% CO_2 atmosphere. The cells were seeded in a 24-well plate at a cell density of 50,000 cells/well and a volume of culture medium of 500 µL per well. The cells were allowed to proliferate until confluency in a humidified cell incubator set at 37 °C and 5% CO_2.

- Tyrosinase enzyme activity and pigmentation assays

B16-F10 cells were exposed to different treatments (500 µL/well of 24-well plate) for 48 h. The different treatment conditions, with three replicates per treatment condition (n = 3), were tested as follows:

- Control (untreated group);
- 3-isobutyl-1-methyl-xanthine (IBMX) at 100 µM (Sigma–Aldrich, Saint Quentin Fallavier, France);
- IBMX + kojic acid at 200 µM (Sigma–Aldrich);
- IBMX + *Willaertia* lysate at 1×10^{-5}%;
- IBMX + *Willaertia* lysate at 1×10^{-4}%;
- IBMX + *Willaertia* lysate at 1×10^{-3}%.

Melanin content was evaluated on BF16-F10 cells lysed in 1N NaOH (200 μL/well of 24-well plate) and incubated at 80 °C for 1 h via optical density measurement at 405 nm.

Tyrosinase enzyme activity was determined on B16-F10 cells lysed in 1% Triton X-100 (100 μL/well of 24-well plate). The cell lysate (20 μL) was incubated with 2 mM of L-DOPA (100 μL) at 37 °C for 1 h, and the kinetics of formation of dopachrome was monitored at 490 nm.

- Statistical analysis

Statistical significance was determined through analysis of variance (ANOVA single factor). Statistically significant differences were indicated by asterisks as follows: * $p < 0.05$, ** $p < 0.01$, and *** $p < 0.001$).

2.5. Aging Studies on Reconstructed Skins

- Ethical considerations and human cutaneous cell isolation

All cells used to build the reconstructed skins were provided by Labskin Creations, Lyon, Rhone-Alpes, France. Human skin tissue was collected according to the principles of the Declaration of Helsinki, and its use was declared to the French Research Ministry (declaration no DC-2020-4346). Written informed consent was obtained from the donor according to the French bioethical law of 2014 (loi 94–954 du 29 Juillet 1994). Primary cultures of human fibroblasts, keratinocytes, and melanocytes were established from the healthy skin biopsy obtained from an infant donor (<5 years old) or an adult donor (>40 years old). Normal human epidermal keratinocytes (NHEKs), melanocytes (NHEMs), and dermal fibroblasts (NHDFs) were isolated from human skin, as previously described [24].

- Three-dimensional reconstructed aged skin for intrinsic aging study

Three-dimensional (3D) full-thickness reconstructed skin models were obtained by culturing NHDFs (>40 years old) in a scaffold made of collagen, glycosaminoglycans, and chitosan (LabSkin matriX™, Lyon, France) for 21 days under optimized cell culture conditions for ECM neo-synthesis, as previously described [25]. NHEKs were then seeded on the top of the dermal equivalent constructs and raised at the air/liquid interface to allow the formation of the epidermal compartment.

From day 10 until the end of the culture period, the 3D reconstructed skin models were treated in a systemic way with *Willaertia* lysate lot 22-LY-040 containing 7.79% dry matter (w/v) diluted at 0.1×10^{-3}%. The treatments were renewed at every medium renewal each week. Skin equivalent samples harvested on day 42 of total cell culture were immediately fixed in 4% neutral-buffered formalin (Diapath) for 24 h and then embedded in paraffin or in an optimal cutting temperature (OCT) compound and frozen at −80 °C for histological and immunohistological analysis, respectively. For each cell culture condition and analysis, 3D skin equivalents were produced in triplicate.

- Three-dimensional reconstructed young skin for extrinsic aging study

Three-dimensional full-thickness reconstructed skin models were obtained by culturing NHDFs (<5 years old) in a scaffold made of collagen, glycosaminoglycans, and chitosan (LabSkin matriX™) for 21 days under optimized cell culture conditions for ECM neo-synthesis as previously described [25]. NHEKs and NHEMs were then seeded on the top of the dermal equivalent constructs and raised at the air/liquid interface to allow the formation of the epidermal pigmented compartment. For each cell culture condition and analysis, 3D skin equivalents were produced in triplicate.

From day 10 until the end of the culture period, the 3D reconstructed skin models were treated in a systemic way with *Willaertia* lysate diluted at 0.2×10^{-3}%. The treatments were renewed at every medium renewal each week. On day 40, the tissue constructs were exposed to 100 mJ/cm^2 UVB with UV Stratalinker® 1800 (Stratagene, La Jolla, CA, USA). Skin equivalent samples harvested on day 42 of total cell culture were immediately fixed in 4% neutral-buffered formalin (Diapath, Martinengo, Italy) for 24 h and then embedded in paraffin or in an optimal cutting temperature (OCT) compound and frozen at −80 °C for histological and immunohistological analysis, respectively.

- Histological analysis

To evaluate the global cutaneous structure of the samples, hematoxylin phloxine saffron (HPS) staining was performed. For each condition, paraffin sections of 5 μm were cut. After dewaxing and rehydration, the samples were stained with HPS. After rinsing, the samples were dehydrated before being mounted on slides using a hydrophobic mounting medium.

- Immuno-histological analysis

To detect immunofluorescence on the paraffin sections, after the heat-mediated antigen retrieval treatment, non-specific binding was blocked in PBS containing 4% BSA. The sections were then incubated with the primary antibodies of interest diluted in PBS containing 4% BSA overnight at room temperature.

After incubation for 1 h with an AlexaFluor-568-conjugated anti-mouse/rabbit secondary antibody (Molecular Probes, Thermo Fisher Scientific, Waltham, MA, USA), nuclear counterstaining using DAPI was carried out routinely. As a negative control, the primary antibody was replaced by the corresponding IgG class.

For immunohistochemistry on the paraffin sections, after the heat-mediated antigen retrieval treatment, non-specific binding was blocked in PBS containing 4% BSA. The sections were then incubated with the primary antibodies of interest diluted in PBS containing 4% BSA overnight at room temperature. The studied markers were the following:

- Ki67, loricrin, and filaggrin for epidermis;
- Collagen-7 for the dermal–epidermal junction (DEJ);
- Hyaluronic acid for dermis.
- Malonyl dialdehyde (MDA) and carbonylation for antioxidative stress

After incubation for 1 h with EnVision anti-mouse/rabbit-HRP secondary antibody (Dako EnVision+, HRP) and applying DAB+ substrate solution to the sections, the color of the antibody staining was revealed. The slides were counterstained by immersing them in a 25% Harris hematoxylin counterstaining solution. As a negative control, the primary antibody was replaced by the corresponding IgG class.

- Image acquisition and analysis

The specimens stained in HPS were observed using an Axio Observer A1/D1 optical microscope, and images were captured using AxioCam HRc and ZEN 2 pro (Carl Zeiss SAS, Marly-le-Roi, France). Sixteen-bit images were saved in an uncompressed tagged image file format (tiff). Nine representative images were captured for each condition in the same manner. The immunostained specimens were observed using a Zeiss Axio Observer D1 microscope. Sixteen-bit images were saved. Three representative images were captured for each condition in the same manner (n = 9 in total).

For the markers of interest, positive red-stained tissue areas were automatically detected. The surface area of interest was measured automatically. The data were normalized based on the dermal–epidermal junction length for dermal–epidermal markers and based on the dermal area for dermal markers (for example loricrin/DEJ length).

The data were expressed in percentage of density.

Statistical analysis was performed with R (R Foundation for Statistical Computing, Vienna, Austria), and normality was checked using the Shapiro–Wilk normality test. The normal distribution led to a parametric test. Analysis of variance was performed using an ANOVA test, then the conditions were compared with each other using pairwise *t*-tests as post-hoc tests. The absence of a normal distribution led to a non-parametric test. Analysis of variance was therefore performed using a non-parametric Kruskal–Wallis test, and then the conditions were compared with each other using a non-parametric Wilcoxon–Mann–Whitney test as post hoc tests.

For all data, statistical significance was assessed via a one-way Student's *t*-test, and statistically significant differences were indicated by asterisks as follows: * $p < 0.05$, ** $p < 0.01$, and *** $p < 0.001$.

3. Results

3.1. Preliminary Test to Evaluate the Potential Activity of Willaertia Lysate

Willaertia lysate at two dilutions was tested to evaluate its potential activity on 11 genes involved in the cellular repair process (Table 1).

Table 1. Targeted genes and their function.

Gene Name	Function	References
TLR2	Immunity	[26–28]
TGF-beta	Cellular multiplication, anti-inflammation	[29,30]
RunX1	Cell regeneration regulation, cellular multiplication, cell cohesion, and differentiation	[31–33]
PiwiL1	Cell renewal	[34,35]
SGK1	Blood pressure regulation, UV protection, oxidative stress, cell survival	[36–38]
FOXO1	Stress resistance, antioxidation	[39–42]
BCL2	Regulation of apoptosis	[43–45]
HMMR	Hyaluronic acid receptor, neoplastic process	[46]
COL1A1	Collagen synthesis	[47]
HAS	Hyaluronic acid synthesis	[48–50]
FBL5	Wound healing	[51–53]

A moderate stimulation (1.5- to 1.99-fold increase) was observed in genes related to wound healing (FBL5) and hyaluronic acid synthesis (HAS) at a 0.2% dilution of *Willaertia* lysate (1.6×10^{-2}% dry weight), and a strong stimulation (≥2-fold increase) was noted for genes associated with anti-apoptosis (BCL2), cellular renewal (PIWI1L), UV protection, and antioxidative stress (FOXO1 and SGK1), with maximum expression observed at a 0.2% dilution, whereas the highest expression of the BCL2 gene was achieved with a 0.1% dilution of *Willaertia* lysate (Figure 1).

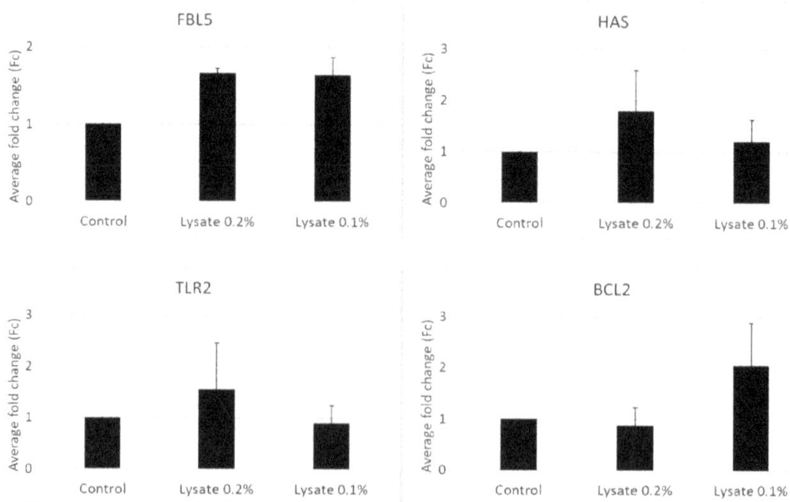

Figure 1. *Cont.*

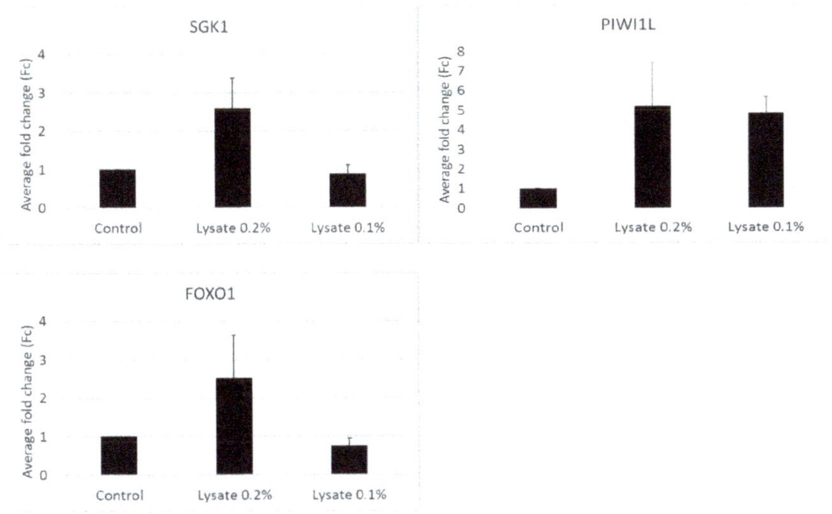

Figure 1. Average fold change in the expression of overexpressed targeted genes.

Therefore, *Willaertia* lysate showed significant potential in enhancing cellular functions linked to aging, including cell protection, cell renewal, UV protection, and antioxidative stress. The findings support the potential efficacy of *Willaertia* lysate in promoting cellular health and longevity.

3.2. Wound Healing Assay

Considering the overexpressed genes, the potential of *Willaertia* lysate in promoting cell migration was evaluated via a scratch test on human dermal fibroblast cells and the human keratinocyte cell line.

Regarding the control curve, the wound closure area increases gradually over time, indicating the occurrence of basal cell migration and natural wound healing processes (Figure 2). NHDF cells treated with 10^{-4}% of *Willaertia* lysate (9.2×10^{-6}% dry weight) exhibit a higher wound closure area compared to the control and the 10^{-5} treated cells (Figure 2). This result demonstrates the dose-dependent efficacy of the lysate in promoting cell migration, thereby improving the wound-healing process.

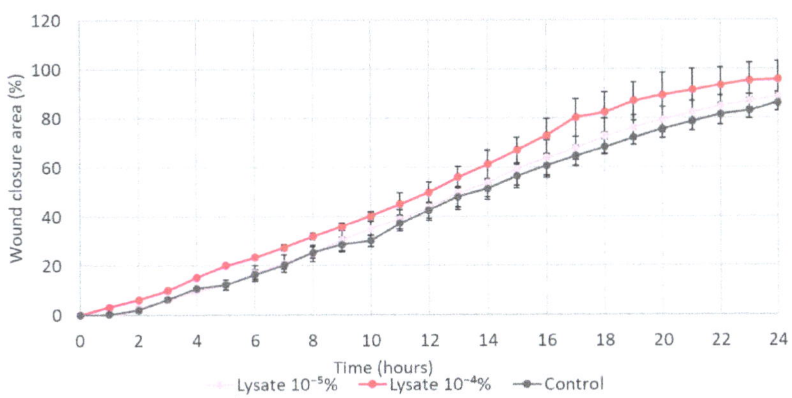

Figure 2. Wound closure area of human dermal fibroblasts treated with two dilutions of *Willaertia* lysate. Results are expressed as mean ± SD (n = 3 biological replicates).

The area under the curve (AUC) is a measure of the total extent of cell migration over time. The treatment of NHDF cells with *Willaertia* lysate solution (10^{-4}% dilution) increased the AUC significantly by 20% ($p < 0.05$) compared to the control group under normal culture conditions, suggesting that it promoted cell migration in a dose-dependent manner (Figure 3).

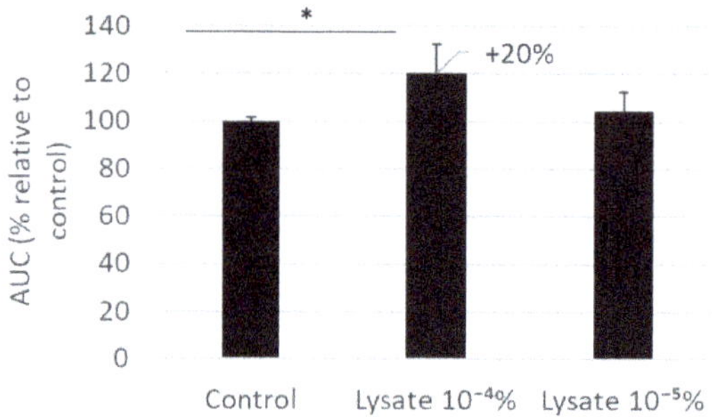

Figure 3. The area under the curve of the closure area of human dermal fibroblasts was treated with two dilutions of *Willaertia* lysate. Results are expressed as mean ± SD (n = 3 biological replicates), * $p < 0.05$.

On human keratinocytes (HaCat cells), the treatment with 10^{-4}% *Willaertia* lysate resulted in an increase in the AUC by +14% compared to the control group. This difference was visualized through image acquisition (Figure 4).

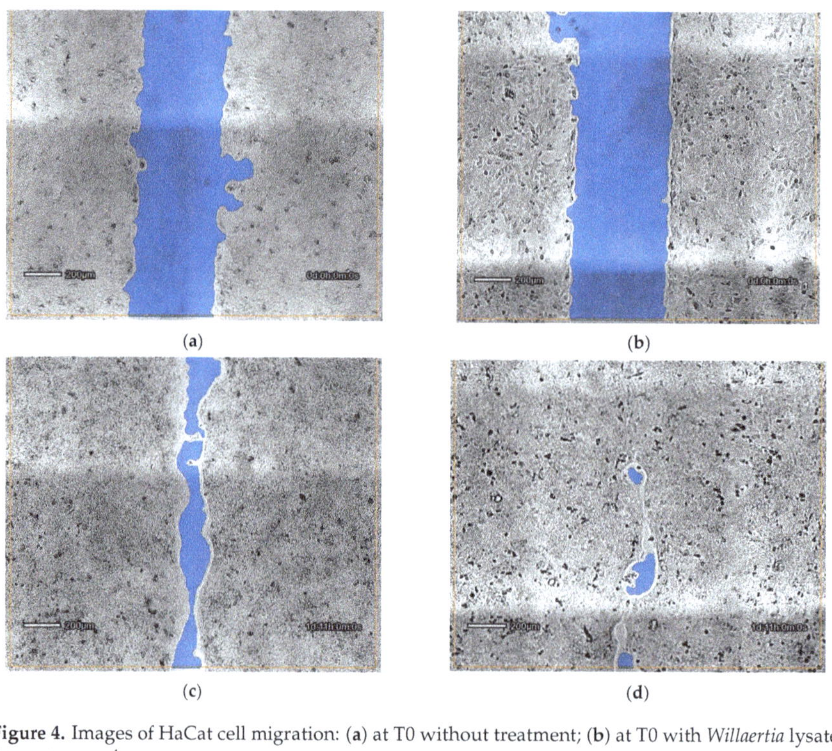

Figure 4. Images of HaCat cell migration: (**a**) at T0 without treatment; (**b**) at T0 with *Willaertia* lysate diluted at 10^{-4}%; (**c**) after 35 h of culturing without treatment; and (**d**) after 35 h of culturing with *Willaertia* lysate diluted at 10^{-4}%. The blue color indicates the cell-free scratch area.

These results demonstrate that *Willaertia* lysate promotes cell migration in both keratinocyte and fibroblast cell types.

3.3. *Pigmentation Assay*

The objective of this assay was to evaluate the potential effect of *Willaertia* lysate on pigmentation using the B16-F10 murine melanoma cell line, both by measuring its capacity to decrease tyrosinase enzyme activity and melanin content.

3.3.1. Tyrosinase Activity

Tyrosinase activity increased significantly by 49% in the presence of IBMX, known to stimulate pigmentation mechanism, compared to the control group. This increase was significantly reduced by 15% with the addition of kojic acid, known to inhibit tyrosinase activity, and by 7%, 11%, and 9% with treatment with *Willaertia* lysate diluted at 10^{-5}%, 10^{-4}%, and 10^{-3}%, respectively (Figure 5). These data suggest that *Willaertia* lysate inhibited the tyrosinase activity in B16-F10 melanocytes when used between 10^{-5}% and 10^{-3}%. It is worth noticing that, at 10^{-4}% dilution, the decrease in tyrosinase activity was not significantly different from that under the kojic acid condition ($p = 0.1$), meaning that the efficacy of the *Willaertia* lysate at this concentration was as good as the effect of kojic acid.

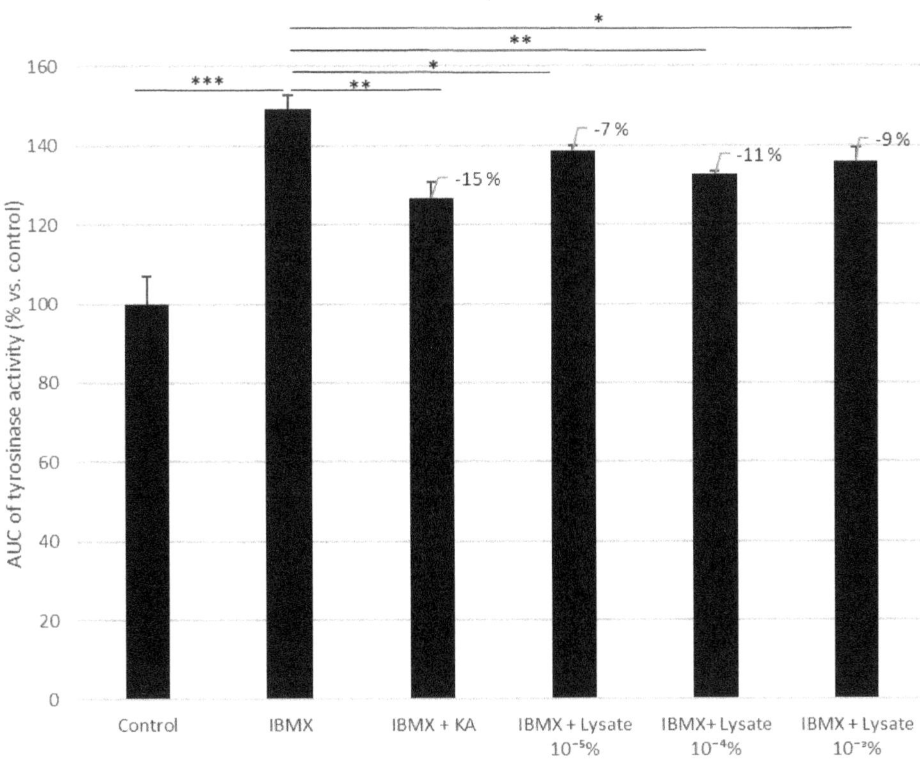

Figure 5. Area under the curve of tyrosinase activity of B16-F10 melanocytes treated with IBMX, IBMX + kojic acid (KA), and three dilutions of *Willaertia* lysate (10^{-5}%, 10^{-4}%, and 10^{-3}%). Results are expressed as mean ± SD (n = 3 biological replicates), * $p < 0.05$, ** $p < 0.01$, *** $p < 0.001$.

3.3.2. Melanin Synthesis

In B16-F10 cells treated with IBMX and kojic acid, the melanin content decreased significantly by 24% (Figure 6) compared to the IBMX group ($p < 0.001$). This confirmed the effect of kojic acid on melanin synthesis, thereby validating the pigmentation assay as kojic acid represented the positive control. These results clearly show that *Willaertia* lysate induced a concentration-dependent decrease in melanin synthesis in B16-F10 melanocytes. It significantly decreased the melanin content by 12% ($p < 0.05$) and 22% ($p < 0.01$) at 10^{-4}% and 10^{-3}% dilutions, respectively.

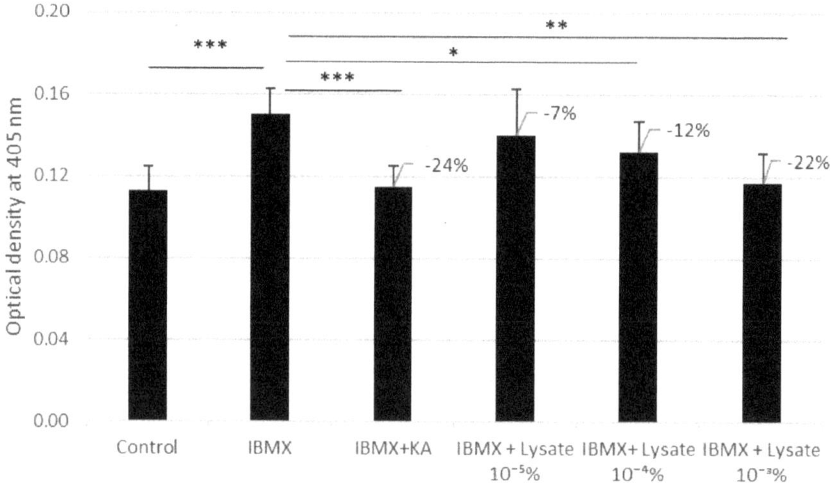

Figure 6. Concentration of melanin in B16-F10 melanocytes treated with IBMX, IBMX + kojic acid (KA), and three dilutions of *Willaertia* lysate (10^{-5}%, 10^{-4}%, and 10^{-3}%). Results are expressed as mean ± SD (n = 3 biological replicates), * $p < 0.05$, ** $p < 0.01$, *** $p < 0.001$.

The treatment of murine melanoma B16-F10 cells with *Willaertia* lysate showed a concentration-dependent decrease in both tyrosinase activity and melanin content. These two combined effects indicated that *Willaertia* lysate was effective in reducing pigmentation.

3.4. Intrinsic Aging Study

The objective of this investigation was to evaluate the effects of *Willaertia* lysate on the histological characteristics of 3D reconstructed aged skin models, more precisely on different specific cutaneous markers using immunohistological techniques.

3.4.1. Histological Analysis of 3D Young vs. 3D Aged Untreated and Treated Skin Models

The global structure of the reconstructed aged skin model was consistent at both the epidermal and dermal levels, showing a pluristratified epidermis that was well differentiated from the basal layer to the *stratum corneum*, and well anchored on top of the dermal equivalent filled with fibroblasts and a neosynthesized extracellular matrix (ECM). The dermal–epidermal junction (DEJ) seemed cohesive and regular. However, compared with the young condition (Figure 7a), aging had an impact on both dermal and epidermal compartments. The extracellular matrix was less dense and abundant, and it was disorganized, impacting tissue elasticity and hydration. The epidermal thickness decreased with fewer layers of living cells. Moreover, the terminal differentiation was disrupted, with thinner *stratum granulosum* and *stratum corneum*, thereby impacting barrier function (Figure 7b).

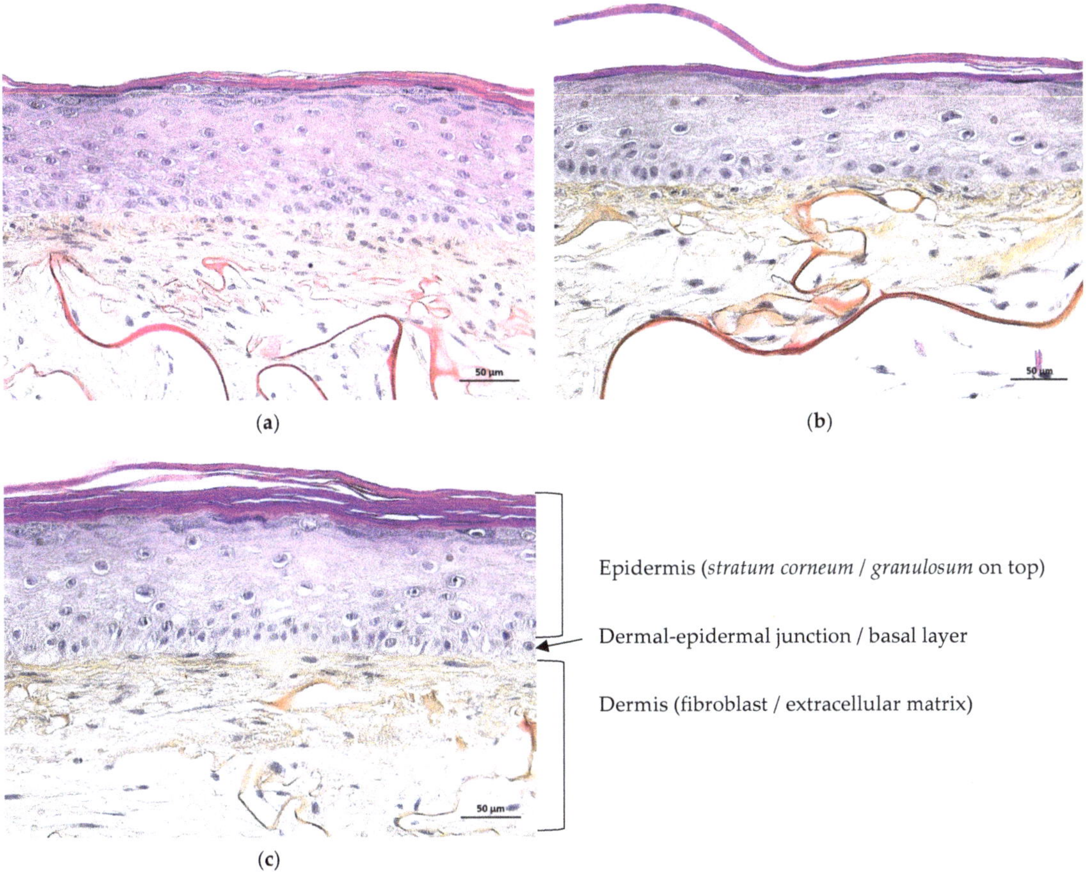

Figure 7. Histological analysis of 3D reconstructed skins: (**a**) untreated young skin; (**b**) untreated aged skin; and (**c**) aged skin treated with *Willaertia* lysate diluted at 0.1×10^{-3}%.

The histological observations showed some beneficial effects of *Willaertia* lysate on the 3D reconstructed aged skin model. Compared with the untreated aged condition, the treatment with *Willaertia* lysate diluted at 0.1×10^{-3}% (7.8×10^{-6}% dry weight) increased the epidermal thickness. Moreover, there was a recovery of terminal differentiation after the treatment compared with the untreated condition. Concerning the dermal compartment, the *Willaertia* lysate treatment improved the quality and abundance of the ECM (Figure 7c).

3.4.2. Immuno-Histological Analysis of Specific Markers in 3D Young Untreated vs. 3D Aged Untreated and Treated Skin Models

To gain a deeper understanding of these effects, further investigations were carried out to decipher more precisely the cellular mechanisms underlying the effects of *Willaertia* lysate at both the epidermal and dermal levels. To conduct such investigations and confirm the histological observations, different cutaneous markers were selected to be explored by means of immuno-histological analysis. The selected markers were Ki67 (Figure 8a), filaggrin (Figure 8b), and loricrin (Figure 8c) for the epidermis; collagen-7 (COL VII) (Figure 8d) for the dermo-epidermal junction (DEJ); and hyaluronic acid (HA) for the dermis (Figure 8e).

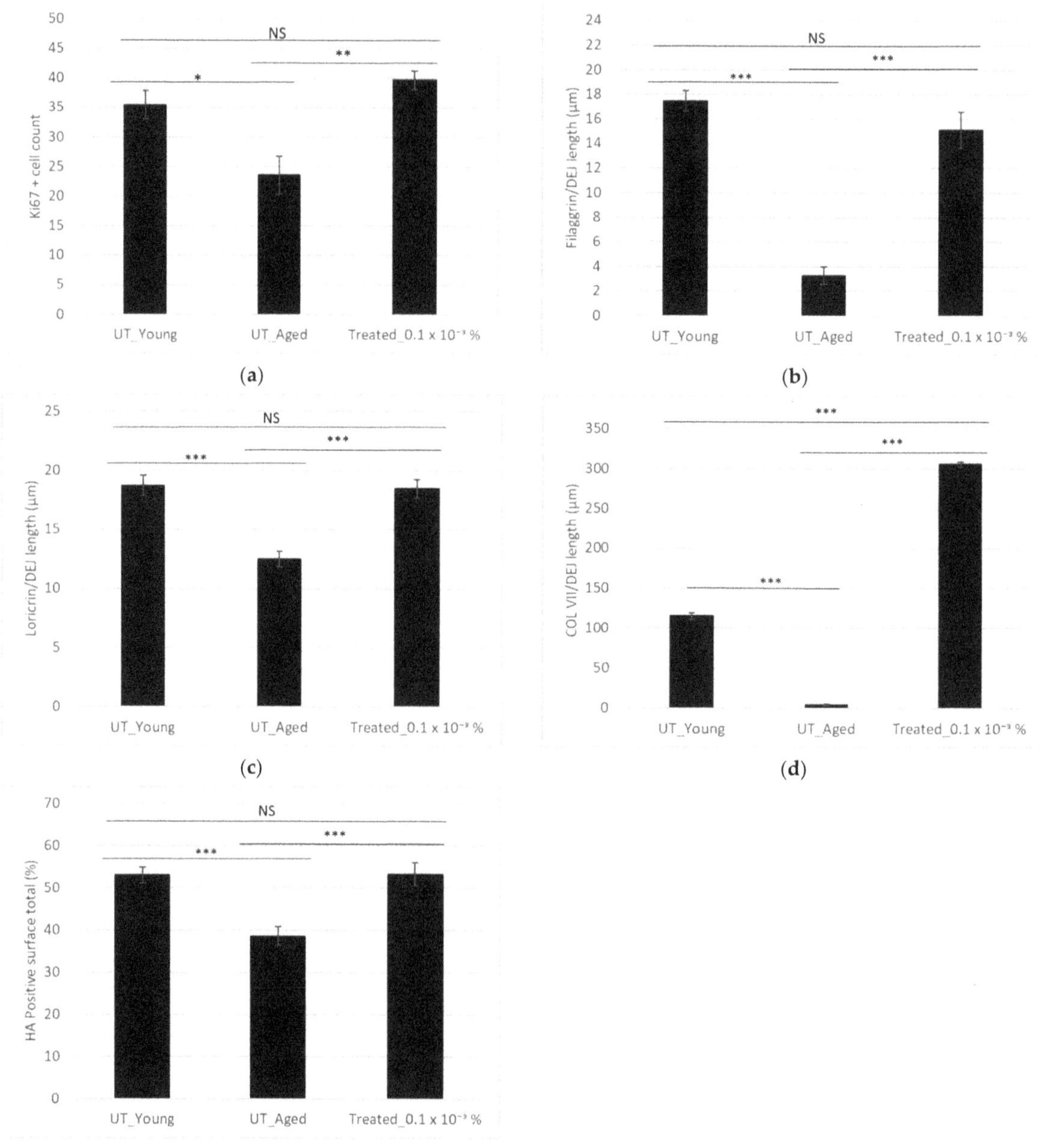

Figure 8. Immuno-histological analysis of 3D reconstructed skins via image automated quantification analysis of (**a**) Ki67, (**b**) filaggrin, (**c**) loricrin, (**d**) collagen-7 (COL VII), and (**e**) hyaluronic acid (HA). *** $p < 0.001$, ** $p < 0.01$, * $p < 0.05$, NS: non-significant, UT: untreated. There were 3 experimental replicates and 3 image acquisition replicates for each marker; thus, 9 images were quantified per experimental condition.

Compared with the young untreated model, the expression levels of the three main cutaneous markers of the skin barrier—Ki67, loricrin, and filaggrin—were significantly decreased in the aged untreated model by −34%, −33%, and −81%, respectively. Compared with the young untreated model, COL VII, and HA were also significantly decreased in the aged untreated model by −96% and −27%, respectively. These results were expected and validate the experiment (Figure 8).

Compared with the aged untreated control, the treatment with 0.1×10^{-3}% of *Willaertia* lysate resulted in a significant increase in Ki67, filaggrin, loricrin, COL VII, and HA by +68%, +361%, +48%, +7170%, and +38%, respectively, in the 3D reconstructed aged skin model. The expression level of these cutaneous markers returned to a similar level to that of the untreated young skin model, except for COL VII, which increased significantly by 64% (Figure 8).

3.5. Extrinsic Aging Study

The objective of this investigation was to evaluate the effects of *Willaertia* lysate on different specific cutaneous markers in 3D UV-irradiated reconstructed young skin using immuno-histological techniques.

3.5.1. Histological Analysis of 3D Unirradiated Untreated Young Skin vs. 3D Irradiated Untreated and Irradiated Treated Young Skins

The global structure of the UVB-exposed reconstructed skin was consistent at both the epidermal and dermal levels, showing a pluristratified epidermis that was well differentiated from the basal layer to the *stratum corneum* and well anchored on top of the dermal equivalent filled with fibroblasts and a neosynthesized ECM. The DEJ seemed cohesive and regular (Figure 9a). However, compared with the unexposed condition, UVB stress had an impact on both the dermal and epidermal compartments. The extracellular matrix was less dense and abundant and was disorganized, impacting tissue elasticity and hydration. The extracellular matrix was also disorganized. The epidermal thickness decreased with fewer layers of living cells, impacting barrier function (Figure 9b). Compared with the untreated and UVB-exposed condition, the treatment with *Willaertia* lysate diluted at 0.2×10^{-3}% (1.6×10^{-5}% dry weight) improved ECM quality and abundance. Moreover, the treatment induced an increase in epidermal thickness (Figure 9c).

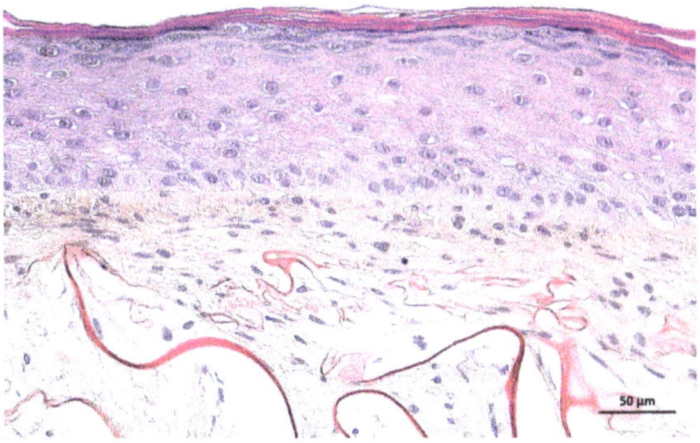

(a)

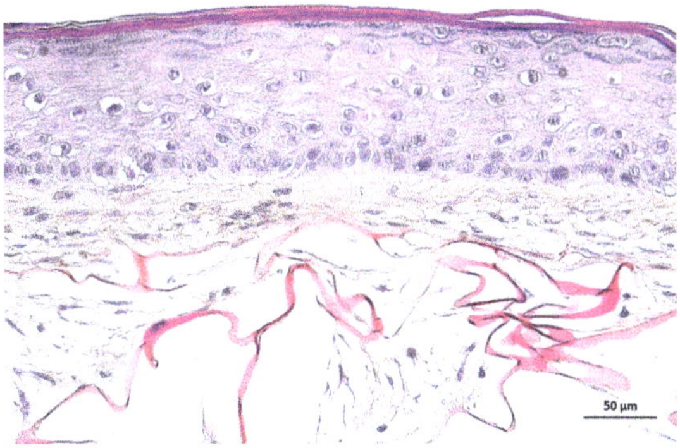

(b)

Figure 9. *Cont.*

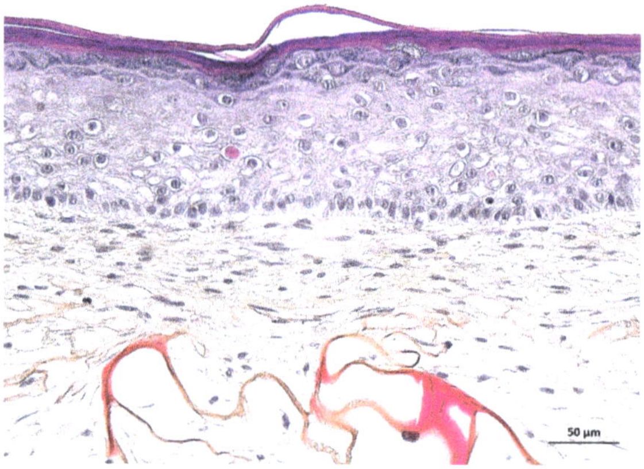

(c)

Figure 9. Histological analysis of 3D reconstructed skins: (**a**) untreated unirradiated young skin, (**b**) untreated irradiated young skin, and (**c**) irradiated young skin treated with *Willaertia* lysate diluted at 0.2×10^{-3}%.

3.5.2. Immuno-Histological Analysis of Specific Markers in 3D Unirradiated Untreated Young Skin vs. 3D Irradiated Untreated and Irradiated Treated Young Skins

To gain a deeper understanding of these effects, further investigations were carried out to decipher more precisely the cellular mechanisms underlying the effects of *Willaertia* lysate on the epidermis with filaggrin (Figure 10a) and loricrin (Figure 10b) as the markers; on the DEJ with laminin-332 (Figure 10c) and COL VII (Figure 10d) as the markers; and on the dermis with HA (Figure 10e), collagen-1 (COL I) (Figure 10f), and elastin (Figure 10g) as the markers.

Compared with the untreated and unirradiated control, significant decreases in filaggrin, loricrin, laminin-332, COL VII, total HA, COLI, and elastin expression by −56%, −19%, −12%, −30%, −17%, −14%, and −19%, respectively, were observed after UVB irradiation. These results were expected. Compared with the untreated and UVB-irradiated control, significant increases in filaggrin, loricrin, laminin-332, COL VII, HA, COLI, and elastin expression by +57%, +27%, +28%, +32%, +31%, +28%, and +69%, respectively, were observed with the addition of the *Willaertia* lysate treatment diluted at 0.2×10^{-3}% (Figure 10).

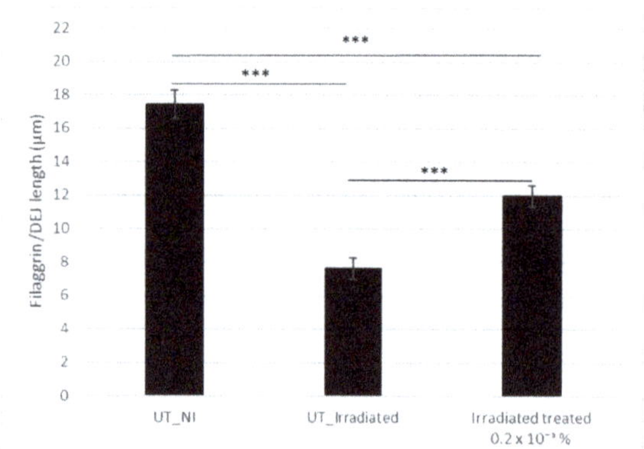

(a)

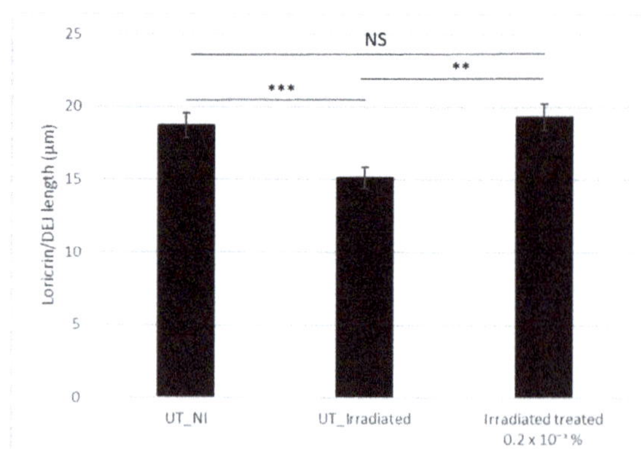

(b)

Figure 10. *Cont.*

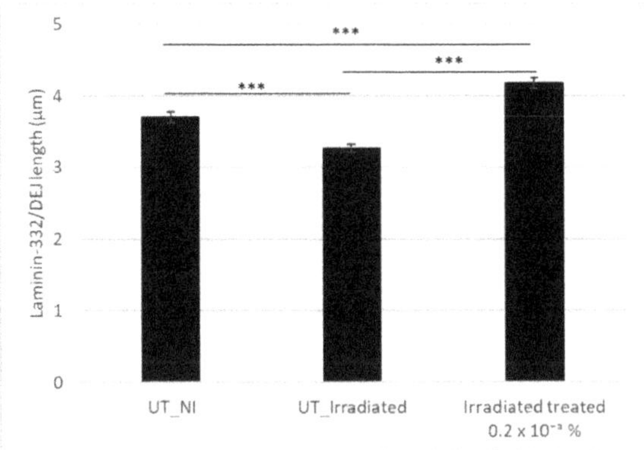

(c)

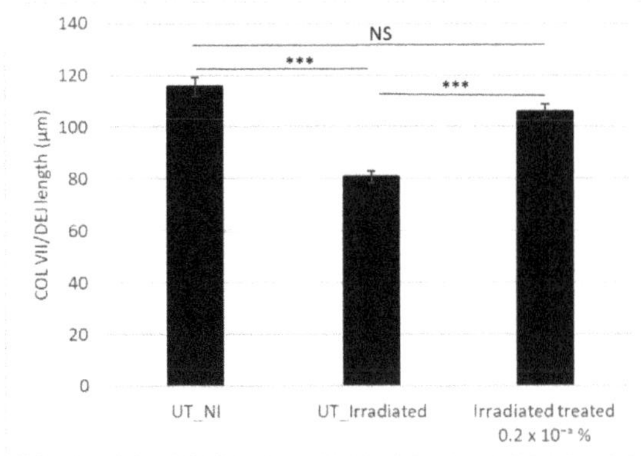

(d)

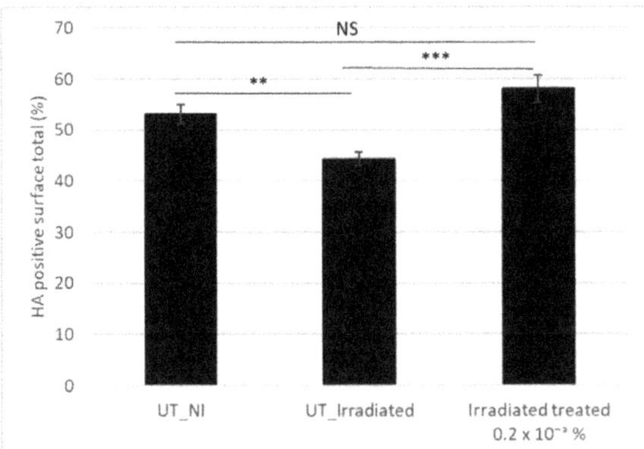

(e)

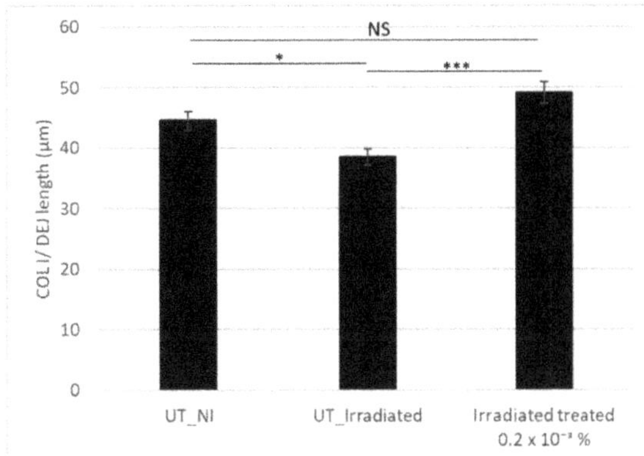

(f)

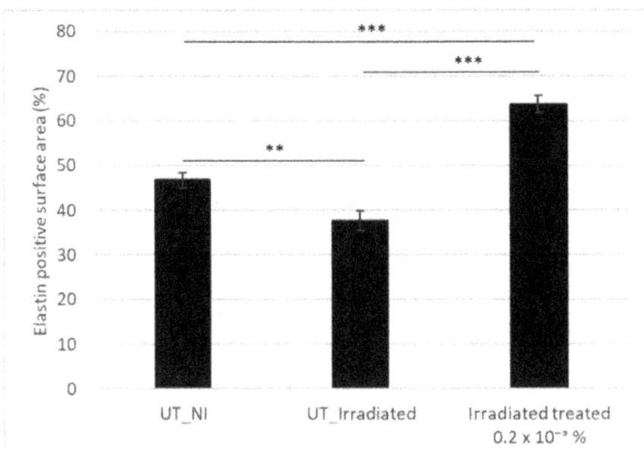

(g)

Figure 10. Immuno-histological analysis of 3D reconstructed young skins through image automated quantification analysis of: (**a**) filaggrin, (**b**) loricrin, (**c**) laminin-332, (**d**) COL VII, (**e**) hyaluronic acid (HA), (**f**) COL I, and (**g**) elastin. *** $p < 0.001$; ** $p < 0.01$; * $p < 0.05$; NS: non-significant; UT: untreated; NI: unirradiated. There were 3 experimental replicates and 3 image acquisition replicates for each marker; thus, 9 images were quantified per experimental condition.

It is interesting to note that the expression level of loricrin, COL VII, HA, and COL I returned to a similar level as that of the untreated and unirradiated control, and that the expression level of laminin-332 and elastin increased significantly by +13% and +36%, respectively.

Finally, considering UV exposure, the level of oxidative stress was also evaluated by determining the level of lipid peroxidation via malondialdehyde (MDA) measure (Figure 11a) and protein carbonylation (Figure 11b). Compared with the untreated and unirradiated control, a significant increase of +682% and +26% in MDA and protein carbonylation expression, respectively, was measured after UVB irradiation of the untreated young skin. Compared with the untreated and UVB irradiated control, a significant decrease of −86% and −25% in MDA and protein carbonylation expression, respectively, was observed with the addition of the *Willaertia* lysate treatment diluted at 0.2×10^{-3}%. The level of MDA expression after treatment with *Willaertia* lysate was similar to that of the untreated and unirradiated control, while protein carbonylation decreased significantly by 5%.

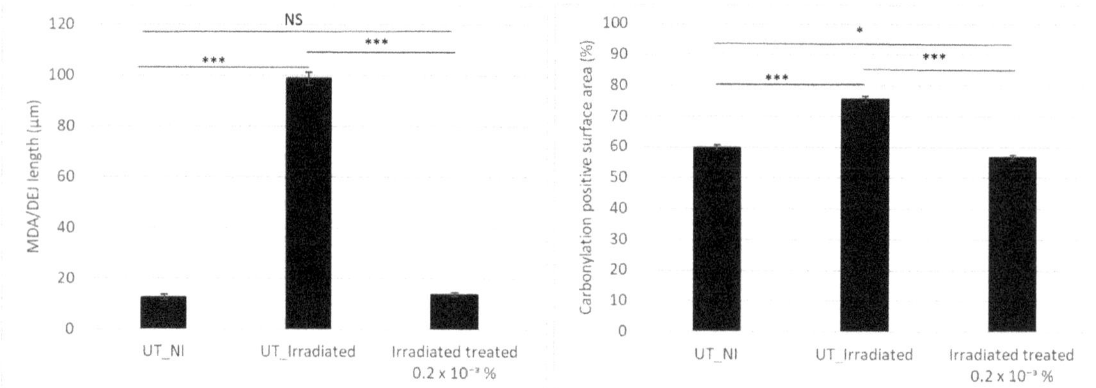

(a) (b)

Figure 11. Immuno-histological analysis of 3D reconstructed young skins through image automated quantification analysis of (**a**) MDA and (**b**) carbonylation. *** $p < 0.001$; * $p < 0.05$; NS: non-significant; UT: untreated; NI: unirradiated. There were 3 experimental replicates and 3 image acquisition replicates for each marker; thus, 9 images were quantified per experimental condition.

An additional in vitro test was conducted to evaluate the antioxidative properties of *Willaertia* lysate. Strong activity was observed at 15 mg/mL dry weight with 83% inhibition, and the IC_{50} was determined to correspond to a concentration of 8.02 mg/mL. In comparison, *Moringa oleifera* seed oil, known for its antioxidant properties [54] has an IC_{50} of 121.9 mg/mL, and thus fifteen less fold efficient than *Willaertia* lysate.

To conclude, we demonstrated that treatment of the 3D reconstructed skin model exposed to UVB with *Willaertia* lysate improved the quality and abundance of the ECM and increased the epidermal thickness. By stimulating epidermal markers such as loricrin and filaggrin, the *Willaertia* lysate treatment demonstrated its protective effects on the skin barrier; by stimulating DEJ markers such as COL VII and laminin-332, and dermal markers such as COL I, elastin, and hyaluronic acid, the *Willaertia* lysate treatment demonstrated its efficacy in improving skin firmness, elasticity, architecture, and hydration.

Moreover, the antioxidant activity of the *Willaertia* lysate treatment was demonstrated through an efficient reduction in the quantity of DPPH radicals and its efficacy in reducing oxidative stress after UV exposure was shown by the decreases in exposome markers such as protein carbonylation and lipid peroxidation expression. This complete set of coherent results demonstrated the global protective efficacy of *Willaertia* lysate against the effects of photoaging.

4. Discussion

Thermal spring waters have long been revered for their medicinal and wellness properties, showing benefits in treating a variety of conditions related to rheumatology, gastroenterology, pulmonology, and dermatology. To date, twelve samples of thermal spring water have been recognized to have potential as active ingredients in cosmetic formulations [55,56]. To the best of our knowledge, this is the first study to investigate an ingredient sourced from the thermal spring waters of Aix-les-Bains. These thermal spring waters are unique because of their origin from two different boreholes: Chevalley and Reine Hortense. The Chevalley borehole reaches a depth of 2200 m with a temperature of 71 °C, while the Reine Hortense borehole extends to 1100 m with a temperature of 38 °C. Although nearly identical in composition, the waters from these two boreholes were blended, with 36% sourced from Reine-Hortense and 34% from Chevalley. Since Roman times, the thermal spring waters of Aix les Bains have been renowned for their richness in trace elements, including calcium bicarbonate, sulfate, and sulfur. These mineral-rich waters are recognized for their therapeutic virtues, particularly in alleviating and treating rheumatic, respiratory, dermatological and circulatory conditions [57,58]. However, the mineral composition of these waters alone does not fully explain their positive effects on atopic skin. Therefore, it is worth exploring the interactions between the biological components of thermal waters and the skin, an area that has been relatively under-researched.

The living microbiota present in thermal spring waters, referred to as "hydrobiome", is increasingly recognized for its therapeutic effects on skin health. The bioactive compounds extracted from the hydrobiome help prevent skin aging by enhancing protection against UV exposure, strengthening the skin barrier function, maintaining homeostasis of skin defenses, repairing damaged skin, promoting wound healing, and reducing uneven skin pigmentation [59]. *Willaertia* lysate demonstrates all of these properties. In this study, it was demonstrated to promote cell migration in primary cultures of human dermal fibroblasts and keratinocytes, decrease the tyrosinase activity and melanin content of murine melanoma cells, and improve the quality and abundance of the extracellular matrix in aged skins to a similar level as that in young skins by stimulating epidermal markers such as Ki67, loricrin, and filaggrin, dermo-epidermal junction markers such as COLVII, and dermal markers such as hyaluronic acid. Hyaluronic acid (HA) is known to play a key role during the aging process by maintaining skin moisture. Treatment of cultured cells with *Willaertia* lysate induced a 1.8-fold increase in the expression of HAS gene responsible for HA synthesis at a dilution of 0.2%, corresponding to $1.6 \times 10^{-2}\%$ dry weight of the active substance. Comparatively, retinol at 0.1% can induce a six-fold increase in HAS gene expression [60]. The expression stimulation with *Willaertia* lysate was confirmed by a significant increase of 38% in HA content in the 3D aged skin model treated with $0.1 \times 10^{-3}\%$ of lysate, corresponding to the $9.2 \times 10^{-5}\%$ dry weight of the active substance, compared with the untreated aged skin. The HA content of the treated aged skin was not significantly different from that of the young skin. Therefore, even if the gene stimulation by *Willaertia* lysate is less efficient than retinol, it is sufficient to restore the skin's properties, and the *Willaertia* lysate concentration used is ten-fold lower than the retinol concentration to obtain a similar result. Finally, based on the provided evidence, *Willaertia* lysate appears to promote cell migration by creating a favorable environment for cell movement through extracellular matrix (ECM) remodeling. The lysate's ability to improve the quality and abundance of the ECM in the 3D skin model suggests that it can create a more favorable environment for cell migration.

In the case of the reconstructed young skin exposed to UVB, we demonstrated that the *Willaertia* lysate treatment improved the quality and abundance of the extracellular matrix, and increased the epidermal thickness. By stimulating epidermal markers such as loricrin and filaggrin, the *Willaertia* lysate treatment demonstrated its protective effects on the skin barrier; by stimulating dermo-epidermal junction markers, such as COLVII and laminin-332, and dermal markers, such as COL I, elastin, and HA, the *Willaertia* lysate treatment demonstrated its efficacy in improving skin firmness, elasticity, architecture, and hydration. Moreover, the efficacy of this treatment in reducing oxidative stress after UV exposure was shown by the decreases in the levels of exposome markers such as protein carbonylation and lipid peroxidation. The exposome is a concept encompassing the totality of exposures to environmental factors, i.e., non-genetic factors, that a human organism undergoes from conception to the end of its life. Thus, the reduction in DPPH radicals and markers of oxidative damage (protein carbonylation and lipid peroxidation) indicates that *Willaertia* lysate possesses direct antioxidant properties.

This complete set of coherent results demonstrated the global protective efficacy of *Willaertia* lysate against the effects of (photo)aging and its potential for well-aging and longevity. The next step will consist of conducting a clinical study to confirm the benefits of *Willaertia* lysate for human skin care.

5. Conclusions

This is the first study to report the potential application of a non-viable form of *Willaertia magna* C2c Maky (lysate) in the field of cosmetics as a natural and biosourced postbiotic active ingredient. Significant statistical results may be partly due to natural biological variations; however, the diverse models used and the various experiments conducted all show the same trend toward efficiency. Our study's findings on the anti-aging effects of *Willaertia* lysate, a postbiotic isolated from French Alps thermal spring waters, may offer a partial explanation for the broader skin benefits attributed to thermal waters. The positive outcomes observed in our research suggest that the unique components and properties of these natural waters, including potential microbial contributions, could contribute to the therapeutic effects reported in historical and contemporary studies.

Author Contributions: Conceptualization, S.T.; methodology, M.D.S., J.R., K.L., M.B., A.S., A.C. and H.O.-P.; formal analysis, M.D.S., J.R., M.B., B.Q. and H.O.-P.; writing—original draft preparation, S.T.; writing—review and editing, M.D.S. and H.O.-P.; supervision, S.T. All authors have read and agreed to the published version of the manuscript.

Funding: This research received no external funding. The company Amoéba supported all expenses.

Institutional Review Board Statement: This study uses anonymized skin samples obtained as a byproduct from Lyon University Hospital (Hospice Civils de Lyon, France) or Sterlab (Vallauris, France). French Law No. 2004–801 on bioethics did not require the study to be reviewed or approved by an ethics committee because of the anonymized nature and by-product status of the samples. All the human skin samples used in this study were obtained with the explicit informed consent of the patients in compliance with the French Ethics Committee of the French Research Ministry (Declaration no. DC-2020-4346 delivered to LabSkin Creations, Lyon, France; Declaration no. DC-2018-3312 delivered to PKDERM, Grasse, France; Declaration AC-2018-3159 delivered to Sterlab, Vallauris, France).

Informed Consent Statement: Informed consent was obtained from all subjects involved in the study, patients who are minors are from their parents or relatives/guardians.

Data Availability Statement: The raw data supporting the conclusions of this study will be made available by the authors upon request.

Acknowledgments: The authors are very grateful to the production department of Amoéba for the *Willaertia* lysate production.

Conflicts of Interest: Authors Morgan Dos Santos, Julie Rorteau and Kilian Laho were employed by the company LabSkin Creations; LabSkin Creations, Lyon, France. Hanan Osman-Ponchet and Manon Barthe were employed by the company Laboratoires PKDERM; Laboratoires PKDERM, Grasse, France. The remaining authors were employed by the company Amoéba; Amoéba, Chassieu, France. The authors declare the funders had no role in the design of the study; in the collection, analyses, or interpretation of data; in the writing of the manuscript; or in the decision to publish the results.

References

1. Gilchrest, B.A. *Skin and Aging Processes;* CRC Press: Boca Raton, FL, USA, 1984; ISBN 978-0-8493-5472-4.
2. Kligman, A.M. Early Destructive Effect of Sunlight on Human Skin. *JAMA* **1969**, *210*, 2377–2380. [CrossRef] [PubMed]
3. Martin, K.I.; Glaser, D.A. Cosmeceuticals: The New Medicine of Beauty. *Mo. Med.* **2011**, *108*, 60–63. [PubMed]
4. Dhaliwal, S.; Rybak, I.; Ellis, S.R.; Notay, M.; Trivedi, M.; Burney, W.; Vaughn, A.R.; Nguyen, M.; Reiter, P.; Bosanac, S.; et al. Prospective, Randomized, Double-Blind Assessment of Topical Bakuchiol and Retinol for Facial Photoageing. *Br. J. Dermatol.* **2019**, *180*, 289–296. [CrossRef] [PubMed]
5. Fathalla, Z.; Shoman, M.E.; Barakat, H.S.; Al Fatease, A.; Alamri, A.H.; Abdelkader, H. Cyclodextrins and Amino Acids Enhance Solubility and Tolerability of Retinoic Acid/Tretinoin: Molecular Docking, Physicochemical, Cytotoxicity, Scratch Assay, and Topical Gel Formulations Investigation. *Pharmaceutics* **2024**, *16*, 853. [CrossRef]
6. Chaudhuri, R.K.; Bojanowski, K. Bakuchiol: A Retinol-like Functional Compound Revealed by Gene Expression Profiling and Clinically Proven to Have Anti-Aging Effects. *Int. J. Cosmet. Sci.* **2014**, *36*, 221–230. [CrossRef]
7. Malinauskiene, L.; Linauskiene, K.; Černiauskas, K.; Chomičiene, A. Bakuchiol-A New Allergen in Cosmetics. *Contact Dermat.* **2019**, *80*, 398–399. [CrossRef]
8. Gibielle, C.; Bousseksou, L.; Guéhenneux, S.; Vié, K. In a Preliminary Study on Human Subjects, a Cosmetic Cream Containing a Harungana Madagascariensis Plant Extract Induces Similar Anti-Aging Effects to a Retinol-Containing Cream. *Clin. Cosmet. Investig. Dermatol.* **2023**, *16*, 1051–1058. [CrossRef]
9. Gupta, P.L.; Rajput, M.; Oza, T.; Trivedi, U.; Sanghvi, G. Eminence of Microbial Products in Cosmetic Industry. *Nat. Prod. Bioprospect.* **2019**, *9*, 267–278. [CrossRef]
10. Janjetovic, Z.; Slominski, A.T. Promising Functions of Novel Vitamin D Derivatives as Cosmetics: A New Fountain of Youth in Skin Aging and Skin Protection. *Cosmetics* **2024**, *11*, 37. [CrossRef]
11. Luangpraditkun, K.; Pimjuk, P.; Phimnuan, P.; Wisanwattana, W.; Wisespongpand, C.; Waranuch, N.; Viyoch, J. Anti-Aging Properties of Cannabis Sativa Leaf Extract against UVA Irradiation. *Cosmetics* **2024**, *11*, 45. [CrossRef]
12. Gavrilescu, M.; Chisti, Y. Biotechnology—A Sustainable Alternative for Chemical Industry. *Biotechnol. Adv.* **2005**, *23*, 471–499. [CrossRef] [PubMed]
13. Murray, P.M.; Moane, S.; Collins, C.; Beletskaya, T.; Thomas, O.P.; Duarte, A.W.F.; Nobre, F.S.; Owoyemi, I.O.; Pagnocca, F.C.; Sette, L.D.; et al. Sustainable Production of Biologically Active Molecules of Marine Based Origin. *New Biotechnol.* **2013**, *30*, 839–850. [CrossRef] [PubMed]
14. Pérez-Rivero, C.; López-Gómez, J.P. Unlocking the Potential of Fermentation in Cosmetics: A Review. *Fermentation* **2023**, *9*, 463. [CrossRef]
15. Kim, J.; Lee, Y.I.; Mun, S.; Jeong, J.; Lee, D.-G.; Kim, M.; Jo, H.; Lee, S.; Han, K.; Lee, J.H. Efficacy and Safety of Epidermidibacterium Keratini EPI-7 Derived Postbiotics in Skin Aging: A Prospective Clinical Study. *Int. J. Mol. Sci.* **2023**, *24*, 4634. [CrossRef]
16. Duarte, M.; Oliveira, A.L.; Oliveira, C.; Pintado, M.; Amaro, A.; Madureira, A.R. Current Postbiotics in the Cosmetic Market—An Update and Development Opportunities. *Appl. Microbiol. Biotechnol.* **2022**, *106*, 5879–5891. [CrossRef]
17. Dey, R.; Mameri, M.R.; Trajkovic-Bodennec, S.; Bodennec, J.; Pernin, P. Impact of Inter-Amoebic Phagocytosis on the L. Pneumophila Growth. *FEMS Microbiol. Lett.* **2020**, *367*, fnaa147. [CrossRef]
18. Hasni, I.; Chelkha, N.; Baptiste, E.; Mameri, M.R.; Lachuer, J.; Plasson, F.; Colson, P.; La Scola, B. Investigation of Potential Pathogenicity of Willaertia Magna by Investigating the Transfer of Bacteria Pathogenicity Genes into Its Genome. *Sci. Rep.* **2019**, *9*, 18318. [CrossRef]
19. De Jonckheere, J.F.; Dive, D.G.; Pussard, M.; Vickerman, K. Willaertia Magna Gen. Nov., Sp. Nov.(Vahlkampfiidae), a Thermophilic Amoeba Found in Different Habitats. *Protistologica* **1984**, *20*, 5–13.
20. Robinson, B.S.; Christy, P.E.; De Jonckheere, J.F. A Temporary Flagellate (Mastigote) Stage in the Vahlkampfiid Amoeba Willaertia Magna and Its Possible Evolutionary Significance. *Biosystems* **1989**, *23*, 75–86. [CrossRef]
21. Fuentes-Herrera, P.B.; Herrera-Cabrera, B.E.; Martínez-Ayala, A.L.; Zamilpa, A.; Delgado-Alvarado, A. Content and Yield of L-DOPA and Bioactive Compounds of Broad Bean Plants: Antioxidant and Anti-Inflammatory Activity In Vitro. *Plants* **2023**, *12*, 3918. [CrossRef]
22. Rahman, M.M.; Islam, M.B.; Biswas, M.; Khurshid Alam, A.H.M. In Vitro Antioxidant and Free Radical Scavenging Activity of Different Parts of Tabebuia Pallida Growing in Bangladesh. *BMC Res. Notes* **2015**, *8*, 621. [CrossRef] [PubMed]
23. Brand-Williams, W.; Cuvelier, M.E.; Berset, C. Use of a Free Radical Method to Evaluate Antioxidant Activity. *LWT-Food Sci. Technol.* **1995**, *28*, 25–30. [CrossRef]
24. Germain, L.; Rouabhia, M.; Guignard, R.; Carrier, L.; Bouvard, V.; Auger, F.A. Improvement of Human Keratinocyte Isolation and Culture Using Thermolysin. *Burns* **1993**, *19*, 99–104. [CrossRef]
25. Dos Santos, M.; Metral, E.; Boher, A.; Rousselle, P.; Thepot, A.; Damour, O. In Vitro 3-D Model Based on Extending Time of Culture for Studying Chronological Epidermis Aging. *Matrix Biol.* **2015**, *47*, 85–97. [CrossRef]
26. de Oliviera Nascimento, L.; Massari, P.; Wetzler, L. The Role of TLR2 in Infection and Immunity. *Front. Immunol.* **2012**, *3*, 79. [CrossRef]
27. Khanmohammadi, S.; Rezaei, N. Role of Toll-like Receptors in the Pathogenesis of COVID-19. *J. Med. Virol.* **2021**, *93*, 2735–2739. [CrossRef]
28. Takeda, K.; Akira, S. TLR Signaling Pathways. *Semin. Immunol.* **2004**, *16*, 3–9. [CrossRef]

29. Koshak, A.E.; Algandaby, M.M.; Mujallid, M.I.; Abdel-Naim, A.B.; Alhakamy, N.A.; Fahmy, U.A.; Alfarsi, A.; Badr-Eldin, S.M.; Neamatallah, T.; Nasrullah, M.Z.; et al. Wound Healing Activity of Opuntia Ficus-Indica Fixed Oil Formulated in a Self-Nanoemulsifying Formulation. *Int. J. Nanomed.* **2021**, *16*, 3889–3905. [CrossRef]
30. Shive, C.; Pandiyan, P. Inflammation, Immune Senescence, and Dysregulated Immune Regulation in the Elderly. *Front. Aging* **2022**, *3*, 840827. [CrossRef]
31. Abbasi, S.; Sinha, S.; Labit, E.; Rosin, N.L.; Yoon, G.; Rahmani, W.; Jaffer, A.; Sharma, N.; Hagner, A.; Shah, P.; et al. Distinct Regulatory Programs Control the Latent Regenerative Potential of Dermal Fibroblasts during Wound Healing. *Cell Stem Cell* **2020**, *27*, 396–412.e6. [CrossRef]
32. Johnson, K.; Zhu, S.; Tremblay, M.S.; Payette, J.N.; Wang, J.; Bouchez, L.C.; Meeusen, S.; Althage, A.; Cho, C.Y.; Wu, X.; et al. A Stem Cell-Based Approach to Cartilage Repair. *Science* **2012**, *336*, 717–721. [CrossRef] [PubMed]
33. Umansky, K.B.; Gruenbaum-Cohen, Y.; Tsoory, M.; Feldmesser, E.; Goldenberg, D.; Brenner, O.; Groner, Y. Runx1 Transcription Factor Is Required for Myoblasts Proliferation during Muscle Regeneration. *PLoS Genet.* **2015**, *11*, e1005457. [CrossRef] [PubMed]
34. Xue, H.; Zha, M.; Tang, Y.; Zhao, J.; Du, X.; Wang, Y. Research Progress on the Extraction and Purification of Anthocyanins and Their Interactions with Proteins. *Molecules* **2024**, *29*, 2815. [CrossRef] [PubMed]
35. Zeng, A.; Li, H.; Guo, L.; Gao, X.; McKinney, S.; Wang, Y.; Yu, Z.; Park, J.; Semerad, C.; Ross, E.; et al. Prospectively Isolated Tetraspanin+ Neoblasts Are Adult Pluripotent Stem Cells Underlying Planaria Regeneration. *Cell* **2018**, *173*, 1593–1608.e20. [CrossRef]
36. Busjahn, A.; Aydin, A.; Uhlmann, R.; Krasko, C.; Bähring, S.; Szelestei, T.; Feng, Y.; Dahm, S.; Sharma, A.M.; Luft, F.C.; et al. Serum- and Glucocorticoid-Regulated Kinase (SGK1) Gene and Blood Pressure. *Hypertension* **2002**, *40*, 256–260. [CrossRef]
37. Jiang, D.; Fu, C.; Xiao, J.; Zhang, Z.; Zou, J.; Ye, Z.; Zhang, X. SGK1 Attenuates Oxidative Stress-Induced Renal Tubular Epithelial Cell Injury by Regulating Mitochondrial Function. *Oxid. Med. Cell. Longev.* **2019**, *2019*, 2013594. [CrossRef]
38. Mason, J.A.; Cockfield, J.A.; Pape, D.J.; Meissner, H.; Sokolowski, M.T.; White, T.C.; Valentín López, J.C.; Liu, J.; Liu, X.; Martínez-Reyes, I.; et al. SGK1 Signaling Promotes Glucose Metabolism and Survival in Extracellular Matrix Detached Cells. *Cell Rep.* **2021**, *34*, 108821. [CrossRef]
39. Daitoku, H.; Kaneko, Y.; Yoshimochi, K.; Matsumoto, K.; Araoi, S.; Sakamaki, J.; Takahashi, Y.; Fukamizu, A. Nontranscriptional Function of FOXO1/DAF-16 Contributes to Translesion DNA Synthesis. *Mol. Cell. Biol.* **2016**, *36*, 2755–2766. [CrossRef]
40. Dang, R.; Yang, M.; Cui, C.; Wang, C.; Zhang, W.; Geng, C.; Han, W.; Jiang, P. Activation of Angiotensin-Converting Enzyme 2/Angiotensin (1-7)/Mas Receptor Axis Triggers Autophagy and Suppresses Microglia Proinflammatory Polarization via Forkhead Box Class O1 Signaling. *Aging Cell* **2021**, *20*, e13480. [CrossRef]
41. Graves, D.T.; Milovanova, T.N. Mucosal Immunity and the FOXO1 Transcription Factors. *Front. Immunol.* **2019**, *10*, 2530. [CrossRef]
42. Link, W. Introduction to FOXO Biology. In *FOXO Transcription Factors. Methods in Molecular Biology*; Humana: New York, NY, USA, 2019; Volume 1890, pp. 1–9. [CrossRef]
43. Korsmeyer, S.J. BCL-2 Gene Family and the Regulation of Programmed Cell Death. *Cancer Res.* **1999**, *59*, 1693s–1700s. [CrossRef] [PubMed]
44. Ruvolo, P.P.; Deng, X.; May, W.S. Phosphorylation of Bcl2 and Regulation of Apoptosis. *Leukemia* **2001**, *15*, 515–522. [CrossRef] [PubMed]
45. Siddiqui, W.A.; Ahad, A.; Ahsan, H. The Mystery of BCL2 Family: Bcl-2 Proteins and Apoptosis: An Update. *Arch. Toxicol.* **2015**, *89*, 289–317. [CrossRef] [PubMed]
46. He, Z.; Mei, L.; Connell, M.; Maxwell, C.A. Hyaluronan Mediated Motility Receptor (HMMR) Encodes an Evolutionarily Conserved Homeostasis, Mitosis, and Meiosis Regulator Rather than a Hyaluronan Receptor. *Cells* **2020**, *9*, 819. [CrossRef]
47. Wang, S.; Xiong, Y.; Chen, J.; Ghanem, A.; Wang, Y.; Yang, J.; Sun, B. Three Dimensional Printing Bilayer Membrane Scaffold Promotes Wound Healing. *Front. Bioeng. Biotechnol.* **2019**, *7*, 348. [CrossRef]
48. Liu, L.; Liu, Y.; Li, J.; Du, G.; Chen, J. Microbial Production of Hyaluronic Acid: Current State, Challenges, and Perspectives. *Microb. Cell Factories* **2011**, *10*, 99. [CrossRef]
49. Siiskonen, H.; Oikari, S.; Pasonen-Seppänen, S.; Rilla, K. Hyaluronan Synthase 1: A Mysterious Enzyme with Unexpected Functions. *Front. Immunol.* **2015**, *6*, 43. [CrossRef]
50. Skandalis, S.S.; Karalis, T.; Heldin, P. Intracellular Hyaluronan: Importance for Cellular Functions. *Semin. Cancer Biol.* **2020**, *62*, 20–30. [CrossRef]
51. Claus, S.; Fischer, J.; Mégarbané, H.; Mégarbané, A.; Jobard, F.; Debret, R.; Peyrol, S.; Saker, S.; Devillers, M.; Sommer, P.; et al. A p.C217R Mutation in Fibulin-5 from Cutis Laxa Patients Is Associated with Incomplete Extracellular Matrix Formation in a Skin Equivalent Model. *J. Investig. Dermatol.* **2008**, *128*, 1442–1450. [CrossRef]
52. Halper, J.; Kjaer, M. Basic Components of Connective Tissues and Extracellular Matrix: Elastin, Fibrillin, Fibulins, Fibrinogen, Fibronectin, Laminin, Tenascins and Thrombospondins. *Adv. Exp. Med. Biol.* **2014**, *802*, 31–47. [CrossRef]
53. Papke, C.L.; Yanagisawa, H. Fibulin-4 and Fibulin-5 in Elastogenesis and beyond: Insights from Mouse and Human Studies. *Matrix Biol.* **2014**, *37*, 142–149. [CrossRef] [PubMed]
54. Athikomkulchai, S.; Tunit, P.; Tadtong, S.; Jantrawut, P.; Sommano, S.R.; Chittasupho, C. Moringa Oleifera Seed Oil Formulation Physical Stability and Chemical Constituents for Enhancing Skin Hydration and Antioxidant Activity. *Cosmetics* **2021**, *8*, 2. [CrossRef]

55. Figueiredo, A.C.; Rodrigues, M.; Mourelle, M.L.; Araujo, A.R.T.S. Thermal Spring Waters as an Active Ingredient in Cosmetic Formulations. *Cosmetics* **2023**, *10*, 27. [CrossRef]
56. Merial-Kieny, C.; Castex-Rizzi, N.; Selas, B.; Mery, S.; Guerrero, D. Avène Thermal Spring Water: An Active Component with Specific Properties. *J. Eur. Acad. Dermatol. Venereol.* **2011**, *25*, 2–5. [CrossRef] [PubMed]
57. Station Thermale d'Aix Les Bains—Thermes Chevalley | ValVital. Available online: https://www.valvital.fr/Nos-stations-thermales/Aix-les-Bains-Station-thermale-d-Aix-les-Bains (accessed on 19 August 2024).
58. Cure Thermale Rhumatologie & Voies Respiratoires à Aix-les-Bains. Available online: https://www.thermes-aixlesbains.com/ (accessed on 19 August 2024).
59. Mourelle, M.L.; Gómez, C.P.; Legido, J.L. Hydrobiome of Thermal Waters: Potential Use in Dermocosmetics. *Cosmetics* **2023**, *10*, 94. [CrossRef]
60. Li, W.-H.; Wong, H.-K.; Serrano, J.; Randhawa, M.; Kaur, S.; Southall, M.D.; Parsa, R. Topical Stabilized Retinol Treatment Induces the Expression of HAS Genes and HA Production in Human Skin in Vitro and in Vivo. *Arch. Dermatol. Res.* **2017**, *309*, 275–283. [CrossRef]

cosmetics

MDPI

Article

Anti-Aging Effects of Low-Molecular-Weight Collagen Peptide Supplementation on Facial Wrinkles and Skin Hydration: Outcomes from a Six-Week Randomized, Double-Blind, Placebo-Controlled Trial

Juan Antonio Carrillo-Norte [1,*], Baldomero García-Mir [2], Lluis Quintana [3], Bruno Buracchio [3] and Rafael Guerrero-Bonmatty [4]

1 Department of Medical and Surgical Therapeutics, Division of Clinical Pharmacology, School of Medicine and Health Sciences, University of Extremadura, 06006 Badajoz, Spain
2 Department of Dermatology, Don Benito-Villanueva Hospital, 06400 Don Benito, Spain
3 Viscofan BioEngineering, Viscofan DE GmbH, 69469 Weinheim, Germany; quintana@bio.viscofan.com (L.Q.); buracchio@bio.viscofan.com (B.B.)
4 Department of Nursing, School of Nursing, University of Extremadura, 06800 Mérida, Spain; rafael.bonmatty@gmail.com
* Correspondence: carrillo@unex.es

Abstract: In recent decades, there has been a rising demand for anti-aging interventions aimed at postponing or potentially reversing indicators of skin aging. The use of collagen-based nutraceutical supplements has gained popularity as they have shown promise in enhancing skin health and reducing signs of aging. The aim of this randomized, placebo-controlled, blinded study was to investigate the effects of 2.5 g COLLinstant® LMW, a novel cosmeceutical containing low-molecular-weight (≤1000 Da) collagen peptides, on skin aging and health. The trial was conducted with 80 healthy women aged 30 years and older. They received a daily oral dose of either the food supplement (n = 40) or placebo (n = 40) for six weeks. Skin assessment was performed based on validated objective methods, such as Visioface 1000D (skin wrinkling), cutometry (elasticity and fatigue), and corneometry (skin hydration) at baseline (T0) and at week 6 (T6). After 6 weeks, participants that received collagen had significant improvements in biometric skin wrinkle parameters from baseline, with a reduction in volume by 46%, in area by 44%, and in depth by 9%, along with a greater increase in skin moisturization (by 34%) than those in the placebo group ($p < 0.001$). The food supplement did not significantly modify skin firmness or fatigue and had only slight beneficial effects on skin elasticity. The investigational product was well tolerated. The observed effects aligned closely with the subjective assessments reported by study participants. The study provides substantiated evidence supporting the efficacy of low-molecular-weight collagen peptides in restoring altered skin biometric parameters, as objectively assessed. Thus, regular supplementation with this nutraceutical may contribute to achieving smoother and more radiant skin.

Keywords: low-molecular-weight collagen; anti-aging; skin health; facial wrinkles; moisturization; nutraceutical; randomized controlled trial

1. Introduction

The skin, the body's largest organ, plays a crucial role in numerous functions. Beyond its sensory capacity, it acts as a dynamic interface between the internal and external environments, enabling the continuous adaptation and acclimatization of an organism throughout its lifespan [1,2].

Collagen, a protein primarily synthesized by fibroblasts in the connective tissues, constitutes the most abundant component of the extracellular matrix of the skin, representing over 75% of the dry weight of a young and healthy human dermis [3,4]. This protein is characterized by a triple-helix structure formed by the repetition of glycine every third residue, with proline and hydroxyproline occupying the remaining positions [5,6]. These amino acids are essential in the structure of collagen. Fibroblasts, as connective tissue cells within the dermis, play a pivotal role in the synthesis and organization of the collagen matrix. These cells exhibit sensitivity to both physical and chemical stimuli, which can trigger fibroblast activation and proliferation. Chemical stimuli operate via a key–lock mechanism, wherein small ligands bind to receptors located on the fibroblast extracellular membrane, leading to their activation. On the other hand, physical stimuli directly influence the interactions between collagen and fibroblasts [3,4].

Thus, collagen is a cornerstone in the skin's extracellular matrix, essential for maintaining structural integrity and physiological functions. It retains water and plays a pivotal role in maintaining the skin's smoothness, firmness, and resilience [4]. This enables the skin to effectively respond to the ever-changing onslaught of environmental stressors [7].

The aging process of the skin is an ongoing phenomenon, marked by a gradual decline in structural and physiological functions, often exacerbated by environmental factors and dermatological disorders [8,9].

Aging precipitates a decline in collagen synthesis within mature skin, attributed to diminished activity of enzymes involved in collagen post-translational processing, a reduction in the population of collagen-synthesizing fibroblasts, and a decrease in skin vasculature [10–12]. Consequently, the skin's biomatrix begins to deteriorate as the collagen scaffold loses its strength and stability [4,13,14].

Extrinsic factors such as sunlight exposure, smoking, pollution, alcohol consumption, an unbalanced diet, and stress-related micronutrient deficiencies expedite this process, contributing to the process of collagen loss associated with aging skin [4,14–16]. The consequences of aging on the skin are manifold. It undergoes regressive changes, losing its integrity and becoming progressively thin and dry, leading to a compromised ability to retain enough moisture. Additionally, the reduction in dermal thickness and elasticity over time manifests as lines and wrinkles, further contributing to the visible signs of aging [17].

The psychosocial impact of skin aging has spurred the demand for effective interventions, including topical creams, injectable fillers, and collagen supplements [2,18].

While topically applied collagen from skincare products such as creams, lotions, and serums often fails to penetrate the deeper layers of the skin [2,19], injectable fillers, such as hyaluronic-acid-based products, can be costly and may cause adverse effects, including bruising, swelling, and infection, among others [20].

On the other hand, orally administered hydrolyzed collagen has gained significant popularity in recent years as a safe and cost-effective option to enhance skin health and maintain a youthful appearance [4,21–24]. It is available in various formulations, including gels, liquids, capsules, and powder, making them easy to incorporate into daily routines.

The use of low-molecular-weight collagen hydrolysate has been shown to be a promising and effective strategy to improve skin hydration and elasticity, thereby counteracting the changes associated with aging [2,4,25]. However, a comprehensive understanding and further research are paramount to substantiate these claims and optimize collagen use in dermatological applications.

Therefore, COLLinstant® LMW is a novel cosmeceutical that distinguishes itself by containing a high proportion of low-molecular-weight ($\leq$1000 Da) glycine- and proline-rich bioactive peptides.

This product is the result of a tailored enzymatic digestion of native bovine collagen (type I and III) that preserves the collagen-specific sequence Gly-Pro-Hyp (GPH), yielding a low-molecular-weight (LMW) hydrolysate enriched with this active tripeptide, which appears to exert beneficial effects in various tissues, including skin, muscles, joints, and bones [26–28].

Unlike standard-weight collagen, which is degraded in the gastrointestinal tract, low-molecular-weight collagen (LMW) peptides, such as GPH and PH, easily cross the intestinal barrier via the PEPT1 transporter and remain intact throughout the gastrointestinal pathway [26,29,30]. These bioactive components are highly stable and facilitate rapid and efficient uptake, enhancing bioavailability and overall efficacy by providing essential building blocks and stimulating the synthesis of new collagen, elastin, and hyaluronic acid [27,28].

The objective of the present study is to assess the effectiveness of daily supplementation with COLLinstant® LMW over a period of six weeks in ameliorating visible signs of aging. This includes assessing its impact on skin wrinkle reduction, as well as its potential to enhance skin elasticity and moisturization. COLLinstant® LMW was administered orally in a single-center, randomized, double-blind, placebo-controlled clinical trial. A secondary objective involves comparing skin improvement, assessing product satisfaction, and monitoring adverse events among middle-aged female volunteers.

2. Materials and Methods

2.1. Study Design and Ethical Aspects

This was a 6-week, prospective, randomized, placebo-controlled, double-blind, monocentric study performed at GALA Laboratories in Don Benito–Villanueva (Badajoz, Spain). This study is listed on the ClinicalTrials.gov registry (NCT06321770).

Participants were individually randomized (1:1 ratio) to a strategy of receiving either COLLinstant® LMW (collagen group) or a placebo regimen and were followed up with for 6 weeks. Subjects were instructed to follow the administration guidelines provided in the manufacturer's package for both product regimens and, if necessary, investigator guidance.

The investigation was performed according to the ethical guidelines detailed in the Declaration of Helsinki (amendment of the 64th General Assembly, Fortaleza, Brazil, October 2013) and the national regulations of Spain, and in full compliance with the applicable principles of good clinical practice (GCP) and International Council for Harmonisation (ICH) of Technical Requirements for Pharmaceuticals for Human Use [31]. The trial protocol was approved (code 075/2022) by the Clinical Research Ethics Committee at the University Hospital of Cáceres (Cáceres, Spain) and written informed consent was obtained from all subjects prior to any study procedures being initiated.

2.2. Investigational Product

The preparation under study was COLLinstant® LMW (Viscofan DE GmbH, Weinheim, Germany), an oral food supplement based on bovine bioactive hydrolyzed (type I and III) collagen peptides.

Following ICH-GCP requirements and applicable local regulations [31], the investigational product and placebo were formulated as powder-containing sachets for oral suspension that were identical in appearance and odor.

Each sachet of the investigational product (COLLinstant® LMW) contained 2.5 g low-molecular-weight hydrolyzed collagen peptides. Detailed information of each ingredient is shown in Table 1. The placebo did not contain any nutrients.

Table 1. Ingredient composition of the investigational product and the placebo.

Supplement Facts	Sachet Type			
	Investigational Product (COLLinstant® LMW)		Placebo	
	Amount	%DV	Amount	%DV
Manufacturer	14 sachets per container		14 sachets per container	
Serving size	1 sachet (3.13 g)		1 sachet (3.13 g)	
Calories	11.30	0.6%	10.80	0.54%
Fat	0.004 g	0.005%	0	0%
Protein	2.7 g	5.4%	0	0%

Table 1. *Cont.*

Supplement Facts	Sachet Type			
	Investigational Product (COLLinstant® LMW)		Placebo	
	Amount	%DV	Amount	%DV
Sodium	37.40 mg	1.62%	37.40 mg	1.62%
Total carbohydrate	0.126 g	0.046%	2.7 g	0.98%
Low MW collagen peptides	2.5 g	+		
Other ingredients				
Lemon flavor	467 mg	+	467 mg	+
Anhydrous citric acid	150 mg	+	150 mg	+
Sucralose	8.5 mg	+	8.5 mg	+
Stevia (97%)	7.1 mg	+	7.1 mg	+
Maltodextrin			2.5 g	+

% Daily values (DV) are based on a 2000 Calorie diet, + daily values (DV) not established for the ingredient.

2.3. Study Subjects

We recruited a total of 80 women (aged 30–65 years) with phototypes I–IV (Fitzpatrick scale) [32] who were mentally and physically healthy, had a BMI 20.0–29.9 kg/m^2, and displayed visible signs of natural and photoaging on their face (crow's feet) rated from moderate to severe [33].

The Fitzpatrick scale is a numerical classification system for human skin color, ranging from type I to type VI, based on the amount of melanin in the skin. This classification informs the skin's susceptibility to burns and its ability to tan [32,34].

During the screening phase, participants met all inclusion and exclusion criteria. Subjects were excluded if they were pregnant or lactating, had acute or chronic skin diseases or dermatological disorders, used natural health supplements for skin improvement within one month prior to the study commencement, followed a low-protein diet, had planned or unavoidable UV radiation exposure, had tattoos on or near the test area, used systemic corticosteroids or applied topical alpha hydroxy acids near the test site within four weeks of enrollment, used topical medications near the test area within six weeks of enrollment, underwent Botulinum toxin A (Botox) treatment or filler injection (collagen, hyaluronic acid, etc.) near the test sites within two years of enrolment, were cognitively impaired and/or unable to provide informed consent, or had any other condition that, in the medical investigator's opinion, might adversely affect the individual's ability to complete the study or its measures or pose significant risk to the individual.

2.4. Study Schedule and Biometric Evaluation

All participants (test and placebo groups) were instructed to consume the content of one sachet daily, in the morning, on an empty stomach for 6 weeks. The product was required to be dissolved in at least 100 mL of water, juice or other liquid.

Participants agreed to refrain from prolonged exposure to ultraviolet (UV) radiation, consuming any similar dietary supplements, and using any skincare treatments such as face masks, packs, and massages. They were also not permitted to apply topical cosmetics except those provided during the study. This restriction was enforced for a 2-week wash-out period before the study began and continued throughout the 6-week study period to maintain consistent skin conditions. Each participant visited the research center twice for assessment: once prior to intake of the study formulation at baseline (T0), and again 6 weeks after intake of the study formulation (T6), for efficacy measurements and safety evaluations.

For all women participating in the study, skin parameters were assessed at baseline (T0), and biometric changes were also evaluated after 6 weeks of treatment with the products (T6).

Measurement of skin wrinkling parameters (volume, area, and depth) was evaluated at the crow's feet region and changes were analyzed and digitally photographed in all patients by a VisioFace® 1000D (equipped with a high-resolution reflex camera) [35].

According to Mödinger et al. [36], subjects underwent acclimatization for a minimum of 30 min in the air-conditioned measurement room set at a temperature of 21 ± 1 °C and a relative humidity of 50 ± 5%.

Skin elasticity at the crow's feet region was quantified using a Cutometer® dual MPA 580 (Courage + Khazaka), a non-invasive instrument designed to assess skin biomechanical properties [37]. This device evaluates skin elasticity by applying negative force that mechanically deforms the skin. The operational principle involves using a probe with negative pressure (450 mbar) to suction the skin, drawing the test area into the probe's aperture. A non-contact optical measurement system then determines the depth of skin penetration. The evaluated parameters (R0, R2, R5, R7, and R9) are key indicators used to characterize skin biomechanical properties following suction force application [38–44].

R0 represents the final distension of the initial curve, reflecting the passive response of the skin to suction force and correlating with skin firmness. It is calculated as the difference between the highest point of amplitude at the end of the suction phase and the baseline reading (R0 = Uf). The R2 parameter is related to the gross elasticity/viscoelasticity, representing the skin's resistance to mechanical suction force relative to its ability to recover (R2 = Ua/Uf). R5, referred to as net elasticity, signifies the ratio of elastic deformation during suction to rapid recovery during relaxation (R5 = Ur/Ue), indicating higher skin elasticity with increasing values. R7 is related to biological elasticity, quantifying the immediate elastic recovery within the first 0.1 s compared to the total deformation after suction (R7 = Ur/Uf) and can be interpreted as another marker of elasticity, with aging causing its reduction. Lastly, R9 denotes residual deformation at the conclusion of the measurement cycle, reflecting skin fatigue following repeated suction (R9 = R3 − R0) [37,42,43,45,46]. The measurements were carried out in triplicate.

Furthermore, stratum corneum hydration was assessed at each study visit using the electrical capacitance method with a Corneometer® CM 825 (Courage + Khazaka, Cologne, Germany). A minimum of five measurements were taken per measurement area at four distinct locations (middle forehead, both cheekbones, and the chin area), and the average value was used for subsequent analysis [36,44,46].

2.5. Safety and Self-Reported Measures

Safety was evaluated based on adverse reactions reported by the patients during the treatment period. Adverse events were documented at the final visit following six weeks of treatment.

At the end of the treatment, the volunteers filled out questionnaires to subjectively assess their perception of different parameters such as efficacy, organoleptic properties, tolerability, and satisfaction since the last time they took the product. The Spanish version of the Treatment Satisfaction Questionnaire with Medication (TSQM) [47] and a 3-point Likert scale with the items dissatisfied, slightly satisfied, and very satisfied were used.

2.6. Statistical Methods

Statistical analyses were performed according to the principles of ICH (International Council for Harmonization) guideline E9(R1) on "Statistical Principles for Clinical Trials" [48] using IBM® SPSS® Statistics for Windows (version 27.0) and Jeffreys's Amazing Statistics Program (JASP) (version 0.17.1) computer software. The analysis of the distribution and normality tests of the variables were carried out using the Shapiro–Wilk tests.

All measured data are presented as mean ± SD (standard deviation). Categorical variables were summarized using the number and percentages of patients in each category.

Skin parameters (wrinkling, elasticity, and hydration) were evaluated descriptively at baseline (T0) and after 6 weeks of collagen supplementation (T6). The efficacy was assessed based on relative changes in these parameters, calculated as the differences between means at T0 and T6.

Graphical representations of the results were generated using box-and-whisker plots. The box spans from the 25th to the 75th percentile of the data. Whiskers extend from the edges of the box to the minimum and maximum values that are not considered outliers. Outliers, defined as values more than 1.5 times the interquartile range from the box, are depicted as symbols beyond the whiskers.

The between-group comparison (placebo group vs. verum group) was carried out to determine population homogeneity at baseline (T0) and treatment-related differences after 6 weeks (at T6) were analyzed by means of the Mann–Whitney U test, a non-parametric test applied to two independent samples. Comparisons between categorical variables were performed using the Chi-square test.

The within-group mean changes of skin parameters between the initial (T0) and final (T6) visits were evaluated by the non-parametric Wilcoxon signed-rank test for paired data. The threshold of statistical significance was set, in all cases, at a value of $p < 0.05$.

3. Results

3.1. CONSORT (Consolidated Standards of Reporting Trials) Flowchart of the Controlled Interventional Trial

Eighty women aged 30 to 60 years were included in the statistical analysis. No subjects were excluded during screening or throughout the study, and no protocol violations occurred. Compliance during the trial was excellent; therefore, all volunteers ($n = 80$) who were screened for eligibility and requested to participate completed the study protocol, and could therefore be analyzed. The flow of subjects through the controlled intervention trial is depicted in a diagram according to CONSORT guidelines (Figure 1) [49].

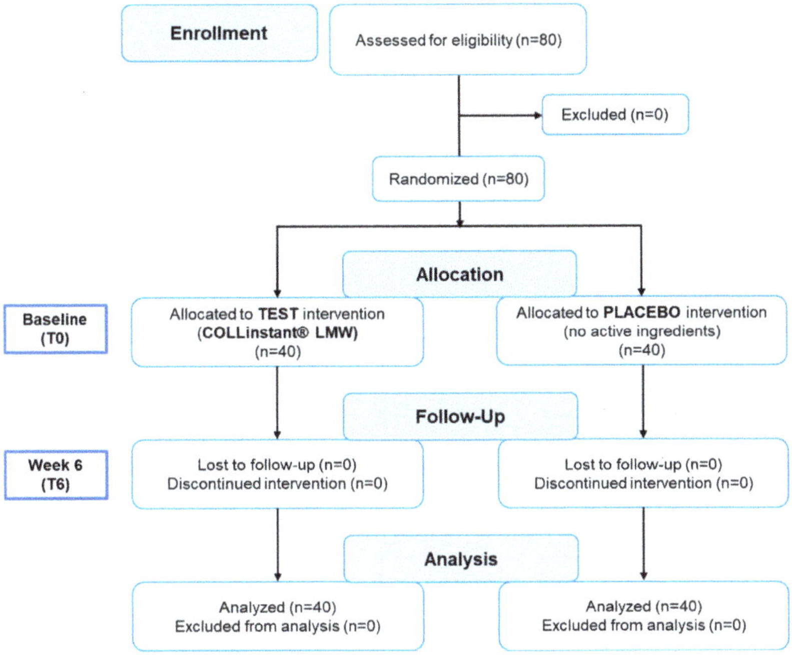

Figure 1. Flow chart of subjects' recruitment, randomization, and follow-up.

3.2. Characteristics of the Population

Participants (n = 80) were randomized at baseline (T0) and allocated to the verum group (n = 40) or the placebo group (n = 40). Demographic and general features of the volunteers did not show any significant difference between the test product and the placebo group at baseline (Table 2).

Table 2. Demographics and general characteristics of women randomly allocated to the placebo and the test group.

		Placebo (n = 40)	Collagen (n = 40)	p Value
Age, years		47 ± 7.7	45 ± 7.1	0.22
Height (m)		1.68 ± 0.06	1.69 ± 0.07	0.68
Weight (kg)		69.1 ± 9.13	68.5± 9.27	0.59
BMI (kg/m^2)		24.5 ± 2.7	24.1 ± 2.6	0.71
Skin type, n (%)	Normal	30 (75)	27 (67.5)	0.27
	Sensitive	8 (20)	12 (30)	
	Dry	0 (0)	1 (2.5)	
	Oiled	2 (5)	0 (0)	
Skin phototype, (Fitzpatrick [a]), n (%)	II	0 (0)	3 (7.5)	0.20
	III	36 (90)	34 (85)	
	IV	4 (10)	3 (7.5)	

Data are presented as means ± SD, except where otherwise indicated. [a] Fitzpatrick scale [32].

The investigational treatment group included 40 women who received the bioactive collagen peptide-based food supplement orally.

The mean age in the placebo group (n = 40) was 47 ± 7.7 years (age range between 32 and 60 years) and in the experimental group (n = 40) 45 ± 7.1 years (age range between 30 and 58 years), with no significant differences between the two groups (p = 0.22).

The predominant skin type in both groups (placebo and experimental) was "normal skin" (67.5% in the experimental group vs. 75% in the placebo group), followed by "sensitive skin" (30% in the experimental group vs. 20% in the placebo group), with no significant differences between the two groups (p = 0.27).

Among all participants, the most represented Fitzpatrick skin classification was phototype III (slightly brown skin and brown hair), which was predominant in both groups (90% in the placebo group; 85% in the verum group). No significant differences were detected between the two groups for this dermatological parameter (p = 0.20).

Homogeneity tests between groups revealed no significant differences for the mean initial skin parameters at baseline (T0) between the placebo group and the collagen group (Table 3).

Table 3. Skin biometric parameters at baseline (T0).

		Placebo (n = 40)	Collagen (n = 40)	
	Parameter	Mean (SD)	Mean (SD)	p Value *
Skin wrinkling	Volume (px^3)	48.50 (35.2)	58.77 (47.7)	0.52
	Area (px^2)	4.13 (2.6)	4.53 (3.1)	0.72
	Depth (px)	11.95 (2.7)	11.50 (2.4)	0.45
Elasticity	R0 (mm)	0.40 (0.09)	0.38 (0.10)	0.25
	R2 (%)	53.28 (15.7)	53.14 (11.6)	0.89
	R5 (%)	48.59 (17.4)	48.16 (13.9)	0.96
	R7 (%)	34.03 (14.5)	32.1 (10.2)	0.95
	R9 (mm)	0.07 (0.03)	0.07 (0.02)	0.27
Hydration (AU)		54.77 (9.4)	55.5 (8.6)	0.52

* p values for intergroup comparison were determined by Mann–Whitney U test.

3.3. Skin Wrinkling

The descriptive analysis of skin wrinkling biometric parameters (volume, area, and depth) of the facial crow's feet region before intake of the product (at T0) and after 6 weeks of intake (at T6) is summarized in Table 4. The mean value of three determinations was used for analysis. At baseline, the mean skin wrinkling biometric parameters were similar between both groups of treatment (Table 3). Regarding the intraindividual comparison from baseline (T0–T6), all biometric parameters (volume, area, and depth) considerably decreased at T6 after 6 weeks of treatment in the collagen group, whereas they remained unchanged in the placebo group during the same period (Table 4).

Table 4. Skin crow's feet parameters (volume, area, and depth) assessed by VisioFace® 1000D.

		Placebo (n = 40)		Collagen (n = 40)		
Parameter	**Time-Point**	**Mean (SD)**	***p* Value ***	**Mean (SD)**	***p* Value ***	**Test/Placebo *p* Value †**
Volume (px^3)	Baseline Week 6	48.50 (35.2)) 50.30 (40.4)	0.13	58.77 (47.7) 30.24 (24.4)	0.001	0.01
Area (px^2)	Baseline Week 6	4.13 (2.6) 4.14 (2.5)	0.92	4.53 (3.1) 2.60 (2.1)	0.001	0.001
Depth (px)	Baseline Week 6	11.95 (2.7) 12.25 (2.6)	0.23	11.50 (2.4) 10.35 (2.1)	0.001	0.001

* p value for within-group comparison (pre-treatment vs. post-treatment; T0–T6) (Wilcoxon signed-rank test).
† p value for intergroup comparison (experimental vs. placebo group) at T6, after 6 weeks of treatment (Mann–Whitney U test).

At baseline (T0), no significant differences were found for the mean skin wrinkle volume at the crow's feet region between the placebo group (48.50 px^3) and the collagen group (58.77 px^3) (Table 3). After 6 weeks of intake of the study products, at T6, the mean volume of wrinkles significantly decreased from baseline (T0–T6) by −45.9% [−90.8%–(−2.9%)] in the collagen group (58.77 vs. 30.24 px^3; $p < 0.001$) but remained relatively unchanged by 0.32% (−90.4%–20.9%) in the placebo group (48.50 vs. 50.30 px^3) (Table 4 and Figure 2). The difference between groups (placebo vs. collagen) at T6 proves to be significant ($p < 0.01$) in favor of the test product (Table 4 and Figure 2).

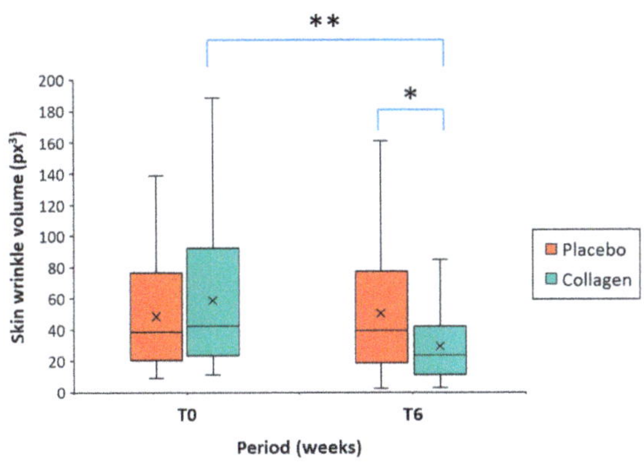

Figure 2. Box plot representing skin wrinkle volume at crow's feet region before (T0) and after intake of the product for 6 weeks (T6) in the group of women receiving a low-molecular-weight (LMW) collagen preparation or placebo (n = 40/group). The mean value (x) is also represented. The asterisks indicate statistically significant differences in the intergroup comparison (* $p < 0.01$; ** $p < 0.001$).

Starting at baseline from similar initial mean wrinkle areas of 4.13 px^2 (placebo group) and 4.53 px^2 (collagen group) (Table 3), after an intake of 6 weeks of either placebo or collagen, the area of skin wrinkles at the crow's feet region significantly decreased from baseline (T0–T6) by −43.8% [−94.4%–(−2.6%)] (4.53 vs. 2.60 px^2; $p < 0.001$) in the collagen group, but the intraindividual difference was almost negligible at 1.66% (−24.6%–41.7%) in the placebo group (4.13 vs. 4.14 px^2) (Table 4 and Figure 3). At T6, the intergroup comparison (placebo group vs. collagen group) of the mean wrinkle area proved to be significant ($p < 0.001$) in favor of the test product (Table 4 and Figure 3).

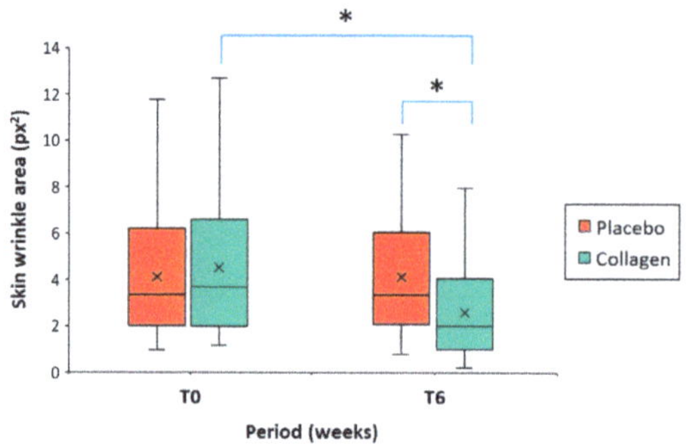

Figure 3. Box plot representing skin wrinkle area at crow's feet region before (T0) and after intake of the product for 6 weeks (T6) in the group of women receiving a low-molecular-weight (LMW) collagen preparation or placebo (n = 40/group). The mean value (x) is also represented. The asterisk (*) indicates statistically significant differences in the intergroup comparison (* $p < 0.001$).

Table 3 shows that mean wrinkle depth values were similar at baseline of 11.95 px (placebo) and 11.50 px (collagen group). However, in line with improvements in volume and area, skin wrinkle depth decreased during intake of the investigational product. As shown in Table 4 and Figure 4, at T6, the depth of wrinkles significantly decreased from baseline (T0–T6) by −9.0% (−40.0%–0.0%) (11.50 vs. 10.35 px; $p < 0.001$) in the collagen group. On the other hand, the intraindividual variation was non-significant and limited to 3.3% (−33.3%–40.0%) (11.95 vs. 12.25 px) in the placebo group. The differences between the groups at T6 (placebo group vs. collagen group) of the mean wrinkle depth proved also to be highly significant ($p < 0.001$) in favor of the test product (Table 4 and Figure 4).

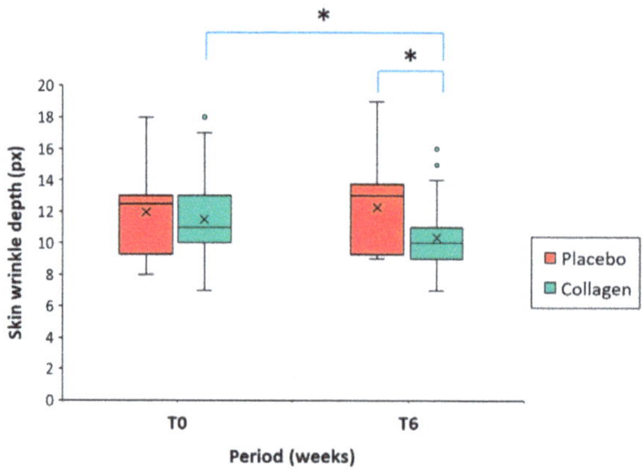

Figure 4. Box plot representing skin wrinkle depth at crow's feet region before (T0) and after intake of the product for a 6-week period (T6) in the group of women receiving a low-molecular-weight (LMW) collagen preparation or placebo (n = 40/group). The mean value (x) is also represented. The asterisk (*) indicates statistically significant differences in the intergroup comparison ($p < 0.001$). ° represent outliers.

Figure 5 illustrates photographs of three volunteers showcasing the improvement in the appearance of skin wrinkles in the crow's feet areas from the initial visit at T0, following the consumption of the investigational product COLLinstant® LMW for six weeks. The skin assessment was conducted using objective and validated methods (Visioface 1000D). In volunteer no. 13, there was a percentage change in volume, area, and depth of −59.18%, −51.91%, and −15.4%, respectively. For volunteer no. 60, the corresponding changes were −64.8%, −46.2%, and −40.0%, and for volunteer no. 70, the changes were −37.1%, −31.4%, and −7.2%, respectively.

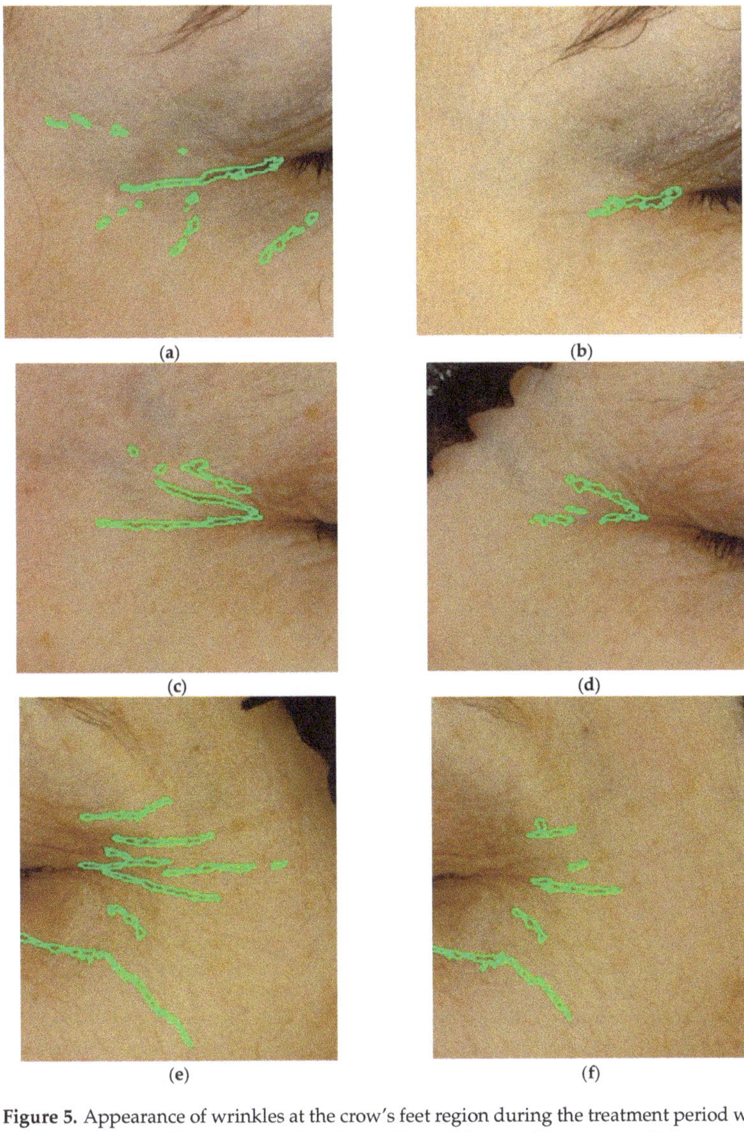

Figure 5. Appearance of wrinkles at the crow's feet region during the treatment period with the investigational product. Photographs of the frown lines in three volunteers, no. 13 (top), no. 60 (middle), and no. 70 (bottom), at baseline (images (**a**,**c**,**e**) and at T6 after 6 weeks of oral supplementation with the test product (images (**b**,**d**,**f**). The green lines provide a visual guide to researchers to precisely observe and analyze skin features. Data obtained from Visioface® 1000D.

3.4. Skin Elasticity

The descriptive analysis of skin elasticity obtained from the Cutometer® device before intake of the product (at T0) and after 6 weeks of intake (at T6) is summarized in Table 5. The mean value of three determinations at the crow's feet region was used for analysis. At baseline (T0), the skin elasticity parameters were similar between both groups of treatment (Table 3).

Table 5. Mechanical characteristics of the skin through the analysis of elasticity parameters (mean ± SD) at the crow's feet region, before intake of the study products at baseline (T0) and after 6 weeks of treatment (T6) as assessed by Cutometer®.

		Placebo (n = 40)		Test Group (n = 40)		
Elasticity Parameter	**Time-Point**	**Mean (SD)**	***p* Value ***	**Mean (SD)**	***p* Value ***	**Test/Placebo *p* Value †**
R0 (mm)	Baseline	0.402 (0.09)		0.380 (0.09)		
	Week 6	0.362 (0.09)	0.001	0.329 (0.07)	0.001	0.07
R2 (%)	Baseline	53.28 (15.7)		53.14 (11.66)		
	Week 6	51.31 (18.1)	0.31	51.86 (13.7)	0.68	0.53
R5 (%)	Baseline	48.59 (17.5)		48.16 (13.9)		
	Week 6	45.84 (17.6)	0.15	48.54 (15.5)	0.95	0.39
R7 (%)	Baseline	34.03 (14.5)		32.09 (10.2)		
	Week 6	31.65 (13.7)	0.09	32.27 (13.2)	0.65	0.63
R9 (mm)	Baseline	0.073 (0.03)		0.070 (0.02)		
	Week 6	0.066 (0.02)	0.14	0.065 (0.02)	0.13	0.41

* p value for within-group comparison (pre-treatment vs. post-treatment; T0–T6). † p value for intergroup comparison (experimental vs. placebo group) after 6 weeks of treatment at T6.

Results showed that skin firmness (R0) significantly increased in both groups of treatment. Compared to baseline (T0), the assessment of total elongation and skin firmness (R0) showed statistically significant lower values ($p < 0.001$) after 6 weeks in both the treatment group receiving placebo and the collagen group (Table 5). At the end of the study, at T6, we could not demonstrate significant differences in the R0 parameter between both group means (Table 5).

The percentage of variation in mean skin firmness (R0) from baseline was −8.6% (−50.8–24.1%) in the placebo and −10.7% (−61.3–29.0%) in the group that received the food supplement.

The assessment of the remaining skin elasticity parameters (R2, R5, R7, and R9) showed a moderate improvement, but the differences were not statistically significant (Table 5).

At T6, the assessment of the rest of skin elasticity parameters (R2, R5, R7, and R9) obtained with the Cutometer® device did not show any statistical significance between groups (Table 5), indicating that the food supplement did not show an improvement in most skin elasticity parameters in our study.

3.5. Skin Hydration

The descriptive analysis of skin hydration before intake of the product (at T0) and after 6 weeks of intake (at T6) is summarized in Table 6. The mean value of five determinations at four different locations (middle forehead, right and left cheek, and chin) was used for analysis. At baseline (T0), the skin hydration values were similar between both groups of treatment, 54.77 AU (placebo group) vs. 55.50 AU (collagen group) (Table 3).

Table 6. Skin hydration values (mean ± SD) before intake of the study products at baseline (T0) and after 6 weeks of treatment (T6).

		Placebo (n = 40)		Collagen (n = 40)		
	Time-Point	**Mean (SD)**	***p* Value ***	**Mean (SD)**	***p* Value ***	**Test/Placebo *p* Value †**
Skin hydration (AU)	Baseline	54.77 (9.4)		55.50 (8.6)		
	Week 6	55.03 (9.5)	0.79	74.13 (11.9)	0.001	0.001

* p value for within-group comparison (pre-treatment vs. post-treatment; T0–T6). † p value for intergroup comparisons (experimental vs. placebo group) at T6, after 6 weeks of treatment. AU indicates arbitrary units.

A corneometric methodology corroborated the improvement in skin hydration in the volunteers who received the test product. Compared to baseline (T0), a statistically significant increase in skin hydration by +34.4% (9.2–98.0%) (55.50 AU vs. 74.13 AU, $p < 0.001$) was detected in the volunteers who received the investigational product for 6 weeks (T0–T6) (Table 6). On the other hand, the percentage change in skin hydration was non-significant and limited to +1.0% (−30.3–17.6%) (54.77 vs. 55.03 AU) in the placebo group (Table 6 and Figure 6). The differences between the groups at T6 (placebo group vs. collagen group) of the mean hydration values proved to be highly significant ($p < 0.001$) in favor of the test product (Table 6 and Figure 6).

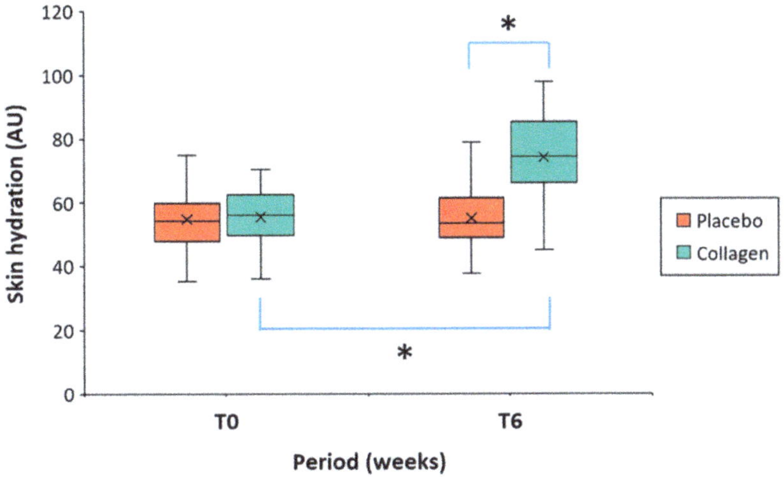

Figure 6. Boxplot representing skin hydration before (T0) and after intake of the product for a 6-week period (T6) in the group of women receiving a low-molecular-weight (LMW) collagen preparation or placebo (n = 40/group). The mean value (x) is also represented. The asterisk (*) indicates statistically significant differences in the intergroup comparison ($p < 0.001$).

3.6. Overall Assessment of the Efficacy

Compared with placebo, the efficacy of the test product is summarized in Figure 7. The differences between the relative changes (T0–T6) in skin wrinkle parameters (volume, area, and depth), skin elasticity, and skin hydration are illustrated. There was a statistically significant improvement ($p < 0.001$) in the skin wrinkle parameters and skin hydration in the verum group. Regarding skin elasticity, the food supplement had slight beneficial effects on skin elasticity parameters but we could not demonstrate statistical significance.

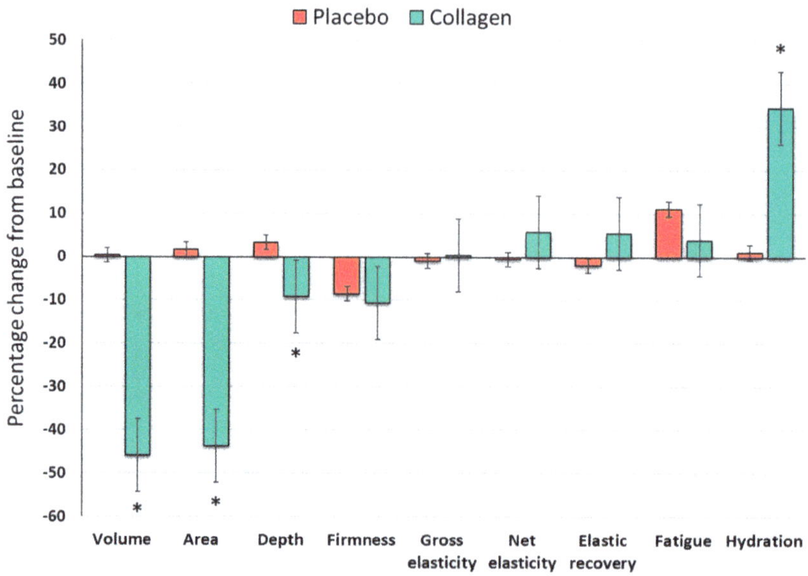

Figure 7. Percentage change in biometric skin parameters between the baseline visit and the end of the 6 weeks of the interventional period (T0–T6) in the placebo group (red bars) and in the group taking the investigational product (green bars). Error bars indicate the standard error of the mean. The asterisk (*) indicates $p < 0.001$ for intergroup comparison (placebo group vs. collagen group).

3.7. Safety and Subjective Rating

During the study intervention, the collagen supplement did not cause any side effects and proved to be safe and well tolerated during the entire period of application.

Over two thirds of the volunteers rated the overall effectiveness of COLLinstant® LMW as good, a survey result that was matched by the ratings of the treating physicians. Over 90% of the volunteers in the experimental group indicated a high degree of satisfaction with the ability of the supplement to improve skin hydration or wrinkles, confirming the robust effect of the treatment. In addition, 87.5% of the volunteers in this group were highly satisfied with the brief time period it takes for the supplement to start showing its effects, underlining the significance of obtaining quick results with a collagen food supplement (Table 7).

Table 7. Results of the survey conducted by the principal investigators with the volunteers who received the food supplement.

	Very Satisfied	Slightly Satisfied	Dissatisfied
Effectiveness (%)	76.2	21.3	2.5
Tolerability (%)	88.8	10.0	1.25
Acceptability of the product (%)	88.8	10.0	1.25

4. Discussion

The current body of evidence indicates that the administration of hydrolyzed collagen can significantly improve skin conditions and mitigate signs of aging, thereby establishing it as a popular and promising nutraceutical for skin rejuvenation and anti-aging interventions [4,50,51].

Upon oral intake of collagen hydrolysates, small bioactive peptides, such as the tripeptide GPH, are readily absorbed intact across the intestinal barrier [3]. These peptides exhibit high bioavailability and are detectable in human blood shortly after consumption [52,53]. Subsequently, they are distributed to various tissues, with notable accumulation in the skin, where they remain at elevated levels and persist longer (up to 14 days) compared to other tissue [30]. Within the skin, GPH is further hydrolyzed into the bioactive dipeptide PH [27].

In light of these considerations, COLLinstant® LMW, a novel cosmeceutical containing a unique composition of bioactive glycine- and proline-rich peptides, was selected for the clinical trial due to its potential to enhance the product's efficacy (Table 1). The study specifically aimed to assess the efficacy, safety, and tolerability of 2.5 g COLLinstant® LMW over a 6-week period by evaluating various skin parameters.

Our observations suggest that the low-molecular-weight peptides in the investigational product are efficiently absorbed and biologically active, leading to a significant reduction in several skin wrinkle biomarkers such as volume, area, and depth in the crow's feet region. Additionally, compared to the placebo, increased skin hydration and a moderate improvement in skin elasticity were observed. However, we were unable to demonstrate differences in skin elasticity parameters between the control group and the investigational group. The effects were exclusively attributed to COLLinstant® LMW, as the subjects did not receive any other form of cosmetic treatment.

The importance of dosage and duration of the treatment appear as crucial factors to be considered. In contrast to the higher dosages and longer durations recommended for standard-molecular-weight collagen hydrolysates to achieve clear beauty or health benefits [25,54] notably, a daily intake of 2.5 g COLLinstant® LMW for 6 weeks was sufficient to yield beneficial effects on some biometric parameters of skin health in this study.

Regarding skin elasticity, our findings indicated a positive trend (as measured by the Cutometer) in the group treated with the investigational product. However, the observed differences did not reach statistical significance when compared with the placebo group.

Kim et al. [55] reported significant improvement in skin hydration from baseline as early as 6 weeks after intake of 1 g of low-molecular-weight collagen peptides in the verum group. However, benefits in skin wrinkling and elasticity were observed later, after 12 weeks of treatment. Similarly, in another study, the intake of 1.65 g of low-molecular-weight bioactive peptides led to a significant increase in skin hydration and a reduction in skin desquamation after 4 weeks of administration, as well as reduced wrinkling after 12 weeks and increased elasticity after 8 weeks [56].

In light of previous studies on orally based collagen intake from various sources, significant improvements in skin hydration, elasticity, and wrinkling were generally observed following 12 weeks of collagen supplementation [2,57,58]. Thus, to achieve efficacy in enhancing skin elasticity through hydrolyzed collagen supplementation, a treatment duration exceeding 8 weeks appears to be necessary [51]. However, beneficial effects on skin hydration might be obtained earlier.

COLLinstant® LMW is orally administered, making it easy to incorporate into daily routines. Nevertheless, not all sources of hydrolyzed collagen are equally effective, and further studies are needed to determine the optimal source and therapeutic duration for combating skin aging. Different types of collagen used in these supplements, such as fish, porcine, chicken, and bovine collagen, may exhibit varying effects based on their source [2,4,51,59].

Within the scope of skincare and nutraceuticals, hydrolyzed bovine collagen has become a popular ingredient. COLLinstant® LMW is the first bovine collagen hydrolysate on the market, providing a beneficial amino acid profile that acts as both a building material and a stimulator for the synthesis of new collagen, elastin, and hyaluronic acid in the skin as corroborated by various studies [51]. The remarkable improvements in skin parameters observed in this investigation may be attributed to the substantial similarity between the collagen peptides derived from the bovine collagen complex and those naturally present in human collagen. Specifically, the hydrolysis of bovine collagen produces bioactive short-chain peptides that closely match the amino acid profile of human collagen type I, as well as elastin and hyaluronic acid in the skin [4,50,60–63]. This chemical resemblance likely contributes to the efficacy of the supplementation in improving skin health and appearance.

During the aging process, the dysregulation of extracellular matrix turnover, particularly the degradation of collagen fibers by matrix metalloproteinases (MMPs) and other proteases, is often a pivotal molecular event [64]. Consequently, the skin undergoes structural and degenerative changes such as dehydration and loss of elasticity, resulting in dry, loose skin with the appearance of furrows or wrinkles [64–66]. According to Oesser et al. [67], the skin health benefits observed in our study are likely attributable to changes in protein turnover and the restoration of collagen synthesis within the dermal stratum of the skin. This was notably evidenced by decreases in the volume, area, and depth of skin wrinkles following oral supplementation with low-molecular-weight collagen peptides.

The efficacy of bioactive collagen di- and tripeptides in enhancing skin health appears to be attributed to several key mechanisms. First, these compounds have the ability to induce collagen expression through the mitogen-activated protein kinase 38 (p38 MAPK) pathway [68]. Second, they increase the availability of essential free amino acids, which promotes the synthesis of collagen and elastin fibers. Third, they have the potential to stimulate fibroblasts to produce collagen and hyaluronic acid [3,4,69–72].

The beneficial effects of collagen supplementation were validated not only through objective testing methods but also by the subjective assessments provided by the volunteers. Additionally, consistent with findings from other studies [55,57], the tested product was demonstrated to be safe, with no adverse reactions reported.

5. Conclusions

Exploring its potential as a natural intervention to uphold skin health and combating signs of aging, low-molecular-weight collagen peptides (COLLinstant® LMW) were investigated in a randomized, placebo-controlled clinical trial, following a daily oral supplementation of 2.5 g of these peptides.

LMW collagen hydrolysate, which contains a high concentration of these characteristic peptides, offers several advantages over standard-molecular-weight products, including quicker and more efficient peptide uptake, higher bioavailability, and enhanced stability and efficacy, particularly in the skin.

The study confirmed the efficacy of these nutrients in restoring altered skin biometric parameters, as objectively assessed. We observed a significant reduction in wrinkles, a considerable increase in skin hydration, and a modest increase in skin elasticity. Regular supplementation may contribute to achieving smoother and more radiant skin.

Ongoing supplementation has the potential to enhance skin texture by providing essential building materials in this tissue, stimulating the synthesis of new collagen, elastin, and hyaluronic acid. Importantly, the collagen supplementation regimen was found to be devoid of any adverse effects, proving to be safe and well-tolerated throughout the entire administration period.

Author Contributions: Conceptualization, J.A.C.-N., L.Q., B.B. and R.G.-B.; methodology, J.A.C.-N., B.G.-M., L.Q., B.B. and R.G.-B.; validation, J.A.C.-N., B.G.-M. and R.G.-B.; formal analysis, J.A.C.-N.; investigation, J.A.C.-N. and B.G.-M.; resources, B.G.-M., L.Q. and B.B.; data curation, J.A.C.-N.; writing—original draft preparation, J.A.C.-N. and R.G.-B.; supervision, J.A.C.-N., B.G.-M. and B.B.; project administration, J.A.C.-N.; funding acquisition, L.Q. and B.B. All authors have read and agreed to the published version of the manuscript.

Funding: This research was funded by Nutraresearch S.L. (Barcelona, Spain), the company that manufactures the active collagen peptides, and Viscofan BioEngineering (Viscofan DE GmbH, Weinheim, Germany), the company owning IP rights of COLLinstant® LMW, the nutraceutical product used in this study. The APC was funded by Nutraresearch S.L. (Barcelona, Spain).

Institutional Review Board Statement: The study was conducted in accordance with the Declaration of Helsinki and approved by the Clinical Research Ethics Committee at the Cáceres University Hospital, Cáceres, Spain (no. 075/2022).

Informed Consent Statement: Written informed consent has been obtained from the patients to publish this paper. Informed consent was obtained from all subjects involved in the study. Written informed consent has been obtained from the patients to publish this paper.

Data Availability Statement: The original contributions presented in the study are included in the article, further inquiries can be directed to the corresponding author.

Conflicts of Interest: L.Q. and B.B. are full-time employees of Viscofan BioEngineering (Viscofan DE GmbH, Weinheim, Germany). The remaining authors (J.A.C.-N., B.G.-M., and R.G.-B.) declare no conflicts of interest. The sponsors had no influence in the design of the study; in the collection, analyses, or interpretation of data; in the writing of the manuscript; or in the decision to publish the results.

References

1. Pereira, E.S.P.; Barros Langen, S.; Fidelis, M.C.; Pereira, M.O.; Costa, A. Skin and Menopause. In *Skin, Mucosa and Menopause*; Farage, M.A., Miller, K.W., Woods, N.F., Maibach, H.I., Eds.; Springer: Berlin/Heidelberg, Germany, 2015; pp. 15–23, ISBN 978-3-662-44079-7.
2. Evans, M.; Lewis, E.D.; Zakaria, N.; Pelipyagina, T.; Guthrie, N. A Randomized, Triple-Blind, Placebo-Controlled, Parallel Study to Evaluate the Efficacy of a Freshwater Marine Collagen on Skin Wrinkles and Elasticity. *J. Cosmet. Dermatol.* **2021**, *20*, 825–834. [CrossRef] [PubMed]
3. Sibilla, S.; Godfrey, M.; Brewer, S.; Budh-Raja, A.; Genovese, L. An Overview of the Beneficial Effects of Hydrolysed Collagen as a Nutraceutical on Skin Properties: Scientific Background and Clinical Studies. *Open Nutraceuticals J.* **2015**, *8*, 29–42. [CrossRef]
4. Bolke, L.; Schlippe, G.; Gerß, J.; Voss, W. A Collagen Supplement Improves Skin Hydration, Elasticity, Roughness, and Density: Results of a Randomized, Placebo-Controlled, Blind Study. *Nutrients* **2019**, *11*, 2494. [CrossRef] [PubMed]
5. Shoulders, M.D.; Raines, R.T. Collagen Structure and Stability. *Annu. Rev. Biochem.* **2009**, *78*, 929–958. [CrossRef] [PubMed]
6. Li, P.; Wu, G. Roles of Dietary Glycine, Proline, and Hydroxyproline in Collagen Synthesis and Animal Growth. *Amino Acids* **2018**, *50*, 29–38. [CrossRef] [PubMed]
7. Quan, T.; Fisher, G.J. Role of Age-Associated Alterations of the Dermal Extracellular Matrix Microenvironment in Human Skin Aging: A Mini-Review. *Gerontology* **2015**, *61*, 427–434. [CrossRef] [PubMed]
8. Tobin, D.J. Introduction to Skin Aging. *J. Tissue Viability* **2017**, *26*, 37–46. [CrossRef] [PubMed]
9. He, X.; Gao, X.; Guo, Y.; Xie, W. Research Progress on Bioactive Factors against Skin Aging. *Int. J. Mol. Sci.* **2024**, *25*, 3797. [CrossRef] [PubMed]
10. Calleja-Agius, J.; Muscat-Baron, Y.; Brincat, M.P. Skin Ageing. *Menopause Int.* **2007**, *13*, 60–64. [CrossRef] [PubMed]
11. Castelo-Branco, C.; Duran, M.; González-Merlo, J. Skin Collagen Changes Related to Age and Hormone Replacement Therapy. *Maturitas* **1992**, *15*, 113–119. [CrossRef] [PubMed]
12. Frantz, C.; Stewart, K.M.; Weaver, V.M. The Extracellular Matrix at a Glance. *J. Cell Sci.* **2010**, *123*, 4195–4200. [CrossRef]
13. Sato, K. The Presence of Food-Derived Collagen Peptides in Human Body-Structure and Biological Activity. *Food Funct.* **2017**, *8*, 4325–4330. [CrossRef]
14. Krutmann, J.; Bouloc, A.; Sore, G.; Bernard, B.A.; Passeron, T. The Skin Aging Exposome. *J. Dermatol. Sci.* **2017**, *85*, 152–161. [CrossRef]
15. Lee, D.E.; Huh, C.S.; Ra, J.; Choi, I.D.; Jeong, J.W.; Kim, S.H.; Ryu, J.H.; Seo, Y.K.; Koh, J.S.; Lee, J.H.; et al. Clinical Evidence of Effects of Lactobacillus Plantarum HY7714 on Skin Aging: A Randomized, Double Blind, Placebo-Controlled Study. *J. Microbiol. Biotechnol.* **2015**, *25*, 2160–2168. [CrossRef] [PubMed]
16. Nistico, S.P.; Silvestri, M.; Zingoni, T.; Tamburi, F.; Bennardo, L.; Cannarozzo, G. Combination of Fractional CO_2 Laser and Rhodamine-Intense Pulsed Light in Facial Rejuvenation: A Randomized Controlled Trial. *Photobiomodul. Photomed. Laser Surg.* **2021**, *39*, 113–117. [CrossRef]
17. Varani, J.; Dame, M.K.; Rittie, L.; Fligiel, S.E.G.; Kang, S.; Fisher, G.J.; Voorhees, J.J. Decreased Collagen Production in Chronologically Aged Skin: Roles of Age-Dependent Alteration in Fibroblast Function and Defective Mechanical Stimulation. *Am. J. Pathol.* **2006**, *168*, 1861–1868. [CrossRef]
18. Honigman, R.; Castle, D.J. Aging and Cosmetic Enhancement. *Clin. Interv. Aging* **2006**, *1*, 115–119. [CrossRef] [PubMed]
19. Lee, Y.I.; Lee, S.G.; Jung, I.; Suk, J.; Lee, M.H.; Kim, D.U.; Lee, J.H. Effect of a Topical Collagen Tripeptide on Antiaging and Inhibition of Glycation of the Skin: A Pilot Study. *Int. J. Mol. Sci.* **2022**, *23*, 1101. [CrossRef]
20. Cockerham, K.; Hsu, V.J. Collagen-Based Dermal Fillers: Past, Present, Future. *Facial Plast. Surg.* **2009**, *25*, 106–113. [CrossRef] [PubMed]
21. Lordan, R. Dietary Supplements and Nutraceuticals Market Growth during the Coronavirus Pandemic—Implications for Consumers and Regulatory Oversight. *PharmaNutrition* **2021**, *18*, 100282. [CrossRef] [PubMed]
22. Martinez, M.J.; Dixit, D.; White, M.W.; Rieder, E.A. Motivations for Seeking Cosmetic Enhancing Procedures of the Face: A Systematic Review. *Dermatol. Surg.* **2023**, *49*, 278–282. [CrossRef] [PubMed]

23. Pogačnik, T.; Žmitek, J.; Hristov, H.; Keršmanc, P.; Butina, M.R.; Žmitek, K. The Effect of a 12-Week Dietary Intake of Food Supplements Containing Collagen and MSM on Dermis Density and Other Skin Parameters: A Double-Blind, Placebo-Controlled, Randomised Four-Way Study Comparing the Efficacy of Three Test Products. *J. Funct. Foods* **2023**, *110*, 105838. [CrossRef]
24. Lupu, M.; Gradisteanu Pircalabioru, G.; Chifiriuc, M.; Albulescu, R.; Tanase, C. Beneficial Effects of Food Supplements Based on Hydrolyzed Collagen for Skin Care (Review). *Exp. Ther. Med.* **2019**, *20*, 12–17. [CrossRef] [PubMed]
25. de Miranda, R.B.; Weimer, P.; Rossi, R.C. Effects of Hydrolyzed Collagen Supplementation on Skin Aging: A Systematic Review and Meta-Analysis. *Int. J. Dermatol.* **2021**, *60*, 1449–1461. [CrossRef] [PubMed]
26. Sontakke, S.B.; Jung, J.H.; Piao, Z.; Chung, H.J. Orally Available Collagen Tripeptide: Enzymatic Stability, Intestinal Permeability, and Absorption of Gly-Pro-Hyp and Pro-Hyp. *J. Agric. Food Chem.* **2016**, *64*, 7127–7133. [CrossRef] [PubMed]
27. Yazaki, M.; Ito, Y.; Yamada, M.; Goulas, S.; Teramoto, S.; aki Nakaya, M.; Ohno, S.; Yamaguchi, K. Oral Ingestion of Collagen Hydrolysate Leads to the Transportation of Highly Concentrated Gly-Pro-Hyp and Its Hydrolyzed Form of Pro-Hyp into the Bloodstream and Skin. *J. Agric. Food Chem.* **2017**, *65*, 2315–2322. [CrossRef] [PubMed]
28. Choi, S.Y.; Ko, E.J.; Lee, Y.H.; Kim, B.G.; Shin, H.J.; Seo, D.B.; Lee, S.J.; Kim, B.J.; Kim, M.N. Effects of Collagen Tripeptide Supplement on Skin Properties: A Prospective, Randomized, Controlled Study. *J. Cosmet. Laser Ther.* **2014**, *16*, 132–137. [CrossRef] [PubMed]
29. Yamamoto, S.; Hayasaka, F.; Deguchi, K.; Okudera, T.; Furusawa, T.; Sakai, Y. Absorption and Plasma Kinetics of Collagen Tripeptide after Peroral or Intraperitoneal Administration in Rats. *Biosci. Biotechnol. Biochem.* **2015**, *79*, 2026–2033. [CrossRef] [PubMed]
30. Mari, W.K.; Muneshige, S.; Shin, K.; Yasuki, T.; Hideyuki, S.; Fumiki, M.; Hitoshi, S.; Yuji, F.; Michio, K. Absorption and Effectiveness of Orally Administered Low Molecular Weight Collagen Hydrolysate in Rats. *J. Agric. Food Chem.* **2010**, *58*, 835–841. [CrossRef]
31. International Council for Harmonisation of Technical Requirements for Pharmaceuticals for Human Use. *ICH Harmonised Guideline, Integrated addendum to ICH E6(R1): Guideline for Good Clinical Practice E6(R2)*; European Medicines Agency: London, UK, 2016.
32. Gupta, V.; Sharma, V.K. Skin Typing: Fitzpatrick Grading and Others. *Clin. Dermatol.* **2019**, *37*, 430–436. [CrossRef]
33. Carruthers, A.; Carruthers, J.; Hardas, B.; Kaur, M.; Goertelmeyer, R.; Jones, D.; Rzany, B.; Cohen, J.; Kerscher, M.; Flynn, T.C.; et al. A Validated Grading Scale for Crow's Feet. *Dermatol. Surg.* **2008**, *34*, S173–S178. [CrossRef] [PubMed]
34. Czajka, A.; Kania, E.M.; Genovese, L.; Corbo, A.; Merone, G.; Luci, C.; Sibilla, S. Daily Oral Supplementation with Collagen Peptides Combined with Vitamins and Other Bioactive Compounds Improves Skin Elasticity and Has a Beneficial Effect on Joint and General Wellbeing. *Nutr. Res.* **2018**, *57*, 97–108. [CrossRef] [PubMed]
35. Bazargan, A.S.; Shemshadi, M.; Ziaeifar, E.; Taheri, A.; Roohaninasab, M.; Goodarzi, A.; Mirhashemi, M. Evaluation of Effectiveness of Tranexamic Acid as Mesotherapy in Improvement of Periorbital Wrinkling in a Trial Study. *J. Cosmet. Dermatol.* **2023**, *22*, 2548–2552. [CrossRef] [PubMed]
36. Mödinger, Y.; Schön, C.; Vogel, K.; Brandt, M.; Bielfeldt, S.; Wilhelm, K.P. Evaluation of a Food Supplement with Collagen Hydrolysate and Micronutrients on Skin Appearance and Beauty Effects: A Randomized, Double-Blind, PlaceboControlled Clinical Study with Healthy Subjects. *J. Clin. Cosmet. Dermatol.* **2021**, *5*, 1–5. [CrossRef]
37. Ryu, H.S.; Joo, Y.H.; Kim, S.O.; Park, K.C.; Youn, S.W. Influence of Age and Regional Differences on Skin Elasticity as Measured by the Cutometer®. *Ski. Res. Technol.* **2008**, *14*, 354–358. [CrossRef] [PubMed]
38. Stroumza, N.; Bosc, R.; Hersant, B.; Hermeziu, O.; Meningaud, J.P. Benefits of Using the Cutometer to Evaluate the Effectiveness of Skin Treatments in Plastic and Maxillofacial Surgery. *Rev. Stomatol. Chir. Maxillofac. Chir. Orale.* **2015**, *116*, 77–81. [PubMed]
39. Ohshima, H.; Kinoshita, S.; Oyobikawa, M.; Futagawa, M.; Takiwaki, H.; Ishiko, A.; Kanto, H. Use of Cutometer Area Parameters in Evaluating Age-Related Changes in the Skin Elasticity of the Cheek. *Ski. Res. Technol.* **2013**, *19*, e238–e242. [CrossRef] [PubMed]
40. Granger, C.; Aladren, S.; Delgado, J.; Garre, A.; Trullas, C.; Gilaberte, Y. Prospective Evaluation of the Efficacy of a Food Supplement in Increasing Photoprotection and Improving Selective Markers Related to Skin Photo-Ageing. *Dermatol. Ther.* **2020**, *10*, 163–178. [CrossRef] [PubMed]
41. Saito, M.; Tanaka, M.; Misawa, E.; Yao, R.; Nabeshima, K.; Yamauchi, K.; Abe, F.; Yamamoto, Y.; Furukawa, F. Oral Administration of Aloe Vera Gel Powder Prevents Uvb-Induced Decrease in Skin Elasticity via Suppression of Overexpression of Mmps in Hairless Mice. *Biosci. Biotechnol. Biochem.* **2016**, *80*, 1416–1424. [CrossRef] [PubMed]
42. Salomão Calixto, L.; Picard, C.; Savary, G.; Campos, P.M.B.G.M. Skin Characterization and Immediate Effects of Different Dermocosmetic Treatments in French and Brazilian Skin. *J. Cosmet. Dermatol.* **2020**, *19*, 466–472. [CrossRef] [PubMed]
43. Abbas, D.B.; Lavin, C.V.; Fahy, E.J.; Griffin, M.; Guardino, N.; King, M.; Chen, K.; Lorenz, P.H.; Gurtner, G.C.; Longaker, M.T.; et al. Standardizing Dimensionless Cutometer Parameters to Determine In Vivo Elasticity of Human Skin. *Adv. Wound Care* **2022**, *11*, 297–310. [CrossRef] [PubMed]
44. Bianchi, F.M.; Angelinetta, C.; Rizzi, G.; Praticò, A.; Villa, R. Evaluation of the Efficacy of a Hydrolyzed Collagen Supplement for Improving Skin Moisturization, Smoothness, and Wrinkles. *J. Clin. Aesthetic Dermatol.* **2022**, *15*, 48–52.
45. Woo, M.S.; Moon, K.J.; Jung, H.Y.; Park, S.R.; Moon, T.K.; Kim, N.S.; Lee, B.C. Comparison of Skin Elasticity Test Results from the Ballistometer® and Cutometer®. *Ski. Res. Technol.* **2014**, *20*, 422–428. [CrossRef] [PubMed]

46. Agache, P. Measurements of the Human Skin: Why and How? In *Agache's Measuring the Skin: Non-invasive Investigations, Physiology, Normal Constants*; Humbert, P., Fanian, F., Maibach, H.I., Agache, P., Eds.; Springer: Cham, Switzerland, 2017; pp. 5–14. ISBN 978-3-319-32381-7.
47. Atkinson, M.J.; Sinha, A.; Hass, S.L.; Colman, S.S.; Kumar, R.N.; Brod, M.; Rowland, C.R. Validation of a General Measure of Treatment Satisfaction, the Treatment Satisfaction Questionnaire for Medication (TSQM), Using a National Panel Study of Chronic Disease. *Health Qual. Life Outcomes* **2004**, *2*, 12. [CrossRef] [PubMed]
48. International Council for Harmonisation of Technical Requirements for Pharmaceuticals for Human Use. *ICH Harmonised Guideline, Addendum on Estimands and Sensitivity Analysis in Clinical Trials to the Guideline on Statistical Principles for Clinical Trials E9(R1)*; European Medicines Agency: Amsterdam, The Netherlands, 2019.
49. Schulz, K.F.; Altman, D.G.; Moher, D. CONSORT 2010 Statement. *Obstet. Gynecol.* **2010**, *115*, 1063–1070. [CrossRef] [PubMed]
50. Choi, F.D.; Sung, C.T.; Juhasz, M.L.W.; Mesinkovsk, N.A. Oral Collagen Supplementation: A Systematic Review of Dermatological Applications. *J. Drugs Dermatol.* **2019**, *18*, 9–16. [PubMed]
51. Pu, S.Y.; Huang, Y.L.; Pu, C.M.; Kang, Y.N.; Hoang, K.D.; Chen, K.H.; Chen, C. Effects of Oral Collagen for Skin Anti-Aging: A Systematic Review and Meta-Analysis. *Nutrients* **2023**, *15*, 2080. [CrossRef] [PubMed]
52. Ohara, H.; Matsumoto, H.; Ito, K.; Iwai, K.; Sato, K. Comparison of Quantity and Structures of Hydroxyproline-Containing Peptides in Human Blood after Oral Ingestion of Gelatin Hydrolysates from Different Sources. *J. Agric. Food Chem.* **2007**, *55*, 1532–1535. [CrossRef]
53. Proksch, E.; Segger, D.; Degwert, J.; Schunck, M.; Zague, V.; Oesser, S. Oral Supplementation of Specific Collagen Peptides Has Beneficial Effects on Human Skin Physiology: A Double-Blind, Placebo-Controlled Study. *Ski. Pharmacol. Physiol.* **2013**, *27*, 47–55. [CrossRef] [PubMed]
54. Asserin, J.; Lati, E.; Shioya, T.; Prawitt, J. The Effect of Oral Collagen Peptide Supplementation on Skin Moisture and the Dermal Collagen Network: Eevidence from an Ex Vivo Model and Randomized, Placebo-Controlled Clinical Trials. *J. Cosmet. Dermatol.* **2015**, *14*, 291–301. [CrossRef] [PubMed]
55. Kim, D.U.; Chung, H.C.; Choi, J.; Sakai, Y.; Lee, B.Y. Oral Intake of Low-Molecular-Weight Collagen Peptide Improves Hydration, Elasticity, and Wrinkling in Human Skin: A Randomized, Double-Blind, Placebo-Controlled Study. *Nutrients* **2018**, *10*, 826. [CrossRef] [PubMed]
56. Lee, M.; Kim, E.; Ahn, H.; Son, S.; Lee, H. Oral Intake of Collagen Peptide NS Improves Hydration, Elasticity, Desquamation, and Wrinkling in Human Skin: A Randomized, Double-Blinded, Placebo-Controlled Study. *Food Funct.* **2023**, *14*, 3196–3207. [CrossRef] [PubMed]
57. Seong, S.H.; Lee, Y.I.; Lee, J.; Choi, S.; Kim, I.A.; Suk, J.; Jung, I.; Baeg, C.; Kim, J.; Oh, D.; et al. Low-Molecular-Weight Collagen Peptides Supplement Promotes a Healthy Skin: A Randomized, Double-Blinded, Placebo-Controlled Study. *J. Cosmet. Dermatol.* **2024**, *23*, 554–562. [CrossRef]
58. Kim, J.; Lee, S.G.; Lee, J.; Choi, S.; Suk, J.; Lee, J.H.; Yang, J.H.; Yang, J.S.; Kim, J. Oral Supplementation of Low-Molecular-Weight Collagen Peptides Reduces Skin Wrinkles and Improves Biophysical Properties of Skin: A Randomized, Double-Blinded, Placebo-Controlled Study. *J. Med. Food* **2022**, *25*, 1146–1154. [CrossRef] [PubMed]
59. Wang, H. A Review of the Effects of Collagen Treatment in Clinical Studies. *Polymers* **2021**, *13*, 3868. [CrossRef]
60. Avila Rodríguez, M.I.; Rodríguez Barroso, L.G.; Sánchez, M.L. Collagen: A Review on Its Sources and Potential Cosmetic Applications. *J. Cosmet. Dermatol.* **2018**, *17*, 20–26. [CrossRef]
61. Garcez Duarte, M. Collagen Supplementation for Health of the Skin, Cartilage and Muscles-Current Myths and Truths. *Ann. Nutr. Metab.* **2017**, *71*, 258.
62. León-López, A.; Morales-Peñaloza, A.; Martínez-Juárez, V.M.; Vargas-Torres, A.; Zeugolis, D.I.; Aguirre-Álvarez, G. Hydrolyzed Collagen-Sources and Applications. *Molecules* **2019**, *24*, 4031. [CrossRef] [PubMed]
63. Farage, M.A.; Miller, K.W.; Maibach, H.I. *Textbook of Aging Skin*; Springer: Berlin/Heidelberg, Germany, 2017; ISBN 978-3-662-47397-9.
64. Quan, T. Molecular Insights of Human Skin Epidermal and Dermal Aging. *J. Dermatol. Sci.* **2023**, *112*, 48–53. [CrossRef] [PubMed]
65. Rostkowska, E.; Poleszak, E.; Wojciechowska, K.; Dos Santos Szewczyk, K. Dermatological Management of Aged Skin. *Cosmetics* **2023**, *10*, 55. [CrossRef]
66. Krutmann, J.; Grether-Beck, S.; Makrantonaki, E.; Schikowski, T. Skin Aging Exposome. *Dermatologie* **2023**, *74*, 657–662. [CrossRef] [PubMed]
67. Oesser, S.; Adam, M.; Babel, W.; Seifert, J. Oral Administration of 14C Labeled Gelatin Hydrolysate Leads to an Accumulation of Radioactivity in Cartilage of Mice (C57/BL). *J. Nutr.* **1999**, *129*, 1891–1895. [CrossRef] [PubMed]
68. Morikiri, Y.; Matsuta, E.; Inoue, H. The Collagen-Derived Compound Collagen Tripeptide Induces Collagen Expression and Extends Lifespan via a Conserved P38 Mitogen-Activated Protein Kinase Cascade. *Biochem. Biophys. Res. Commun.* **2018**, *505*, 1168–1173. [CrossRef] [PubMed]
69. Sato, K.; Asai, T.T.; Jimi, S. Collagen-Derived Di-Peptide, Prolylhydroxyproline (Pro-Hyp): A New Low Molecular Weight Growth-Initiating Factor for Specific Fibroblasts Associated with Wound Healing. *Front. Cell Dev. Biol.* **2020**, *8*, 548975. [CrossRef] [PubMed]
70. Asai, T.T.; Oikawa, F.; Yoshikawa, K.; Inoue, N.; Sato, K. Food-Derived Collagen Peptides, Prolyl-Hydroxyproline (Pro-Hyp), and Hydroxyprolyl-Glycine (Hyp-Hly) Enhance Growth of Primary Cultured Mouse Skin Fibroblast Using Fetal Bovine Serum Free from Hydroxyprolyl Peptide. *Int. J. Mol. Sci.* **2020**, *21*, 229. [CrossRef]

71. Campos, L.D.; Santos Junior, V.d.A.; Pimentel, J.D.; Carregã, G.L.F.; Cazarin, C.B.B. Collagen Supplementation in Skin and Orthopedic Diseases: A Review of the Literature. *Heliyon* **2023**, *9*, e14961. [CrossRef] [PubMed]
72. Hee Shin, J.; Hyang Kim, A.; Woo Lee, H.; Il Kim, J.; Kwang Lee, H. The Beneficial Effects of Collagen Tripeptide on Deep Wrinkling and Skin Moisturization: A Randomized Controlled Trial. *J. Food Nutr. Res.* **2021**, *9*, 508–515. [CrossRef]

MDPI AG
Grosspeteranlage 5
4052 Basel
Switzerland
Tel.: +41 61 683 77 34

Cosmetics Editorial Office
E-mail: cosmetics@mdpi.com
www.mdpi.com/journal/cosmetics

www.ingramcontent.com/pod-product-compliance
Lightning Source LLC
Chambersburg PA
CBHW041946260926
51850CB00018B/974
* 9 7 8 3 7 2 5 8 6 8 7 2 8 *